T0351073

Quantum Atom Optics

The rapid development of quantum technologies has driven a revolution in related research areas such as quantum computation and communication, and quantum materials. The first prototypes of functional quantum devices are beginning to appear. One of the leading platforms to realize such devices are neutral atoms, which allow for the observation of sensitive quantum effects, and have important applications in quantum simulation and matter wave interferometry. This modern text offers a self-contained introduction to the fundamentals of quantum atom optics and atomic many-body matter wave systems. Assuming a familiarity with undergraduate quantum mechanics, this book will be accessible for graduate students and early career researchers moving into this important new field. A detailed description of the underlying theory of quantum atom optics is given, before development of the key quantum technological applications, such as atom interferometry, quantum simulation, quantum metrology, and quantum computing.

Tim Byrnes is Assistant Professor in physics at New York University, Shanghai. He has previously held research positions at the National Institute of Informatics and the University of Tokyo in Japan, where his research has been focused on quantum information, particularly quantum simulation, exciton-polariton physics, and alternative models of quantum computation.

Ebubechukwu O. Ilo-Okeke obtained his PhD from Worcester Polytechnic Institute, USA. In 2017, he was appointed Research Fellow at New York University, Shanghai, where his research is focused on ultra-cold atomic gases, quantum atom optics, and quantum information.

Quantum Atom Optics

Theory and Applications to Quantum Technology

TIM BYRNES
New York University, Shanghai

EBUBECHUKWU O. ILO-OKEKE
New York University, Shanghai

CAMBRIDGE
UNIVERSITY PRESS

CAMBRIDGE
UNIVERSITY PRESS

University Printing House, Cambridge CB2 8BS, United Kingdom

One Liberty Plaza, 20th Floor, New York, NY 10006, USA

477 Williamstown Road, Port Melbourne, VIC 3207, Australia

314–321, 3rd Floor, Plot 3, Splendor Forum, Jasola District Centre, New Delhi – 110025, India

103 Penang Road, #05–06/07, Visioncrest Commercial, Singapore 238467

Cambridge University Press is part of the University of Cambridge.

It furthers the University's mission by disseminating knowledge in the pursuit of education, learning, and research at the highest international levels of excellence.

www.cambridge.org
Information on this title: www.cambridge.org/9781108838597
DOI: 10.1017/9781108975353

© Cambridge University Press 2021

This publication is in copyright. Subject to statutory exception and to the provisions of relevant collective licensing agreements, no reproduction of any part may take place without the written permission of Cambridge University Press.

First published 2021

A catalogue record for this publication is available from the British Library.

ISBN 978-1-108-83859-7 Hardback

Color figures for this publication at www.cambridge.org/quantumatomoptics.

Cambridge University Press has no responsibility for the persistence or accuracy of URLs for external or third-party internet websites referred to in this publication and does not guarantee that any content on such websites is, or will remain, accurate or appropriate.

To our parents, loved ones, children, friends, the colleagues who encouraged and supported us, and the visionaries who dared to pursue their curiosity

Contents

Foreword by Jonathan P. Dowling *page* x
Preface xiii

1 Quantum Many-Body Systems 1
1.1 Introduction 1
1.2 Second Quantization 1
1.3 Fock States 3
1.4 Multimode Fock States 5
1.5 Interactions 7
1.6 References and Further Reading 9

2 Bose–Einstein Condensation 10
2.1 Introduction 10
2.2 Bose and Einstein's Original Argument 10
2.3 Bose–Einstein Condensation for a Grand Canonical Ensemble 14
2.4 Low-Energy Excited States 19
2.5 Superfluidity 23
2.6 References and Further Reading 27

3 The Order Parameter and Gross–Pitaevskii Equation 28
3.1 Introduction 28
3.2 Order Parameter 28
3.3 The Gross–Pitaevskii Equation 30
3.4 Ground State Solutions of the Gross–Pitaevskii Equation 31
3.5 Hydrodynamic Equations 36
3.6 Excited State Solutions of the Gross–Pitaevskii Equation 37
3.7 References and Further Reading 43

4 Spin Dynamics of Atoms 44
4.1 Introduction 44
4.2 Spin Degrees of Freedom 44
4.3 Interaction between Spins 46
4.4 Electromagnetic Transitions between Spin States 48
4.5 The ac Stark Shift 53
4.6 Feshbach Resonances 54
4.7 Spontaneous Emission 55
4.8 Atom Loss 59

4.9 Quantum Jump Method 60
4.10 References and Further Reading 62

5 Spinor Bose–Einstein Condensates 64
5.1 Introduction 64
5.2 Spin Coherent States 64
5.3 The Schwinger Boson Representation 66
5.4 Spin Coherent State Expectation Values 69
5.5 Preparation of a Spin Coherent State 70
5.6 Uncertainty Relations 74
5.7 Squeezed States 77
5.8 Entanglement in Spin Ensembles 82
5.9 The Holstein–Primakoff Transformation 83
5.10 Equivalence between Bosons and Spin Ensembles 84
5.11 Quasiprobability Distributions 86
5.12 Other Properties of Spin Coherent States and Fock States 94
5.13 Summary 96
5.14 References and Further Reading 97

6 Diffraction of Atoms Using Standing Wave Light 99
6.1 Introduction 99
6.2 Theory of Diffraction 99
6.3 Ultra-cold Atom Interaction with Standing Light Wave 101
6.4 Bragg Diffraction by Standing Wave Light 106
6.5 Bragg Diffraction by Raman Pulses 109
6.6 References and Further Reading 114

7 Atom Interferometry 115
7.1 Introduction 115
7.2 Optical Interferometry 115
7.3 BEC Interferometry 116
7.4 References and Further Reading 129

8 Atom Interferometry Beyond the Standard Quantum Limit 130
8.1 Introduction 130
8.2 Two-Component Atomic Bose–Einstein Condensates 131
8.3 Husimi Q-function 136
8.4 Ramsey Interferometry and Bloch Vector 137
8.5 Error Propagation Formula and Squeezing Parameter 140
8.6 Fisher Information 144
8.7 Controlling the Nonlinear Phase per Atom 147
8.8 References and Further Reading 150

9 Quantum Simulation 152
9.1 Introduction 152
9.2 Problem Statement: What is Quantum Simulation? 153
9.3 Digital Quantum Simulation 155

9.4 Toolbox for Analogue Quantum Simulators 157
9.5 Example: The Bose–Hubbard Model 167
9.6 References and Further Reading 173

10 Entanglement Between Atom Ensembles 175
10.1 Introduction 175
10.2 Inseparability and Quantifying Entanglement 175
10.3 Correlation-Based Entanglement Criteria 179
10.4 One-Axis Two-Spin Squeezed States 186
10.5 Two-Axis Two-Spin Squeezed States 191
10.6 References and Further Reading 198

11 Quantum Information Processing with Atomic Ensembles 199
11.1 Introduction 199
11.2 Continuous Variables Quantum Information Processing 199
11.3 Spinor Quantum Computing 204
11.4 Deutsch's Algorithm 208
11.5 Adiabatic Quantum Computing 211
11.6 References and Further Reading 214

References 216
Index 247

Foreword

Having never written a book foreword beforeward, I did the logical thing and Googled "How to write a book foreword." I found the following information: 1. Be honest. 2. Use your unique voice. 3. Discuss your connection to the story and authors. 4. Mimic the style of the book. 5. Sign off. From these tidbits of insight, it sounds more like I am conducting an hour-long radio show and not writing a foreword for a book. Nevertheless, I shall endeavor to address each point in turn.

1. Be honest: When my friends and colleagues Tim Byrnes and Ebubechukwu Ilo-Okeke asked me to write the foreword to *Quantum Atom Optics*, I was initially delighted because I have been an active researcher in this field for over 30 years and I have many tens of publications in the field. Hence, I was a bit flabbergasted when they sent me a draft of the book upon which perusing that I discovered that they had not cited a single paper written by me. Thus, my original draft of this foreword was, "Not too shabby book by what's his face and colleague." Since then I have calmed down a bit, but could not let that one slide. Aside from this unintentional slight, the book is actually very nice. I attributed not referencing my papers to the foolishness of youth and the fact that my papers were published before the authors were born. The field of quantum atom optics is hence quite mature by now, although I'm not dead yet, and such a book is timely and I can't say enough nice things about the authors. The book is an exhaustive primer of a field that has a plethora of applications to a host of quantum technologies — particularly quantum information processing and quantum sensing. It would make a very nice supplementary book to any course on this topic. In truth the book is a compendium of just about everything that is known in quantum atom optics and as such it will be a much-used reference for many years.

2. Use your unique voice: As you can see from the above, my voice is so unique that I often have to translate my anonymous referee reports into Russian and then back into English in order to disguise my writing style. I can assure you that *Quantum Atom Optics* is a much better read than that. It is in fact very well written and delightfully organized in a clear and enticing fashion. Each chapter builds on the previous chapters such that the book builds to a true crescendo at the end, much like Ravel's orchestration of *Bolero*.

3. Discuss your connection to the story and authors: As I mentioned above, apparently unbeknownst to the authors, I have been working in the field of quantum atom optics for much of my career, and this book does an excellent job of covering the field. I have been collaborating with Tim and Ebube on this topic for five years now and we have a series of papers together, which gives me a unique insight into the writing of their book.

4. Mimic the style of the book: Well I can't use my unique voice and mimic the style of the book, now can I? The book was written by two sane and distinguished scientists. I'm more of an extinguished scientist myself.

5. Sign off:

Good bye and good luck,

Jonathan P. Dowling

October 13, 2019
St. George, Louisiana

Authors' comment: We made the unfortunate mistake of sending a preliminary version of the manuscript with incomplete referencing. We hope the current version better captures the relevant literature. Among the many new references added, Prof. Dowling's seminal paper ("Quantum Technology: The Second Quantum Revolution," co-authored by Gerard J. Milburn), where the much quoted term "Second Quantum Revolution" was first coined is now included. This, along with several other brilliantly named papers (e.g., "Quantum Optical Metrology – The Lowdown on High-N00N States"), will live on forever in the field. Rest in peace Jon, you were an absolute inspiration and will be missed by us all.

Preface

This book is an introduction to the field of quantum atom optics, the study of atomic many-body matter wave systems. In many ways, the field mirrors the field of quantum optics – the study of light at its fundamental quantum level. In quantum atom optics, many analogous concepts to quantum optics are encountered. The phenomenon of Bose–Einstein condensation (BEC) can be thought of as the atomic counterpart of lasing. Coherent states of light have analogues with spin coherent states for atoms, which can be visualized with Wigner and Q-functions. In this book, we start from the basic concepts of many-body atomic systems, learn methods for controlling the quantum state of atoms, and understand how such atoms can be used to store quantum information and be used for various quantum technological purposes.

Despite the many analogous concepts, there are also many differences of atoms and photons when treated at the quantum level. Most fundamentally, atoms can be bosons or fermions and possess mass, in contrast to photons. While lasing and BECs both have a macroscopic occupation of bosons, the mechanism that forms the BECs is ultimately at thermal equilibrium, whereas lasing is a nonequilibrium process. Coherent states of light do not conserve photon number, whereas generally atom numbers do not change shot to shot. Another stark difference is that atoms tend to possess much stronger interactions, whereas it is generally difficult to make photons interact strongly with each other. This makes the details of many-body atomic systems often quite different, which often leads to rather different approaches conceptually and theoretically.

One of the recent major developments has been the explosion of interest in the field of quantum information and technologies. Within a span of 20 years, it has turned from a niche field studied by a small community of physicists with various backgrounds in quantum optics, computer science, and foundations of quantum mechanics to a major research field in its own right. Much of the way of thinking in the quantum information community originates from the field of quantum optics. Now, there is great excitement in how quantum systems can be utilized toward new technologies. In this book, we cover several of the promising applications that atomic systems offer, including spin and matter wave interferometry, quantum simulation, and quantum computing.

Several excellent texts already exist in the field. Notable are Pierre Meystre's *Atom Optics* and Daniel Steck's *Quantum and Atom Optics*. These are both excellent comprehensive resources that cover the theory of atom optics at the fundamental level. For BECs, we refer the reader to excellent texts such as those by Pitaevskii and Stringari and by Pethick and Smith. Rather than duplicate these works, we wished to provide a text at the senior undergraduate to junior graduate levels that covers the basic principles that are necessary so that one can get up to speed with the current literature in a simple and straightforward way. While the topics that we cover inevitably are only a brief selection of the extensive achievements in the field, we hope our choices of topics reflect the current interest toward applications of such systems. This book would suit students who wish to obtain the necessary skills for working with many-body atomic systems and have an interest toward quantum technology applications.

1 Quantum Many-Body Systems

1.1 Introduction

In this chapter, we will give a brief overview of the mathematical formalism for describing quantum many-body systems. The types of systems that we will consider in this book will typically involve a large number of identical particles forming a composite quantum state. Such systems are best described using the formalism of second quantized operators. We describe how such operators can be defined starting from single-particle wavefunctions, and how they can be used to build up the full Hilbert space of the composite system. We also introduce the way that interactions between atoms can be described using the formalism, illustrating this with the example of s-wave scattering, a fundamental type of interaction between neutral atoms.

1.2 Second Quantization

First, consider the familiar case that you should already be well acquainted with from elementary quantum mechanics – a single particle trapped in a potential $V(\boldsymbol{x})$. The stationary states $\psi_k(\boldsymbol{x})$ are the solutions of the Schrodinger equation

$$H_0(\boldsymbol{x})\psi_k(\boldsymbol{x}) = E_k\psi_k(\boldsymbol{x}), \tag{1.1}$$

where

$$H_0(\boldsymbol{x}) = -\frac{\hbar^2}{2m}\nabla^2 + V(\boldsymbol{x}). \tag{1.2}$$

Here, m is the mass of the particle, E_k are the energies of the stationary states labeled n, and ∇^2 is the Laplacian. As quantum mechanics teaches us, the wavefunction tells us everything about the system that we would like to know (assuming we are working with pure states). This approach is fine if we want to describe a single particle, but what if we have many particles, all possibly interacting with each other as we have in a quantum many-body system?

One way is, of course, to simply increase the number of labels in the wavefunction and write this as $\psi(\boldsymbol{x}_1, \boldsymbol{x}_2, \ldots, \boldsymbol{x}_N)$. Each particle has its own label; in this case, we have N particles. Then, if we are dealing with bosonic particles, we must impose that the wavefunction is symmetric under particle interchange, or antisymmetric in the case of fermions. While this approach is certainly one way and mathematically

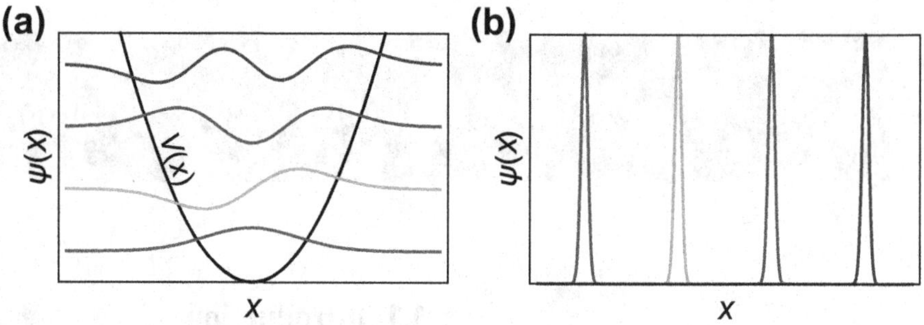

Fig. 1.1 Basis wavefunctions. (a) Wavefunctions $\psi_k(\boldsymbol{x})$ of a quantum harmonic oscillator. (b) Position-basis wavefunctions take the form of delta functions. Color version of this figure available at cambridge.org/quantumatomoptics.

equivalent, using the language of second quantized notation tends to be much more powerful and the method of choice in modern contexts.

In the second quantized approach, one defines an operator that creates or destroys a single particle with an implicit wavefunction. The most intuitive way to define this is in the position basis. The operator corresponding to the creation of one boson at the position $\boldsymbol{x}$ is written $a^\dagger(\boldsymbol{x})$, where the $\dagger$ is Hermitian conjugation. The reverse process of destroying the boson at the position $\boldsymbol{x}$ is written $a(\boldsymbol{x})$. These obey bosonic commutation relations according to

$$[a(\boldsymbol{x}), a^\dagger(\boldsymbol{x}')] = \delta(\boldsymbol{x} - \boldsymbol{x}') \tag{1.3}$$

and

$$[a(\boldsymbol{x}), a(\boldsymbol{x}')] = [a^\dagger(\boldsymbol{x}), a^\dagger(\boldsymbol{x}')] = 0. \tag{1.4}$$

For fermions, one can similarly define creation and destruction operators $c^\dagger(\boldsymbol{x})$ and $c(\boldsymbol{x})$, but this time obeying anticommutation relations

$$\{c(\boldsymbol{x}), c^\dagger(\boldsymbol{x}')\} = \delta(\boldsymbol{x} - \boldsymbol{x}') \tag{1.5}$$

and

$$\{c(\boldsymbol{x}), c(\boldsymbol{x}')\} = \{c^\dagger(\boldsymbol{x}), c^\dagger(\boldsymbol{x}')\} = 0. \tag{1.6}$$

The bosonic operator $a(\boldsymbol{x})$ (and also the fermion operator $c(\boldsymbol{x})$) is only one of an infinity of ways this can be defined. The reason for this is the same as why there are an infinity of ways of defining a basis. The position basis is just the most intuitive way of defining a basis, where the particle is located at the position $\boldsymbol{x}$. We could equally have used a Fourier basis or the complete set of states $\psi_k(\boldsymbol{x})$ (see Fig. 1.1). This can be written explicitly in the following way:

$$a_k = \int d\boldsymbol{x}\, \psi_k^*(\boldsymbol{x}) a(\boldsymbol{x}) \tag{1.7}$$

and

$$a_k^\dagger = \int d\boldsymbol{x}\, \psi_k(\boldsymbol{x}) a^\dagger(\boldsymbol{x}). \tag{1.8}$$

The very neat property of these newly defined operators is that they still have the same commutation relations of bosons that we are familiar with:

$$[a_k, a_l^\dagger] = \delta_{kl} \tag{1.9}$$

and

$$[a_k, a_l] = [a_k^\dagger, a_l^\dagger] = 0. \tag{1.10}$$

We see that, regardless of the basis, bosons always have the same properties. This is one of the reasons why the labels $\boldsymbol{x}$ or n, m are omitted and sometimes even fail to precisely say what the implicit wavefunction of the boson is. It is, however, worth remembering that there is always an implicit wavefunction when defining bosons.

Exercise 1.2.1 Verify (1.9) and (1.10).

Exercise 1.2.2 Show that the fermion version of (1.7),

$$c_k = \int d\boldsymbol{x} \psi_k^*(\boldsymbol{x}) c(\boldsymbol{x}), \tag{1.11}$$

satisfies fermion anticommutation relations

$$\begin{aligned} \{c_k, c_l^\dagger\} &= \delta_{kl}, \\ \{c_k, c_l\} &= \{c_k^\dagger, c_l^\dagger\} = 0. \end{aligned} \tag{1.12}$$

1.3 Fock States

We now have the right mathematical tools to write down a state involving any number of particles. Let's first start with the example that we began the discussion in the previous section – a single particle in a particular stationary state of the Schrodinger equation (1.1). Our starting point is the *vacuum* state, which has absolutely no particles in it at all and which is denoted $|0\rangle$. The vacuum state is defined as the state such that destroying a particle from it gives

$$a_k |0\rangle = 0. \tag{1.13}$$

Then, one particle with wavefunction $\psi_k(\boldsymbol{x})$ is written

$$\begin{aligned} |\psi_k\rangle &= a_k^\dagger |0\rangle \\ &= \int d\boldsymbol{x} \psi_k(\boldsymbol{x}) a^\dagger(\boldsymbol{x}) |0\rangle \\ &= \int d\boldsymbol{x} \psi_k(\boldsymbol{x}) |\boldsymbol{x}\rangle. \end{aligned} \tag{1.14}$$

In the second line above, we see how the wavefunction explicitly comes into the definition of the state, where we used (1.8). In the third line, we defined delta-function localized position eigenstates as

$$|\boldsymbol{x}\rangle = a^\dagger(\boldsymbol{x}) |0\rangle. \tag{1.15}$$

We see that the states can be expanded according to any basis, in the usual way that quantum mechanics allows us to.

Extending this to more than one particle in the way that you can probably already guess: For each particle in the state $\psi_k(\boldsymbol{x})$, we apply a creation operator $a_k^\dagger$. There is one small complication, which is how the normalization factors are defined. Suppose that there are n particles that occupy, say, the ground state. Since we are only talking about the ground state here, let's temporarily drop the k-label, which specifies the state, and write

$$a \equiv a_0. \tag{1.16}$$

Such a state with a particular number of atoms in a state is called a Fock (or number) state, which we would write as

$$|n\rangle = \frac{1}{\sqrt{\mathcal{N}_n}}(a^\dagger)^n|0\rangle. \tag{1.17}$$

Working out the normalization factor is an example of a very routine type of calculation when working with bosonic operators, so it is recommended if you haven't done it before. The essential steps can be done by repeatedly applying the commutation relations (1.9). Since $\langle n|n\rangle = 1$, the normalization factor must be

$$\mathcal{N}_n = \langle 0| \underbrace{aa\dots a}_{n \text{ of these}} \underbrace{a^\dagger a^\dagger \dots a^\dagger}_{n \text{ of these}} |0\rangle. \tag{1.18}$$

The aim is then to take advantage of (1.13) by commuting all the a's to hit the vacuum state ket on the right. Similarly, we can equally commute all the $a^\dagger$'s to the left to hit the vacuum state bra on the left and use the Hermitian conjugate of (1.13):

$$\langle 0|a_k^\dagger = 0. \tag{1.19}$$

For example, taking one of the a's all the way to the right gives

$$\mathcal{N}_n = n\langle 0| \underbrace{aa\dots a}_{n-1 \text{ of these}} \underbrace{a^\dagger a^\dagger \dots a^\dagger}_{n-1 \text{ of these}} |0\rangle. \tag{1.20}$$

Repeating this process, we find that $\mathcal{N}_n = n!$, and so the properly normalized Fock state is

$$|n\rangle = \frac{1}{\sqrt{n!}}(a^\dagger)^n|0\rangle. \tag{1.21}$$

From the commutation relations, it follows that the Fock states are orthonormal:

$$\langle n'|n\rangle = \delta_{nn'}. \tag{1.22}$$

Taking account of this normalization factor, we can use similar methods to work out that applying a creation or destruction operator gives numerical coefficients in addition to increasing or decreasing the number of bosons:

$$a|n\rangle = \sqrt{n}|n-1\rangle \tag{1.23}$$

and

$$a^\dagger|n\rangle = \sqrt{n+1}|n+1\rangle. \tag{1.24}$$

For fermions, it is simpler. We can perform the same logic for fermion operators. Define

$$c \equiv c_0, \tag{1.25}$$

where we have a fermion in the ground state of the states as defined in (1.11). Then, the only Fock states we can define are the vacuum state $|0\rangle$, which satisfies

$$c_k|0\rangle = 0 \tag{1.26}$$

and the one-particle fermion Fock state,

$$|1\rangle_{\mathrm{f}} = c^\dagger|0\rangle, \tag{1.27}$$

where we have labeled the fermion Fock states by a subscript f. We could try to write down a fermion version of (1.21), but from (1.12) we would find that any state with $n \geq 2$ gives zero. This is the Pauli exclusion principle at work – no two fermions can occupy the same state. The creation and destruction operator then simply shifts between these two states:

$$\begin{aligned} c|1\rangle_{\mathrm{f}} &= |0\rangle, \\ c^\dagger|0\rangle &= |1\rangle_{\mathrm{f}}. \end{aligned} \tag{1.28}$$

Exercise 1.3.1 Work through the steps to verify (1.21) and (1.22).

Exercise 1.3.2 Verify (1.23) and (1.24).

Exercise 1.3.3 (a) Show explicitly that the only fermion Fock states are (1.27) and the vacuum. (b) Verify (1.28) using the fermion anticommutation relations (1.12).

1.4 Multimode Fock States

The Fock states that we have written in Section 1.3 are fine if all the particles are in the ground state. This might be the case at zero temperature, or indeed the situation in a Bose–Einstein condensate (BEC), as we will be discussing in later chapters. But more typically, particles will occupy all energy levels, so we need to take into account more than just the ground state. In this case, we can simply just do the same thing as above, for each of the eigenstates labeled by k. The generalized Fock state for bosons is then

$$|n_0, n_1, \ldots, n_k, \ldots\rangle = \prod_k \frac{(a_k^\dagger)^{n_k}}{\sqrt{n_k!}}|0\rangle, \tag{1.29}$$

and similarly, for fermions it is

$$|n_0, n_1, \ldots, n_k, \ldots\rangle_{\mathrm{f}} = \prod_k (c_k^\dagger)^{n_k}|0\rangle. \tag{1.30}$$

Notice we don't need the normalization factors for the fermions because we only ever have $n_k \in \{0, 1\}$.

For a single-particle state, obviously the energy of the state is simply given by the eigenvalue of the Schrodinger equation, E_k. If there are multiple particles involved, how do we write the total energy of a Fock state such as that written above? First, we should generalize the single-particle Schrodinger equation to a multiparticle one, which can be done by simply writing

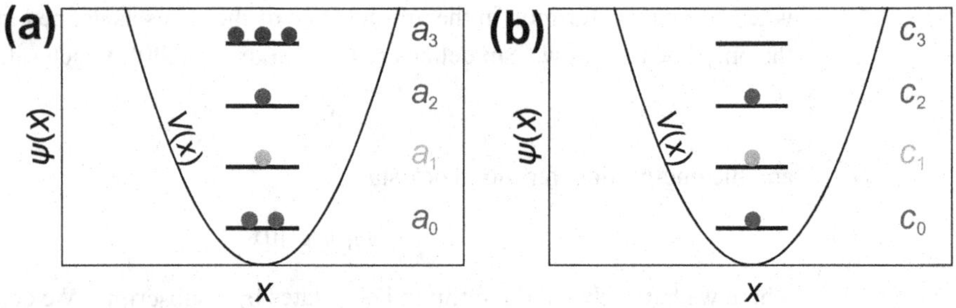

Fig. 1.2 Generalized Fock states for (a) bosons and (b) fermions. The case shown in (b) corresponds to the lowest energy configuration possible for three fermions (Fermi sea). The lowest energy configuration for bosons would correspond to all the particles in the ground state, hence (a) is an excited state for the multiparticle system. Color version of this figure available at cambridge.org/quantumatomoptics.

$$\mathcal{H}_0 = \int d\boldsymbol{x} a^\dagger(\boldsymbol{x}) H_0(\boldsymbol{x}) a(\boldsymbol{x}), \tag{1.31}$$

where $H_0(\boldsymbol{x})$ is given by (1.2). For fermions, this takes the same form, except that the bosonic operators are changed to fermionic ones. We now need the inverse relations of (1.7) and (1.8), given by

$$a(\boldsymbol{x}) = \sum_k \psi_k(\boldsymbol{x}) a_k \tag{1.32}$$

and

$$a^\dagger(\boldsymbol{x}) = \sum_k \psi_k^*(\boldsymbol{x}) a_k^\dagger. \tag{1.33}$$

Substituting this into (1.31), we obtain

$$\mathcal{H}_0 = \sum_k E_k N_k, \tag{1.34}$$

where

$$N_k \equiv a_k^\dagger a_k. \tag{1.35}$$

The operators (1.35) are *number operators* that have Fock states as their eigenstates:

$$N_k |n_0, n_1, \dots, n_k, \dots\rangle = n_k |n_0, n_1, \dots, n_k, \dots\rangle. \tag{1.36}$$

The multiparticle Fock state is thus an eigenstate of the Hamiltonian (1.34):

$$\mathcal{H}_0 |n_0, n_1, \dots, n_k, \dots\rangle = E_{\text{tot}} |n_0, n_1, \dots, n_k, \dots\rangle, \tag{1.37}$$

where the total energy is

$$E_{\text{tot}} = \sum_k E_k n_k. \tag{1.38}$$

We thus have the intuitive result that the total energy of the multiparticle Hamiltonian is the sum of the energies of each particle.

We can picture such Fock states as shown in Fig. 1.2. Each single-particle quantum level is labeled by an index k, and this is occupied by a number n_k, which tells us how many particles are occupied in that level. The total energy of the system is just the sum of the energies of the individual particles. For bosons, we are allowed to fill each level more than once, so the minimum energy is if all the particles are in the ground state. In this case, we would have

$$E_{\text{tot}} = NE_0, \qquad \text{(minimum, bosons)} \tag{1.39}$$

where N is the total number of bosons in the system. For fermions, each level cannot be filled more than once, so the minimum energy will be

$$E_{\text{tot}} = \sum_{k=0}^{N-1} E_k. \qquad \text{(minimum, fermions)} \tag{1.40}$$

Exercise 1.4.1 (a) Derive (1.32) by multiplying it by $\psi_k(\boldsymbol{x}')$ and summing (1.7) over k, using the completeness relation. (b) Derive (1.7) by multiplying it by $\psi_l^*(\boldsymbol{x})$ and integrating (1.32) over x.

Exercise 1.4.2 Verify that (1.34) starting from (1.31).

Exercise 1.4.3 Verify that (1.36) starting from the bosonic commutation relations.

1.5 Interactions

Up to this point, we have not included any interactions between the particles. The energy of the whole system was therefore determined entirely by the sum of the single-particle energies, as seen in (1.38). This is usually called the single-particle or *noninteracting limit*. More realistically, there are interactions present between the particles, due to the presence of some forces – naturally occurring or otherwise – that particles feel between each other. Suppose that the energy between two particles can be written as $U(\boldsymbol{x}, \boldsymbol{y})$, where the location of the first particle is $\boldsymbol{x}$ and the second is $\boldsymbol{y}$. The two-particle time-independent Schrodinger equation would be written in this case as

$$[H_0(\boldsymbol{x}) + H_0(\boldsymbol{y})]\,\psi(\boldsymbol{x}, \boldsymbol{y}) + U(\boldsymbol{x}, \boldsymbol{y})\psi(\boldsymbol{x}, \boldsymbol{y}) = E\psi(\boldsymbol{x}, \boldsymbol{y}). \tag{1.41}$$

The first term on the left-hand side is the single-particle Hamiltonian for the two-particle case, as given in (1.31). The second is the interaction term between the particles. There is only one term because the interaction corresponds to a pair of particles.

To generalize this to any number of particles, we write it in the form

$$\mathcal{H} = \mathcal{H}_0 + \mathcal{H}_I, \tag{1.42}$$

where the interaction Hamiltonian is

$$\begin{aligned}\mathcal{H}_I &= \frac{1}{2}\int dxdy a^\dagger(\boldsymbol{x})a^\dagger(\boldsymbol{y})U(\boldsymbol{x}, \boldsymbol{y})a(\boldsymbol{y})a(\boldsymbol{x}) \\ &= \frac{1}{2}\int dxdy U(\boldsymbol{x}, \boldsymbol{y})n(\boldsymbol{x})(n(\boldsymbol{y}) - \delta(\boldsymbol{x} - \boldsymbol{y})),\end{aligned} \tag{1.43}$$

where we have used the commutation relations (1.3) and (1.4). Equation (1.43) has the simple interpretation of counting all the pairs of particles between the particles

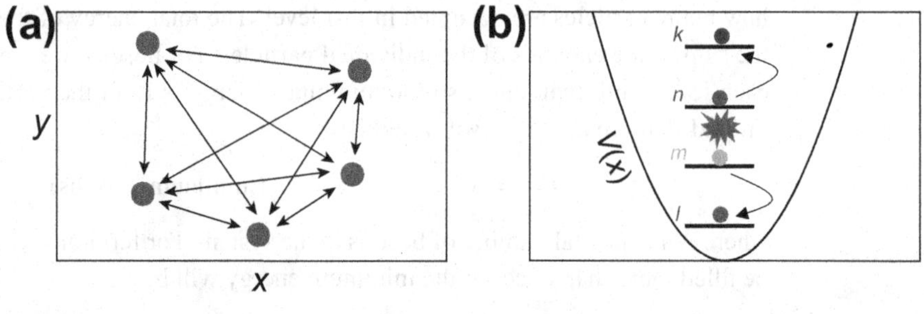

Fig. 1.3 Interactions between particles shown in (a) the position basis and (b) the single-particle eigenstate basis. In (a), for N particles there are $N(N-1)/2$ pairs of interactions. In (b), two particles initially in levels labeled by n and m scatter to other levels k and l. Color version of this figure available at cambridge.org/quantumatomoptics.

(see Fig. 1.3). For an N-particle system, there will be $N(N-1)/2$ pairs of interactions, which are potentially dependent on the positions of the particles $U(\boldsymbol{x},\boldsymbol{y})$.

As was done in the previous sections, it is possible to write the interaction Hamiltonian $\mathcal{H}_I$ in terms of the eigenstates of the single-particle Hamiltonian. This is done by simply substituting (1.32) and (1.33) into (1.43). This gives

$$\mathcal{H}_I = \sum_{klmn} U_{klmn} a_k^\dagger a_l^\dagger a_m a_n, \tag{1.44}$$

where

$$U_{klmn} = \frac{1}{2}\int d\boldsymbol{x} d\boldsymbol{y} \psi_k^*(\boldsymbol{x})\psi_l^*(\boldsymbol{y})U(\boldsymbol{x},\boldsymbol{y})\psi_m(\boldsymbol{x})\psi_n(\boldsymbol{y}). \tag{1.45}$$

In this basis, we find that the interactions are not diagonal. This is in contrast to (1.43), which consists entirely of number operators. This means that for particles that are spatially localized, as shown in Fig. 1.3(a), the interaction Hamiltonian $\mathcal{H}_I$ leaves the state unaffected. In the basis of the eigenstates of the single-particle Hamiltonian, the interaction Hamiltonian $\mathcal{H}_I$ can cause some scattering between states. Specifically, a particle in level n can feel the effect of another particle in level m, and both particles can be scattered to different levels k and l (see Fig. 1.3(b)).

In many cases, it is possible to simplify the form of the interaction. For example, typically we can safely say that the interaction only depends upon the relative position between the particles. This is a reasonable assumption in most experimental situations. The most common type of interaction for BECs is the s-wave interaction, which is a type of contact interaction,

$$U(\boldsymbol{x},\boldsymbol{y}) = U_0\delta(\boldsymbol{x}-\boldsymbol{y}), \tag{1.46}$$

where the interaction energy is

$$U_0 = \frac{4\pi\hbar^2 a_s}{m}. \tag{1.47}$$

Here, a_s is the scattering length, an experimentally measured quantity. In this case, the interaction matrix elements take the form (1.45):

$$g_{klmn} = \frac{2\pi\hbar^2 a_s}{m}\int d\boldsymbol{x} \psi_k^*(\boldsymbol{x})\psi_l^*(\boldsymbol{x})\psi_m(\boldsymbol{x})\psi_n(\boldsymbol{x}). \tag{1.48}$$

Specifically, for the atoms in the ground state, the interaction is

$$g_{0000} = \frac{2\pi\hbar^2 a_s}{m} \int d\boldsymbol{x} |\psi_0(\boldsymbol{x})|^4. \tag{1.49}$$

As we will see in Chapter 2, in a BEC many of the atoms occupy the ground state, and this will be the dominant interaction energy.

For the general interaction (1.48), very often some of the matrix elements will be zero automatically. To see an example of this, consider the simple case where $V(\boldsymbol{x}) = 0$. In this case, the eigenstates are simply plane waves,

$$\psi_{\boldsymbol{k}}(\boldsymbol{x}) = \frac{e^{i\boldsymbol{k}\cdot\boldsymbol{x}}}{\sqrt{V}}, \tag{1.50}$$

where V is a suitable normalization factor. Substituting into (1.48), we find that

$$g_{\boldsymbol{klmn}} = \frac{2\pi\hbar^2 a_s}{mV} \delta(\boldsymbol{n} + \boldsymbol{m} - \boldsymbol{k} - \boldsymbol{l}). \tag{1.51}$$

In this case, the labels $\boldsymbol{k}, \boldsymbol{l}, \boldsymbol{m}, \boldsymbol{n}$ have the interpretation of momentum, and the delta function is a statement of the conservation of momentum between the initial states and the final states. This occurs due to the translational invariance of the potential (1.46). While realistic potentials are not exactly $V(\boldsymbol{x}) = 0$, approximate relations for the conservation of momentum are obeyed in practice.

1.6 References and Further Reading

- Section 1.2: For further details about quantum mechanics and second quantization, see the textbooks [407, 9, 94].
- Sections 1.3, 1.4: Fock states and their algebra in a quantum optics setting are further described in [421, 168, 477].
- Section 1.5: Further details about interactions in Bose–Einstein condensates are provided in [369, 375, 497, 286, 169].

2 Bose–Einstein Condensation

2.1 Introduction

Chapter 1 introduced the formalism for treating many-particle indistinguishable quantum systems. This can describe the wavefunction of any system of bosonic or fermionic particles possibly interacting with each other. In this chapter, we introduce how such a system can undergo Bose–Einstein condensation. The essential feature of a Bose–Einstein condensate (BEC) is the macroscopic occupation of the ground state. The fact that such a state occurs is not completely obvious from the point of view of statistical mechanics – one might naively expect that there is a exponential decay of the probability of occupation following a Boltzmann distribution $p_n \propto \exp(-E_n/k_B T)$, where E_n is the energy of the state, k_B is the Boltzmann constant, and T is the temperature. We first discuss the original argument by Bose and Einstein, showing why such a macroscopic occupation might occur. We then show how macroscopic occupation of the ground state occurs in a grand canonical ensemble and give key results for the condensation temperature and the fraction of atoms in the ground state. We then discuss the effect of interactions on the energy-momentum dispersion relation, which gives rise to the Bogoliubov dispersion relation. This is the key to understanding superfluidity, one of the most astounding features of a quantum fluid.

Bose–Einstein condensation is a expansive subject, and the purpose of this chapter is introduce the minimal amount of background such that one can understand the more modern applications of such systems. For a more detailed discussion of the physics of BECs, we refer the reader to excellent texts such as those by Pitaevskii and Stringari [375] and Pethick and Smith [369].

2.2 Bose and Einstein's Original Argument

To see why macroscopic occupation of the ground state occurs in a system of bosons, we first examine a simple model of N noninteracting two-level particles. To be specific, we can imagine that there is a gas of bosonic atoms and the system is cooled down enough such that the spatial degrees of freedom are all the same for all the particles. As the spatial wavefunction is all the same, all the atoms are indistinguishable from each other, with the exception of the internal states. The two levels are then internal states of the atom, which have an energy E_0 and E_1. Due to the indistinguishable nature of the atoms, it is appropriate to denote them by bosonic creation operators $a_0^\dagger$ and $a_1^\dagger$ for the two states. As we did in Section 1.3, we can write the Fock states corresponding to this system as

$$|k\rangle = \frac{(a_0^\dagger)^{N-k}(a_1^\dagger)^k}{\sqrt{(N-k)!\,k!}}|0\rangle, \tag{2.1}$$

where k is the number of atoms in the state E_1, and the number of atoms in the other state with energy E_0 is $N-k$ because we have N atoms in total. It is easy to see that there is a total of $N+1$ states, because k can be any integer between 0 and N.

For comparison, let's also look at the case when we don't have indistinguishable bosons. Physically, this might correspond to a situation where the system is not quite as cold as the previous case, such that the spatial wavefunctions of the atoms are not the same as each other. In this case, the particles are distinguishable by virtue of some other property of the atoms, such as their velocity or position. Note that the distinguishability has to be merely in principle; no measurement has to be made to ensure this. Let's again look at all the possible states. Since the particles are distinguishable, we can label the states in the jth particle by $|\sigma_j\rangle_j$, where $\sigma_j = 0, 1$ are the two states with energy E_0 and E_1, respectively. The state of the whole system is then

$$|\sigma_1\sigma_2\dots\sigma_N\rangle = |\sigma_1\rangle_1 \otimes |\sigma_2\rangle_2 \otimes \cdots \otimes |\sigma_N\rangle_N. \tag{2.2}$$

Counting the total number of different states, we see a different result from what we saw in (2.1). Since each atom can be in one of two states, the total number of combinations is 2^N.

Let's now look at the energy spectrum for the two cases. For the indistinguishable bosonic case, we can simply use (1.38) to obtain

$$E_{\text{indis}} = E_0(N-k) + E_1 k = NE_0 + k\Delta E, \tag{2.3}$$

where we have defined $\Delta E = E_1 - E_0$. For the distinguishable case, we have

$$\begin{aligned} E_{\text{distin}} &= \sum_{j=1}^{N} E_0(1-\sigma_j) + E_1\sigma_j \\ &= NE_0 + k\Delta E, \end{aligned} \tag{2.4}$$

where we defined $k = \sum_{j=1}^{N} \sigma_j$, which counts the total number of atoms in the excited state E_1.

The expressions (2.3) and (2.4) are exactly the same, which is not surprising considering in both cases we have a set of N two-level atoms with the same energy structure. While both have a variable k, which is the total number of atoms in the excited state, they have a difference in terms of the degeneracy of the states (2.1) and (2.2). In Fig. 2.1, we see the energy spectrum for the whole system. While the indistinguishable case only has one possible state for each k, the distinguishable case has various combinations of spin configurations (2.2) that have the same k. For example, for $N = 5$, there are five states with $k = 1$:

$$|00001\rangle, |00010\rangle, |00100\rangle, |01000\rangle, |10000\rangle. \tag{2.5}$$

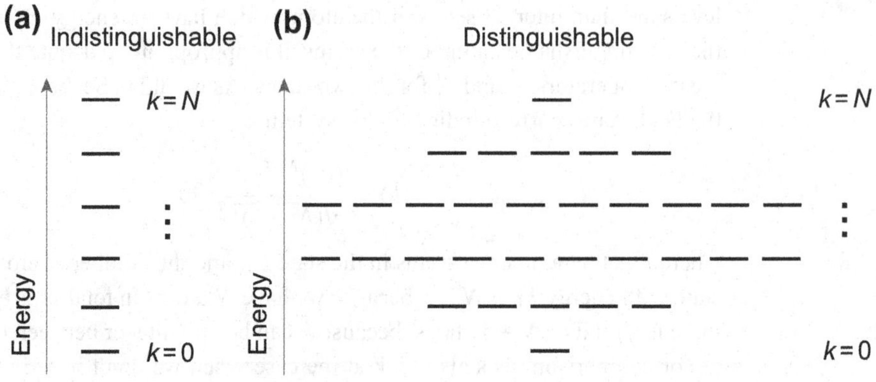

Fig. 2.1 Number of levels for (a) indistinguishable bosonic and (b) distinguishable two-level systems. We show the case $N = 5$.

In general, one can argue from simple combinatorics that the number of states in each sector k is $\binom{N}{k}$, so that we can label the state (2.2) equivalently as

$$|k, d_k\rangle = |\sigma_1\sigma_2 \ldots \sigma_N\rangle, \tag{2.6}$$

where $d_k \in [1, \binom{N}{k}]$.

One may wonder whether there is some kind of correspondence of the states between the distinguishable and indistinguishable cases. The key feature of the bosonic system is that the wavefunction is symmetric under particle interchange. This restriction that indistinguishability imposes greatly reduces the available states that the system can occupy. For example, none of the states (2.5) is symmetric under particle interchange – for each state, interchanging the ordering of the σ_j will give a completely different state. There is, however, a state that is symmetric under interchange:

$$\frac{1}{\sqrt{5}}\left(|00001\rangle + |00010\rangle + |00100\rangle + |01000\rangle + |10000\rangle\right). \tag{2.7}$$

For this state, interchanging the order of the spins still gives the same state. More generally, any state of the form

$$\frac{1}{\sqrt{\binom{N}{k}}} \sum_{d_k=1}^{\binom{N}{k}} |k, d_k\rangle \tag{2.8}$$

is symmetric under particle interchange. This is the only state in the space of states $|k, d_k\rangle$ that has this property – all the phases in the sum (2.7) must be the same. This is thus the state that is equivalent to the bosonic Fock state $|k\rangle$ in (2.1).

This difference in the number of states ultimately is the origin of Bose–Einstein condensation. To see this, let us find the probability of picking a particular atom, and finding it in the ground state for the two cases. Starting with the distinguishable case, first find the probability for the jth atom to be in the ground state using Boltzmann statistics:

$$p_j = \frac{e^{-\beta E_0}}{e^{-\beta E_0} + e^{-\beta E_1}} = \frac{1}{1 + e^{-\beta \Delta E}}, \tag{2.9}$$

where $\beta = 1/k_B T$, k_B is the Boltzmann constant, and T is the temperature. Here, the state of the remaining $N - 1$ atoms does not have any effect on the jth atom. Therefore, if we then measure all N atoms, of course the fraction of atoms that are in the ground state will be

$$f_{\text{distin}} = \frac{1}{1 + e^{-\beta \Delta E}}. \tag{2.10}$$

In this case, we see that the number of atoms N doesn't affect the probability because the particles are independent.

In the indistinguishable boson case, there is no way of simply picking out the jth atom as we did above, since the atoms cannot be distinguished. We can, however, talk about the statistics of the whole system in a well-defined way, since we know the energy structure as shown in Fig. 2.1. The probability of the state (2.1) can be obtained again using Boltzmann statistics:

$$\begin{aligned} p_k &= \frac{e^{-\beta(NE_0 + k\Delta E)}}{\sum_{k=0}^{N} e^{-\beta(NE_0 + k\Delta E)}} \\ &\approx e^{-\beta k \Delta E}(1 - e^{-\beta \Delta E}). \end{aligned} \tag{2.11}$$

where we assumed that N is large for simplicity. This is the distribution of the whole system, but what we are interested in is the probability of *one* atom to be in the ground state. To calculate this, for each state $|k\rangle$, let us count the number of atoms that are in the ground state, which is just equal to $N - k$. On average, one would thus find that

$$\begin{aligned} \langle n_0 \rangle &= \sum_{k=0}^{N} p_k (N - k) \\ &= N - \frac{e^{-\beta \Delta E}}{1 - e^{-\beta \Delta E}}. \end{aligned} \tag{2.12}$$

The fraction of atoms in the ground state is then

$$f_{\text{indis}} = \frac{\langle n_0 \rangle}{N} = 1 - \frac{1}{N} \frac{e^{-\beta \Delta E}}{1 - e^{-\beta \Delta E}}. \tag{2.13}$$

In the limit of $N \to \infty$, we see that the fraction f approaches 1. This means that regardless of the temperature β and energy separation ΔE, the fraction of atoms in the ground state is unity!

The two different results from the fraction of atoms in the ground state (2.10) and (2.13) originates from the different way of counting states for distinguishable and indistinguishable particles. As we see from Fig. 2.1, the main difference is that for the distinguishable case, there are many more levels away from the ground state. In the distinguishable case, due to a state with k excited states having $\binom{N}{k}$ possibilities, this strongly biases the population toward excited states with $k \approx N/2$. Thus, even though the Boltzmann factor $\propto e^{-\beta E}$ exponentially biases the probability toward lower energies, these two effects effectively cancel each other. On the other hand, the indistinguishable case does not have the handicap of the combinatorics. In the limit of $N \to \infty$, the ladder in Fig. 2.1(a) extends indefinitely, and the Boltzmann factor ensures that the population is biased toward the bottom. This makes it much more likely that the ground state is occupied for each atom.

Exercise 2.2.1 Show using a combinatorial argument that for N distinguishable two-level atoms, the total number of states with k atoms in one of the excited states is $\binom{N}{k}$, and hence, verify (2.6).

Exercise 2.2.2 (a) Verify that (2.7) is symmetric under particle interchange. (b) Show explicitly, by evaluating the fidelity or otherwise, that the state (2.7) with a minus sign on one of the terms is not symmetric under particle interchange. (c) Verify that (2.8) is symmetric under particle interchange.

Exercise 2.2.3 For the full system with N distinguishable particles, the probability that the state occupies the state $\sigma_j = 0$ should be equal to $p_{00\ldots0} = \prod_{j=1}^{N} p_j$ in (2.9). This should agree with directly constructing the partition function using (2.6) and (2.4). Check that these two methods give the same answer.

Exercise 2.2.4 (a) Verify (2.11) is true for $N \to \infty$. (b) What is the corresponding result for any N?

2.3 Bose–Einstein Condensation for a Grand Canonical Ensemble

In Section 2.2, we saw how the indistinguishable nature of the particles could affect the occupation of the ground state for a collection of two-level atoms. While this is possibly the simplest example of the macroscopic occupation of the ground state, it is far from a realistic model of Bose–Einstein condensation (BEC) of atoms. In Section 2.2, we did not examine the movement of the atoms at all. In fact, what happens more typically in a BEC experiment is that a collection of atoms are all cooled down to low enough temperatures such that they condense into their ground states. The energy levels that we are talking about here are the motional degrees of freedom, which we can take to be momentum states.

In this section, we will derive BEC for a collection of noninteracting atoms in three dimensions. The argument has a few subtleties, so let's tread carefully and work through it step by step. The argument works in the grand canonical ensemble formulation of statistical mechanics. We will review a few things in the context of bosons and then move on to the main argument showing BEC.

2.3.1 Bose–Einstein Distribution and Chemical Potential

Recall the situation that we had in Fig. 1.2(a): A potential $V(\boldsymbol{x})$ gave rise to a set of levels labeled by k, each of which could be multiply occupied by an integer n_k. The total energy is given by (1.38). According to the Boltzmann statistics, the probability of a particular Fock state configuration being occupied is

$$p(n_0, n_1, \ldots, n_k, \ldots) = \frac{\exp(-\beta \sum_k (E_k - \mu) n_k)}{Z}, \tag{2.14}$$

where Z is the partition function, which plays the role of the normalization factor for the probability distribution. We have also offset the energy by the chemical potential μ, which can for now be viewed as shifting the definition of the zero-point in the energy per particle. The partition function can be evaluated by summing over all possible configurations

$$Z = \sum_{n_0} \sum_{n_1} \cdots \sum_{n_k} \ldots \exp(-\beta \sum_k (E_k - \mu) n_k)$$
$$= \sum_{n_0} \exp(-\beta(E_0 - \mu) n_0) \sum_{n_1} \exp(-\beta(E_1 - \mu) n_1) \ldots . \tag{2.15}$$

Normally, one would like to consider a fixed number of particles, such that $\sum_k n_k = N$. This unfortunately put some restrictions on the summations in (2.15), which makes it difficult to evaluate mathematically. Instead, let's for now consider the unrestricted case, and impose the constraint later. This is the so-called grand canonical distribution, where the particle number is not fixed, and particles can enter or leave the system, very much in the same way as energy being exchanged between the system and reservoir. Evaluating each of the sums by a geometric series, we can write the probability as

$$p(n_0, n_1, \ldots, n_k, \ldots) = \prod_k p_k(n_k), \tag{2.16}$$

where

$$p_k(n_k) = e^{-\beta(E_k - \mu) n_k} (1 - e^{-\beta(E_k - \mu)}) \tag{2.17}$$

is the probability that the kth level is occupied by n_k atoms. The average number of atoms in this level,

$$\bar{n}_k(\mu) = \sum_k p_k n_k = \frac{1}{e^{\beta(E_k - \mu)} - 1}, \tag{2.18}$$

gives the well-known Bose–Einstein distribution.

Since we are using the grand canonical distribution, the number of particles in a given level – and hence the whole system – fluctuates. The number of particles that are in the system is controlled by the chemical potential μ, as can be seen from the Bose–Einstein distribution (2.18). When the chemical potential tends to infinity, $\mu \to -\infty$, we can see from (2.18) that the number of particles in all the levels becomes zero:

$$\lim_{\mu \to -\infty} \bar{n}_k(\mu) = 0. \tag{2.19}$$

The total number of particles in the whole system is also zero in this limit. For the ground state, as the chemical potential approaches E_0 from below, the population diverges:

$$\lim_{\mu \to E_0^-} \bar{n}_0(\mu) = \infty. \tag{2.20}$$

We can thus say that the valid range of the chemical potential is

$$-\infty < \mu < E_0, \tag{2.21}$$

since otherwise we will have a negative population on the state with energy E_0. Between these two extremes, the total particle number

$$\langle N \rangle = \sum_k \bar{n}_k(\mu) \tag{2.22}$$

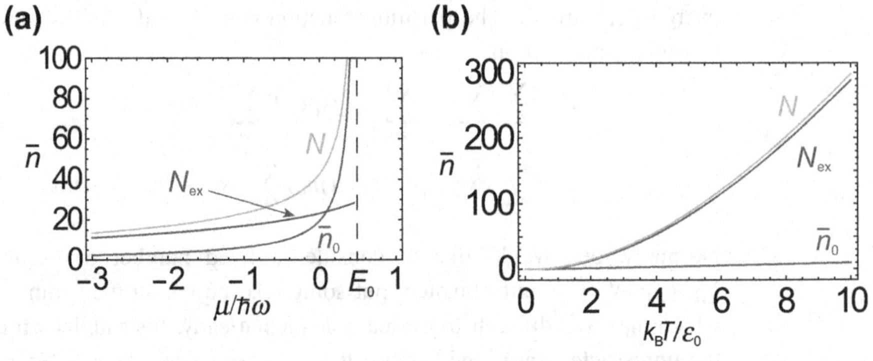

Fig. 2.2 Number of particles occupying the ground state $\bar{n}_0$, all the remaining excited states N_{ex}, and the total particle number $N = \bar{n}_0 + N_{ex}$ according to the Bose–Einstein distribution for (a) a one-dimensional harmonic oscillator with fixed temperature $\beta = \hbar\omega$ and (b) a three-dimensional free particle with fixed chemical potential $\mu = -\epsilon_0$. Color version of this figure available at cambridge.org/quantumatomoptics.

can be swept from 0 to infinity. If we would like to talk about a specific particle number N_0, then we can do this by first finding the chemical potential associated with $\langle N \rangle = N_0$, and then take this to be the chemical potential associated with the system.

The fact that the population in the ground state (2.20) diverges is not particularly surprising, since at this point the chemical potential $\mu = E_0$. But there is something interesting that happens to the remaining levels that is particularly relevant to BEC. If we assume that $E_1 - E_0 > 0$ (that is, that there is a nonzero energy difference between the ground state and first excited state), then there is a maximum to the number of particles that the excited states can contain. This is by virtue of the restricted range that the chemical potential can take (2.21). This maximum is given by

$$N_{ex}^{max} = \sum_{k=1}^{\infty} \bar{n}_k(\mu = E_0) = \sum_{k=1}^{\infty} \frac{1}{e^{\beta(E_k - E_0)} - 1}. \tag{2.23}$$

The specific maximum number depends upon the particular distribution of energies. In Fig. 2.2(a), we show an example for the case of a one-dimensional harmonic oscillator distribution $E_k = \hbar\omega(k + 1/2)$. We see that the ground state population diverges as expected from (2.20), while the population in all the remaining states reaches a maximum according to (2.23). As $\mu \to E_0^-$, the total population N is dominated by the ground state population $\bar{n}_0$.

Exercise 2.3.1 Plot the number of particles occupying the ground state $\bar{n}_0$, all the remaining excited states N_{ex}, and the total particle number $N = \bar{n}_0 + N_{ex}$ versus the chemical potential for a three-dimensional harmonic oscillator, where the bosons have also a spin degree of freedom $\sigma = \pm 1$. The total energy can be modeled by $E_k = \hbar\omega(k_x + k_y + k_z + 3/2) + \mu_B B\sigma$, where μ_B is the Bohr magneton (not to be confused with the chemical potential), and B is the magnetic field. Describe the different cases of $B = 0$, $B > 0$, and $B < 0$.

2.3.2 Bose–Einstein Condensation

In the previous section, we derived the Bose–Einstein distribution, which tells us the average number of bosons for a given energy level E_k for a grand canonical ensemble. As we see from Fig. 2.2, as the chemical potential approaches $\mu \to E_0^-$, the entire population of the bosons resides in the ground state:

$$\lim_{\mu \to E_0^-} \frac{\bar{n}_0}{N} = 1. \tag{2.24}$$

If we consider the signature of Bose–Einstein condensation to be a macroscopic occupation of the ground state, this certainly seems to be what is happening. There is one problem here, which is that this is not exactly what typical experiments do. As the chemical potential is increased, what we are doing is letting more and more particles into the system, and thereby increasing the density. But a more typical situation is that we cool the system to lower and lower temperatures, and at a certain point there is a critical point where macroscopic occupation of the ground state occurs. Equation (2.24) doesn't show any sign of a critical point, so it appears that we haven't quite derived the effect we are seeking yet.

To see Bose–Einstein condensation, it turns out that we must look in three dimensions. As a simple example, let us look at the case of a particle in three dimensions within a box of dimensions $a \times a \times a$. Choosing periodic boundary conditions, the spectrum is

$$E_{k_x k_y k_z} = \epsilon_0 (k_x^2 + k_y^2 + k_z^2), \tag{2.25}$$

where $\epsilon_0 = \frac{h^2}{2ma^2}$ is the energy scale of the Hamiltonian and $k_{x,y,z} \in \{0, \pm 1, \pm 2, \dots\}$. Periodic boundary conditions are chosen for convenience, since this gives a unique ground state with $k_x = k_y = k_z = 0$ at zero energy, but it does not affect the overall results. The ground state occupation number according to (2.18) is

$$\bar{n}_0 = \frac{1}{e^{-\beta\mu} - 1}. \tag{2.26}$$

Meanwhile, the population in the excited states is

$$\begin{aligned} N_{\text{ex}} &= \int d^3k \frac{1}{e^{\beta(\epsilon_0 k_r^2 - \mu)} - 1} \\ &= \left(\frac{\pi}{\epsilon_0 \beta} \right)^{3/2} \text{Li}_{3/2}(e^{\beta\mu}), \end{aligned} \tag{2.27}$$

where $\text{Li}_n(z)$ is the polylogarithm function and $k_r^2 = k_x^2 + k_y^2 + k_z^2$. One may be concerned here that in (2.27) we have included the contribution of the ground state as we have not made any restriction to the integral, as there should be in (2.23). There is, however, nothing to worry about since the ground state automatically has no contribution because in spherical coordinates $d^3k = k_r^2 \sin k_\theta dk_r dk_\theta dk_\phi$, and for the ground state its weight in the integral is zero.

Now let us see whether we see the macroscopic occupation effect as the temperature is lowered. If we calculate the ratio of the ground state to the total population $N = \bar{n}_0 + N_{\text{ex}}$, then we find that

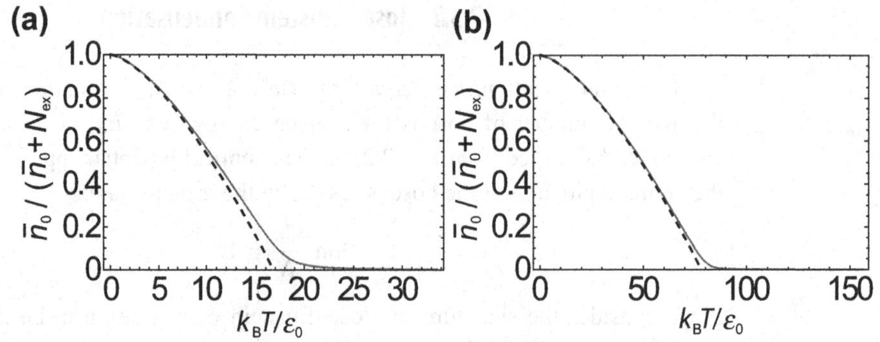

Fig. 2.3 The condensate fraction as a function for temperature for a fixed total number of particles (a) $N = 10^3$ corresponding to $k_B T_c/\epsilon_0 = 16.8$; (b) $N = 10^4$ corresponding to $k_B T_c/\epsilon_0 = 77.9$. Solid lines show a numerical solution of (2.29) for the chemical potential; the dashed lines show the asymptotic result for an infinite number of particles.

$$\lim_{T \to 0} \frac{\bar{n}_0}{N} = 1, \tag{2.28}$$

so we do see that all the particles occupy the ground state as the temperature is lowered. In Fig. 2.2(b), we plot the occupations for the ground state and all the excited states. We see that the basic effect of lowering the temperature is to reduce the population of the excited state. However, we don't see any evidence of a critical temperature where there is a threshold where BEC occurs.

What went wrong? A hint of this can be found in Fig. 2.2(b), where we see that changing the temperature for a fixed chemical potential really has the effect of changing the total population N quite dramatically. What more realistically happens in an experiment is that the particle number should be fixed to a constant as the temperature is lowered. All this happened because we chose to work in the grand canonical ensemble formalism, where particles can freely enter and leave the system.

Let's try again, but this time we vary the chemical potential with the temperature, such that the total number of particles N is fixed. That is, for a particular temperature T, we find the solution with respect to $\mu(T)$ of the equation

$$\bar{n}_0 + N_{\text{ex}} = \frac{1}{e^{-\beta\mu(T)} - 1} + \left(\frac{\pi}{\epsilon_0 \beta}\right)^{3/2} \text{Li}_{3/2}(e^{\beta\mu(T)}) = N. \tag{2.29}$$

Once we have found our function $\mu(T)$ for a fixed N, we can then compare the number of particles in the ground state to the total number, $\bar{n}_0/N$. The results are shown in Fig. 2.3. This time, we do see a sharp transition where the number of particles in the ground state starts to pick up strongly at a particular temperature. As the total number of particles increases, this transition gets sharper and sharper. As before, at $T = 0$ the condensate fraction is 1, but most importantly there is a range of temperatures below which there is a macroscopic occupation of the ground state. This is Bose–Einstein condensation.

Can we estimate what the critical temperature is? We can do this by returning to Fig. 2.2(a) and noticing the crucial property that N_{ex} has a maximum even when the chemical potential is at its largest point, $\mu = E_0$. Also note from Fig. 2.2(b) that the occupation numbers generally decrease with temperature. Suppose, then, that there is a total number of particles that happens to be much bigger than N_{ex}. Then we would surely have a macroscopic population in the ground state, since $\bar{n}_0 = N - N_{ex}$. The temperature at which the ground state population starts to increase strongly can then be estimated by setting $\mu \approx E_0 = 0$ and finding when $N = N_{ex}$. This gives

$$k_B T_c = \frac{h^2}{2\pi m} \left(\frac{n}{\mathrm{Li}_{3/2}(1)} \right)^{2/3}, \tag{2.30}$$

where $n = N/a^3$ is the density of the particles, and $\mathrm{Li}_{3/2}(1) = 2.612$. This can be conveniently rewritten in terms of the thermal de Broglie wavelength,

$$\lambda_T = \frac{h}{\sqrt{2\pi m k_B T}}, \tag{2.31}$$

giving

$$n\lambda_{T_c}^3 = 2.612. \tag{2.32}$$

Since the thermal de Broglie wavelength is the average de Broglie wavelength of particles moving at a particular temperature, this has the physical interpretation that BEC occurs when the matter wave wavelengths start to be comparable to the interparticle distances. We derived the BEC criterion for the particular case of bosons obeying a harmonic oscillator potential. For other geometries, we would generally expect similar results to (2.32), with possibly a different constant on the right-hand side.

We can find an approximate dependence to the ground state population in Fig. 2.3 by again noting that for this temperature range, the chemical potential will be close to the ground state, $\mu \approx E_0 = 0$. The excited state population (2.27) will therefore be

$$N_{ex} \approx N \left(\frac{T}{T_c} \right)^{3/2}, \tag{2.33}$$

where we have used our expression for the critical temperature (2.30). The ground state is then according to $\bar{n}_0 = N - N_{ex}$:

$$\bar{n}_0(T) = N \left(1 - \left(\frac{T}{T_c} \right)^{3/2} \right). \tag{2.34}$$

In practice, the approximation works rather well, as can be seen in Fig. 2.3 by the dashed lines. Strictly speaking, (2.34) is only valid in the limit of infinite density, so the sharp transition is smoothed out in practice.

2.4 Low-Energy Excited States

We have seen that a bosonic gas at low enough temperatures will have a macroscopic occupation of the ground state. Up to this point, we have not included the effect of interactions, as we have discussed in Section 1.5. Typically, the energy due to interactions has a much lower energy than the kinetic energy. Hence, to lowest order, the state of the gas can be approximated by all the bosons occupying the lowest energy state of the Hamiltonian (1.31). In this section, we show the effect of introducing interactions between bosons. We shall see that this has a dramatic effect on the dispersion relation of the bosons, and it is a key reason why BECs have spectacular properties such as superfluidity.

Let us start by writing the full Hamiltonian of the interacting boson gas, assuming the most common situation of an s-wave interaction,

$$H = \sum_{\boldsymbol{k}} E_{\boldsymbol{k}} a_{\boldsymbol{k}}^{\dagger} a_{\boldsymbol{k}} + \frac{U_0}{2V} \sum_{\boldsymbol{k}\boldsymbol{k}'\boldsymbol{q}} a_{\boldsymbol{k}'+\boldsymbol{q}}^{\dagger} a_{\boldsymbol{k}-\boldsymbol{q}}^{\dagger} a_{\boldsymbol{k}} a_{\boldsymbol{k}'}, \tag{2.35}$$

where we have substituted (1.51) into (1.43), and U_0 is the interaction energy as defined in (1.47). We now would like to find the solution of (2.35) in the regime where there is macroscopic occupation of the ground state a_0 and the interactions are weak. Generally, a Hamiltonian that is quadratic in the bosonic annihilation and creation operators $a_{\boldsymbol{k}}$ and $a_{\boldsymbol{k}}^{\dagger}$ can be solved by a bosonic transformation, but since (2.35) is fourth order, it cannot be solved. However, using the fact that we expect most of the bosons to occupy the ground state a_0, we can approximate the above Hamiltonian to second order in bosonic operators, such that it can be solved.

We approximate the interaction part of the Hamiltonian (2.35) by keeping terms only involving the macroscopically occupied state a_0. There is a total of seven such terms. The first such term amounts to setting $\boldsymbol{k} = \boldsymbol{k}' = \boldsymbol{q} = 0$. The other six terms involve setting two of the momentum labels to zero and the other two to nonzero terms. Due to momentum conservation, there are no terms with one or three a_0's. This gives

$$H_{\mathrm{B}} = \sum_{\boldsymbol{k}} E_{\boldsymbol{k}} a_{\boldsymbol{k}}^{\dagger} a_{\boldsymbol{k}} + \frac{U_0}{2V} a_0^{\dagger} a_0^{\dagger} a_0 a_0 + \sum_{\boldsymbol{k}\neq 0} \left[4 a_0^{\dagger} a_{\boldsymbol{k}}^{\dagger} a_{\boldsymbol{k}} a_0 + a_{\boldsymbol{k}}^{\dagger} a_{-\boldsymbol{k}}^{\dagger} a_0 a_0 + a_0^{\dagger} a_0^{\dagger} a_{-\boldsymbol{k}} a_{\boldsymbol{k}} \right]. \tag{2.36}$$

Due to the large population in the ground state, we may replace $a_0 \to e^{i\varphi}\sqrt{N_0}$, where N_0 is the number of atoms in the ground state. Here, φ is the phase that may be present, since a_0 is a non-Hermitian operator and thus can have a complex expectation value. We then have

$$H_{\mathrm{B}} = \sum_{\boldsymbol{k}} E_{\boldsymbol{k}} a_{\boldsymbol{k}}^{\dagger} a_{\boldsymbol{k}} + \frac{U_0}{2V} N_0^2 + \frac{U_0}{2V} \sum_{\boldsymbol{k}\neq 0} \left[4 N_0 a_{\boldsymbol{k}}^{\dagger} a_{\boldsymbol{k}} + N_0 e^{2i\varphi} a_{\boldsymbol{k}}^{\dagger} a_{-\boldsymbol{k}}^{\dagger} + N_0 e^{-2i\varphi} a_{-\boldsymbol{k}} a_{\boldsymbol{k}} \right]. \tag{2.37}$$

The number of atoms in the ground state is related to the total number of atoms by

$$N_0 = N - \sum_{\boldsymbol{k}\neq 0} a_{\boldsymbol{k}}^{\dagger} a_{\boldsymbol{k}}. \tag{2.38}$$

Substituting this into H_{B} and keeping only quadratic terms in bosonic operators, we have

$$\begin{aligned} H_{\mathrm{B}} = {} & E_0 N + \frac{U_0 n N}{2} + \sum_{\boldsymbol{k}\neq 0} (E_{\boldsymbol{k}} - E_0) a_{\boldsymbol{k}}^{\dagger} a_{\boldsymbol{k}} \\ & + \frac{U_0 n}{2} \sum_{\boldsymbol{k}\neq 0} \left[2 a_{\boldsymbol{k}}^{\dagger} a_{\boldsymbol{k}} + e^{2i\varphi} a_{\boldsymbol{k}}^{\dagger} a_{-\boldsymbol{k}}^{\dagger} + e^{-2i\varphi} a_{-\boldsymbol{k}} a_{\boldsymbol{k}} \right], \end{aligned} \tag{2.39}$$

where we have defined the density $n = N/V$.

Now that the Hamiltonian has been approximated to quadratic powers of bosonic operators, it can be diagonalized. The Bogoliubov transformation defines new bosonic operators $b_{\boldsymbol{k}}$ and are related to the existing ones according to

$$\begin{aligned} a_{\boldsymbol{k}} &= e^{i\varphi}(\cosh \xi_{\boldsymbol{k}} b_{\boldsymbol{k}} + \sinh \xi_{\boldsymbol{k}} b^{\dagger}_{-\boldsymbol{k}}) \\ a^{\dagger}_{-\boldsymbol{k}} &= e^{-i\varphi}(\sinh \xi_{\boldsymbol{k}} b_{\boldsymbol{k}} + \cosh \xi_{\boldsymbol{k}} b^{\dagger}_{-\boldsymbol{k}}). \end{aligned} \tag{2.40}$$

The coefficients of the new bosonic operators are chosen such that the operators $a_{\boldsymbol{k}}$ always satisfy bosonic commutations as they should. We have assumed that the coefficients are symmetric: $\xi_{\boldsymbol{k}} = \xi_{-\boldsymbol{k}}$. This ensures that the new operators $b_{\boldsymbol{k}}$ satisfy bosonic commutation relations. Substituting (2.40) into (2.39), we obtain

$$H_{\mathrm{B}} = E_0 N + \frac{U_0 n N}{2} + \sum_{\boldsymbol{k}\neq 0} \Big\{ \big[\delta E_{\boldsymbol{k}} \cosh 2\xi_{\boldsymbol{k}} + U_0 n \sinh 2\xi_{\boldsymbol{k}} \big] b^{\dagger}_{\boldsymbol{k}} b_{\boldsymbol{k}}$$

$$\Big[\frac{\delta E_{\boldsymbol{k}}}{2} \sinh 2\xi_{\boldsymbol{k}} + \frac{gn}{2} \cosh 2\xi_{\boldsymbol{k}}\Big](b_{\boldsymbol{k}} b_{-\boldsymbol{k}} + b^{\dagger}_{\boldsymbol{k}} b^{\dagger}_{-\boldsymbol{k}}) + \delta E_{\boldsymbol{k}} \sinh^2 \xi_{\boldsymbol{k}} + \frac{U_0 n}{2} \sinh 2\xi_{\boldsymbol{k}} \Big\}, \tag{2.41}$$

where $\delta E_{\boldsymbol{k}} = E_{\boldsymbol{k}} - E_0 + U_0 n$. To diagonalize this Hamiltonian, we require setting the off-diagonal terms proportional to $b_{\boldsymbol{k}} b_{-\boldsymbol{k}} + b^{\dagger}_{\boldsymbol{k}} b^{\dagger}_{-\boldsymbol{k}}$ to zero. The condition for this is

$$\coth 2\xi_{\boldsymbol{k}} = -\frac{\delta E_{\boldsymbol{k}}}{U_0 n}. \tag{2.42}$$

Solving for the coefficients themselves, we take the solutions with signs

$$\begin{aligned} \cosh \xi_{\boldsymbol{k}} &= \sqrt{\frac{\delta E_{\boldsymbol{k}}}{2\epsilon_{\boldsymbol{k}}} + \frac{1}{2}} \\ \sinh \xi_{\boldsymbol{k}} &= -\sqrt{\frac{\delta E_{\boldsymbol{k}}}{2\epsilon_{\boldsymbol{k}}} - \frac{1}{2}}, \end{aligned} \tag{2.43}$$

where

$$\epsilon_{\boldsymbol{k}} = \sqrt{(E_{\boldsymbol{k}} - E_0)(E_{\boldsymbol{k}} - E_0 + 2U_0 n)}. \tag{2.44}$$

The diagonalized form of the Hamiltonian then reads

$$H_{\mathrm{B}} = E_0 N + \frac{U_0 n N}{2} + \sum_{\boldsymbol{k}\neq 0} \left\{ \epsilon_{\boldsymbol{k}} b^{\dagger}_{\boldsymbol{k}} b_{\boldsymbol{k}} + \left[\frac{\epsilon_{\boldsymbol{k}} - \delta E_{\boldsymbol{k}}}{2} \right] \right\}. \tag{2.45}$$

Now that the Hamiltonian is diagonalized, we can better understand the effect of the interactions. What (2.45) tells us is that the effect of the interactions is to make a new type of bosonic particle, called a *bogoliubovon*, which is annihilated by the operators $b_{\boldsymbol{k}}$. Since $b^{\dagger}_{\boldsymbol{k}} b_{\boldsymbol{k}}$ is just a number operator for these particles, the eigenstates of (2.45) are simply the Fock states with respect to bogoliubovons,

$$|\boldsymbol{n}_{\boldsymbol{k}}\rangle_b = \prod_{\boldsymbol{k}} \frac{(b^{\dagger}_{\boldsymbol{k}})^{n_{\boldsymbol{k}}}}{\sqrt{n_{\boldsymbol{k}}!}} |\tilde{0}\rangle, \tag{2.46}$$

where we used the same form as (1.29). The difference from (1.29) is that the ground state ($|\tilde{0}\rangle$ in this case) is not the state with no atoms, as it was before. It can explicitly be written as

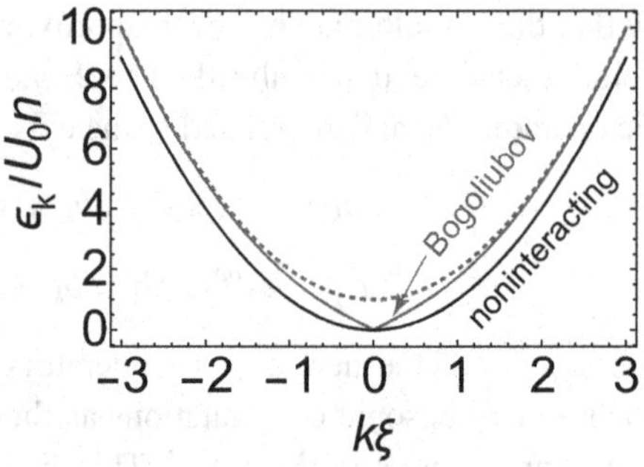

Fig. 2.4 The Bogoliubov dispersion $\epsilon_{\boldsymbol{k}}$ in (2.44). The momentum is in units of the healing length $\xi = \hbar/\sqrt{2mU_0 n}$, and the energy is in units of the interaction strength $U_0 n$. The original noninteracting energy dispersion $E_{\boldsymbol{k}} - E_0 = \frac{\hbar^2 k^2}{2m}$ and the limiting case for large k are also shown for comparison. Color version of this figure available at cambridge.org/quantumatomoptics.

$$|\tilde{0}\rangle = \frac{1}{\cosh \xi_{\boldsymbol{k}}} \exp(e^{2i\varphi} \tanh \xi_{\boldsymbol{k}} a_{\boldsymbol{k}}^\dagger a_{-\boldsymbol{k}}^\dagger)|0\rangle, \tag{2.47}$$

which satisfies

$$b_{\boldsymbol{k}}|\tilde{0}\rangle = 0. \tag{2.48}$$

The lowest energy state has no bogoliubovons and has an energy

$$H_{\mathrm{B}}|\tilde{0}\rangle = \left[E_0 N + \frac{U_0 n N}{2} + \sum_{\boldsymbol{k} \neq 0} \frac{\epsilon_{\boldsymbol{k}} - \delta E_{\boldsymbol{k}}}{2} \right] |\tilde{0}\rangle. \tag{2.49}$$

The state (2.47) in terms of the original $a_{\boldsymbol{k}}$ operators has a nonzero population in the excited state, as can be found by evaluating

$$\langle \tilde{0}| a_{\boldsymbol{k}}^\dagger a_{\boldsymbol{k}} |\tilde{0}\rangle = \sinh^2 \xi_{\boldsymbol{k}}. \tag{2.50}$$

We can interpret this as the interactions causing some of the nonzero momenta states to be excited, thereby lowering the energy of the whole system.

The elementary excitations have an energy dispersion according to (2.44) that is shown in Fig. 2.4. Comparing it to the standard parabolic dispersion, the Bogoliubov dispersion exhibits a linear dispersion at low momenta. Expanding the dispersion (2.44) for small k, we can write

$$\begin{aligned} \epsilon_{\boldsymbol{k}} &\approx \sqrt{2U_0 n (E_{\boldsymbol{k}} - E_0)} \qquad (|k\xi| < 1) \\ &= \sqrt{2} U_0 n \xi |k|, \end{aligned} \tag{2.51}$$

where in the second line we took the original dispersion to be $E_{\boldsymbol{k}} - E_0 = \frac{\hbar^2 k^2}{2m}$, and $\xi = \hbar/\sqrt{2mU_0 n}$ is the healing length of the BEC, which will be discussed more in Section 3.4.5. At larger momenta, the Bogoliubov dispersion again has a parabolic form, which can be approximated by

$$\epsilon_{\boldsymbol{k}} \approx E_{\boldsymbol{k}} - E_0 + U_0 n \qquad (|k\xi| \gg 1). \tag{2.52}$$

The parabolic dispersion is offset by a constant amount $U_0 n$. The approximation works well for momenta $k\xi \gg 1$.

Exercise 2.4.1 Verify that the $b_{\boldsymbol{k}}$ operators satisfy bosonic commutation relations $[b_{\boldsymbol{k}}, b^{\dagger}_{\boldsymbol{k}'}] = \delta_{\boldsymbol{k}\boldsymbol{k}'}$. Hint: First invert the relation (2.40) such that that $b_{\boldsymbol{k}}$ operators are written in terms of the $a_{\boldsymbol{k}}, a^{\dagger}_{-\boldsymbol{k}}$ operators. This is best done by multiplying the two equations by a suitable coefficient and taking advantage of the relation $\cosh^2 x - \sinh^2 x = 1$.

Exercise 2.4.2 Show that (2.47) is the vacuum for the Bogoliubov operators $b_{\boldsymbol{k}}$. Hint: Expand (2.47) as a Taylor series and apply the operator $b_{\boldsymbol{k}}$ written in terms of $a_{\boldsymbol{k}}, a^{\dagger}_{-\boldsymbol{k}}$ operators found in the previous exercise.

Exercise 2.4.3 Verify that the average number of $a_{\boldsymbol{k}}$ particles is (2.50). Try this two ways: First, evaluate it by substituting the state (2.47) into (2.50) and evaluating the Taylor expanded sum. Second, transform the $a_{\boldsymbol{k}}$ operators to $b_{\boldsymbol{k}}$ operators and use the fact that (2.48).

2.5 Superfluidity

The Bogoliubov distribution derived in Section 2.4 gives rise to one of the most remarkable effects seen in BECs: superfluidity. As the name suggests, superfluidity is the phenomenon of a fluid possessing zero viscosity, such that it may flow without resistance. For example, when arranged in a circular geometry, a superfluid can keep flowing around the loop indefinitely. In practice, once the fluid exceeds a particular velocity, the superfluid loses its properties and can no longer flow with zero viscosity. In this section, we derive Landau's criterion for this velocity, which gives insight to the reason why superfluidity occurs.

Consider a BEC that is moving with velocity $\boldsymbol{u}$ in the x-direction. Such a configuration might be prepared by first preparing a BEC such that the atoms condense into a zero-momentum state, and then momentum is added to all the atoms together such that the whole system is moving with velocity u. For simplicity, we assume that the temperature is zero, so there are no thermal excitations and the initial state is exactly in the ground state. Now consider two coordinate frames, one in the laboratory frame F' and another in the moving frame F, which is also moving in the x-direction with velocity u. Thus, with respect to frame F, the laboratory frame F' is moving with velocity $\boldsymbol{v} = -\boldsymbol{u}$ (see Fig. 2.5(a)). The coordinates between the two frames are related according to

$$\boldsymbol{x}' = \boldsymbol{x} + \boldsymbol{u}t, \tag{2.53}$$

where $\boldsymbol{x}, \boldsymbol{x}'$ are the coordinates in the frames F, F', respectively.

Given that in the frame F the dispersion relation is of a form $\epsilon_{\boldsymbol{k}}$, let us work out what the excitation spectrum is in the coordinates of F'. From general considerations of coordinate transformations, we know (see the Box in this section) that this must be

$$\boldsymbol{k}' = \boldsymbol{k} + \frac{m\boldsymbol{u}}{\hbar} \tag{2.54}$$

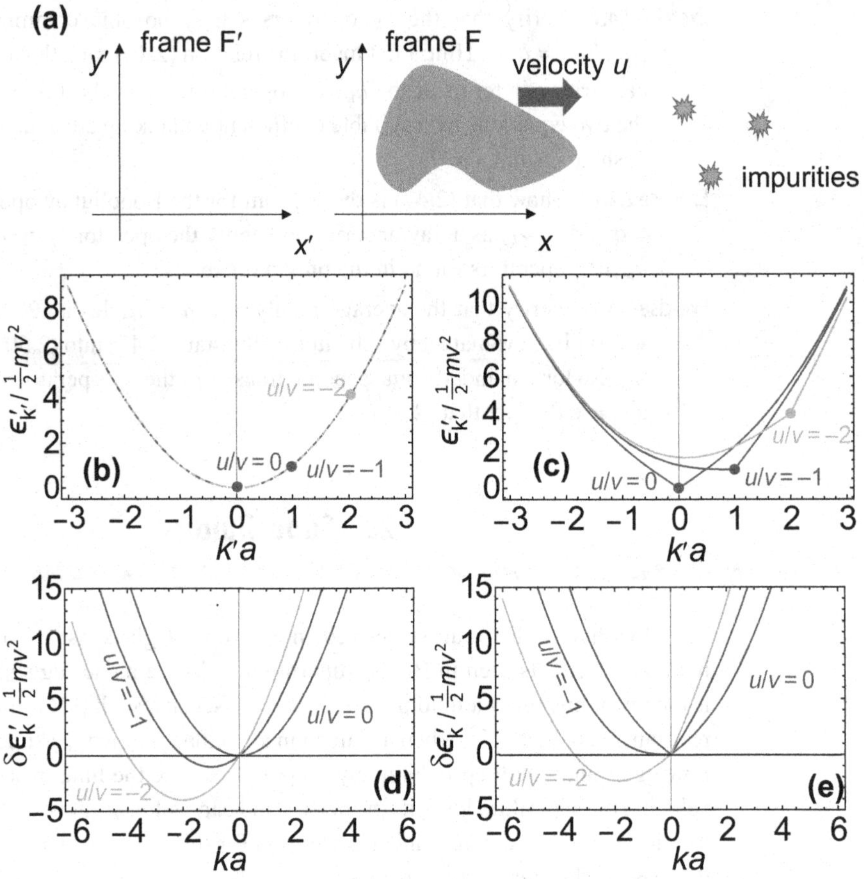

Fig. 2.5 Landau's argument for superfluidity. (a) Coordinate reference frames as defined in the text. Dispersion relations after a Galilean transformation for various velocities u and (b) a noninteracting standard dispersion with $U_0 n = 0$; (c) interacting Bogoliubov dispersion with $U_0 n = mv^2$. The excitation energies defined by (2.59) for the (d) noninteracting and (e) interacting case with the same parameters as (b) and (c), respectively. The length scale as set to be $a = \hbar/mv$, and the energy scale is $\frac{1}{2}mv^2$. The dots indicate the momentum of the zero momentum state $k = 0$ with respect to the frame F. Color version of this figure available at cambridge.org/quantumatomoptics.

$$\epsilon'_{\boldsymbol{k}} = \epsilon_{\boldsymbol{k}} + \hbar \boldsymbol{k} \cdot \boldsymbol{u} + \frac{1}{2} m u^2, \tag{2.55}$$

where $\epsilon'_{\boldsymbol{k}}$ is the energy dispersion in frame F', in terms of the momentum variables $\boldsymbol{k}$ of frame F.

First, let us consider the case of a regular dispersion without interactions $U_0 = 0$. In this case, the dispersion relation in frame F is

$$\epsilon_{\boldsymbol{k}} = \frac{\hbar^2 k^2}{2m}. \tag{2.56}$$

Substituting this relation into (2.54) and (2.55), we can find the dispersion relation in frame F' with the new coordinates, which are

$$\epsilon'_{k'} = \frac{\hbar^2 (k')^2}{2m}, \tag{2.57}$$

which takes exactly the same form. In this frame, the atoms are all moving with momentum

$$\boldsymbol{k}' = \frac{m\boldsymbol{u}}{\hbar}. \tag{2.58}$$

In Fig. 2.5(b), we plot the dispersion of the noninteracting case.

Now suppose that the BEC encounters a few impurities that have fixed positions in the laboratory frame F', and the BEC must flow around these. In a normal fluid ($U_0 = 0$), we expect that such impurities will create excitations such that the atoms no longer all have the momenta given by (2.58). The collision process should scatter many of the atoms to the opposite momenta (e.g., $\boldsymbol{k}' < 0$ for $u > 0$, which also implies $\boldsymbol{k} < 0$), so that eventually the fluid will slow down and finally stop.

Let us look at the energy cost of creating such an excitation in the frame F'. We compare the energy of no excitation (i.e., keeping a particular atom at $\boldsymbol{k} = 0$) and exciting it to some $|\boldsymbol{k}| > 0$. This energy difference, or excitation energy, is

$$\delta\epsilon_{\boldsymbol{k}} = \epsilon'_{\boldsymbol{k}} - \epsilon'_0 = \epsilon_{\boldsymbol{k}} - \epsilon_0 + \hbar \boldsymbol{k} \cdot \boldsymbol{u}. \tag{2.59}$$

A plot of this for the noninteracting case is shown in Fig. 2.5(d) (note that the horizontal axis is with respect to $\boldsymbol{k}$, not $\boldsymbol{k}'$). We see that the excitation energy becomes negative for a range of momenta with $\boldsymbol{k} < 0$, as expected. Since the excitation energy is negative, the system is susceptible to create many excitations with these momenta, which eventually slows and stops the BEC.

If we now repeat the argument for the interacting case, we see completely different behavior. In Fig. 2.5(c), the dispersion relation for the Bogoliubov dispersion derived in Section 2.4 is shown. In this case, the dispersion is not invariant, like in the noninteracting case. In the excitation energy plotted in Fig. 2.5(e), we observe that for the $u/v = -1$ curve, the excitation energy remains positive for all $\boldsymbol{k}$. This means that the system is not susceptible to the creation of many excitations. Without the creation of excitations, the atoms moving with velocity $\boldsymbol{u}$ are not scattered, and the movement of the atoms is unhindered. For the $u/v = -2$ curve, however, we see that the excitation energy is negative, and the system is again susceptible to the creation of excitations that eventually slow the system down.

The criterion for superfluidity is then given by the condition that (2.59) is positive for all momenta:

$$\epsilon_{\boldsymbol{k}} - \epsilon_0 + \hbar \boldsymbol{k} \cdot \boldsymbol{u} > 0. \tag{2.60}$$

As the velocity of the fluid $\boldsymbol{u}$ is increased, superfluidity eventually breaks down, as illustrated above. Let us find what this critical velocity is that marks this boundary. Setting the left-hand side of (2.60) to zero and rearranging, we have

$$u_c = \min_{\boldsymbol{k}} \frac{\epsilon_{\boldsymbol{k}} - \epsilon_0}{\hbar k}, \tag{2.61}$$

where we have taken the minimum of all the momenta that satisfy the boundary condition, in case there are multiple solutions. This is the famous *Landau's criterion*. This predicts the maximum velocity that a fluid can flow in the superfluid phase.

Galilean transformation of the Schrodinger equation

Consider the standard Schrodinger equation in a frame F defined with coordinates $\boldsymbol{x}$ and time t:

$$i\hbar\frac{\partial\psi(\boldsymbol{x},t)}{\partial t} = \left[-\frac{\hbar^2}{2m}\nabla_{\boldsymbol{x}}^2 + V(\boldsymbol{x})\right]\psi(\boldsymbol{x},t), \tag{2.62}$$

where $\nabla_{\boldsymbol{x}}$ indicates the gradient with respect to the variable $\boldsymbol{x}$. Now consider another frame F' that is moving with velocity $\boldsymbol{v}$ with constant velocity. The coordinates in this frame are defined by $\boldsymbol{x}' = \boldsymbol{x} - \boldsymbol{v}t$. Making the transformation, in the new coordinates the Schrodinger equation is written

$$\begin{aligned} i\hbar\frac{\partial\psi(\boldsymbol{x}'+\boldsymbol{v}t,t)}{\partial t} &- i\hbar\boldsymbol{v}\cdot\nabla_{\boldsymbol{x}'}\psi(\boldsymbol{x}'+\boldsymbol{v}t,t) \\ &= \left[-\frac{\hbar^2}{2m}\nabla_{\boldsymbol{x}'}^2 + V(\boldsymbol{x}'+\boldsymbol{v}t)\right]\psi(\boldsymbol{x}'+\boldsymbol{v}t,t), \end{aligned} \tag{2.63}$$

where we used

$$\begin{aligned} \psi(\boldsymbol{x},t) &= \psi(\boldsymbol{x}'+\boldsymbol{v}t,t) \\ \frac{\partial\psi(\boldsymbol{x},t)}{\partial t} &= \frac{\partial\psi(\boldsymbol{x}'+\boldsymbol{v}t,t)}{\partial t} - \boldsymbol{v}\cdot\nabla_{\boldsymbol{x}'}\psi(\boldsymbol{x}'+\boldsymbol{v}t,t) \\ \nabla_{\boldsymbol{x}}\psi(\boldsymbol{x},t) &= \nabla_{\boldsymbol{x}'}\psi(\boldsymbol{x}'+\boldsymbol{v}t,t). \end{aligned} \tag{2.64}$$

In the new frame F', the particle must also obey the Schrodinger equation, which must take the form

$$i\hbar\frac{\partial\psi'(\boldsymbol{x}',t)}{\partial t} = \left[-\frac{\hbar^2}{2m}\nabla_{\boldsymbol{x}'}^2 + V'(\boldsymbol{x}')\right]\psi'(\boldsymbol{x}',t). \tag{2.65}$$

The same physical wavefunction $\psi'(\boldsymbol{x}',t)$ in the frame F' will take a different form from that in the original frame F, since it is described with different variables. For example, the momentum and energy will not be measured to be the same values. We can relate the same wavefunction in the two coordinates using the transformation

$$\psi(\boldsymbol{x}'+\boldsymbol{v}t,t) = e^{i(m\boldsymbol{v}\cdot\boldsymbol{x}'+mv^2t/2)/\hbar}\psi'(\boldsymbol{x}',t). \tag{2.66}$$

One can verify that substitution of (2.66) into (2.63) yields (2.65), where we have defined $V'(\boldsymbol{x}') = V(\boldsymbol{x}'+\boldsymbol{v}t)$.

For example, consider a particle in a plane wave in the frame F with the wavefunction

$$\psi(\boldsymbol{x},t) = e^{i(\boldsymbol{p}\cdot\boldsymbol{x}-Et)/\hbar}. \tag{2.67}$$

Using the formula (2.66), we find that the wavefunction in the frame F' is

$$\psi'(\boldsymbol{x}',t) = e^{i[(\boldsymbol{p}-m\boldsymbol{v})\cdot\boldsymbol{x}'-(E-\boldsymbol{p}\cdot\boldsymbol{v}+mv^2/2)t]/\hbar}. \tag{2.68}$$

Hence, the momentum $\boldsymbol{p}'$ and energy E' in the frame F' is related to the original values as

$$\begin{aligned} \boldsymbol{p}' &= \boldsymbol{p} - m\boldsymbol{v} \\ E' &= E - \boldsymbol{p}\cdot\boldsymbol{v} + \frac{1}{2}mv^2. \end{aligned} \tag{2.69}$$

Exercise 2.5.1 Verify that substitution of (2.66) into (2.63) yields (2.65).

Exercise 2.5.2 For the Bogoliubov dispersion, find the maximum velocity that a superfluid can flow according to Landau's criterion, assuming the linear dispersion (2.51).

2.6 References and Further Reading

- Section 2.1: To learn more about Bose–Einstein condensates in general, see the textbooks [375, 369].
- Section 2.2: The original papers detailing Bose and Einstein's original argument [51, 134].
- Section 2.3: The first experiments showing Bose–Einstein condensation are [15, 110]. Other early experiments are reported in [153, 248, 508, 179, 271]. Reviews and books relating to the topic are given in [17, 375, 369, 180, 101, 295].
- Section 2.4: Bogoliubov's original theory of low-energy excited states of a BEC [49]. Experimental observation [441]. Review articles and books on the topic [375, 295, 369].
- Section 2.5: The original theories of superfluidity [285, 49, 170]. Experimental observation [319]. Review articles and books on the topic [375, 260, 458, 294].

3 The Order Parameter and Gross–Pitaevskii Equation

3.1 Introduction

We have seen that a Bose–Einstein condensate (BEC) can be described by a quantum many-body state of bosons, with a macroscopic occupation of the ground state. While this a mathematically complete framework to describe the system, it is also rather difficult to visualize, since it is inherently involves many particles. In many situations, it is useful to approximately capture the essential physics without having all the details that describe the system. In this respect, the order parameter, and its equation of motion, the Gross–Pitaevskii equation, is a very popular framework to describe a BEC, as it gives a simple way to visualize the state. In this chapter, we describe these and related concepts such as the healing length, vortices, solitons, and hydrodynamic equations.

3.2 Order Parameter

In the previous chapters, we established the essential feature of a BEC, that there is a macroscopic occupation of the ground state. What is the wavefunction of this state? Assuming there are no interactions between the bosons, we can easily write this using the methods that we worked out in Chapter 1. Using the notation of (1.29), we have

$$|\text{BEC}(N)\rangle = |N, 0, \dots\rangle = \frac{(a_0^\dagger)^N}{\sqrt{N!}}|0\rangle. \tag{3.1}$$

Simple enough! Of course, this is an idealization in that we have assumed that all the particles occupy the ground state, which is equivalent to zero temperature. As we have seen in Section 2.4, more realistically there will be some fraction of the bosons occupying the excited states, although most of the bosons will be in the ground state.

Recall that the meaning of a_0 was actually in terms of a single particle wavefunction, which were eigenstates of the potential $V(\boldsymbol{x})$ (see Eq. (1.8)). Thus, in terms of position space, the BEC wavefunction is

$$|\text{BEC}(N)\rangle = \frac{1}{\sqrt{N!}}\left(\int d\boldsymbol{x}\,\psi_0(\boldsymbol{x})a^\dagger(\boldsymbol{x})\right)^N |0\rangle. \tag{3.2}$$

This is obviously N bosons, all with the wavefunction $\psi_0(\boldsymbol{x})$. Since all the bosons have the same wavefunction, it is tempting to define a single macroscopic wavefunction $\Psi(\boldsymbol{x})$, which captures the wavefunction of the whole BEC. This is the idea of the *order parameter*, which in this case would be

$$\Psi(\boldsymbol{x}) = \sqrt{N}\psi_0(\boldsymbol{x}). \tag{3.3}$$

By convention, this is normalized such that when the order parameter is integrated, we obtain the total number of particles in the ground state

$$\int d\boldsymbol{x}|\Psi(\boldsymbol{x})|^2 = N. \tag{3.4}$$

We note that the idea of the order parameter only makes sense when a large number of particles are all occupying the same state, which is not usually what happens in a typical macroscopic state. Taking the example of a general thermal state above the BEC critical temperature, we would expect bosons occupying all kinds of energy states in Fig. 1.2. This means that all the particles would have different wavefunctions $\psi_k(\boldsymbol{x})$, and there would not be any kind of common wavefunction that we can define for the particles. But in this case, since all (or at least most) of the particles have the same wavefunction, we can think of a common giant macroscopic wavefunction for the BEC.

For the case that we wrote above, it was clear already what the answer should be: It is the same wavefunction as the underlying particles. But what if we had a more complicated state, perhaps under more realistic conditions where the particles interact with each other? What would be better is if we could calculate the order parameter given some arbitrary state $|\psi\rangle$. Although the order parameter is a natural idea, it is not completely obvious what the general definition should be. Suppose we tried the most obvious thing, which is to measure the average position of the particles:

$$\langle\psi|a^\dagger(\boldsymbol{x})a(\boldsymbol{x})|\psi\rangle = |\psi_0(\boldsymbol{x})|^2. \tag{3.5}$$

We could take the square root of this to obtain the magnitude of the wavefunction, but we have completely lost all the phase component. Since the phase is an important part of the wavefunction in general, this doesn't really work as a good definition of the order parameter.

To keep the phase information, suppose we define it instead as

$$\Psi(\boldsymbol{x}) = \langle\psi(N-1)|a(\boldsymbol{x})|\psi(N)\rangle, \tag{3.6}$$

where $|\psi(N)\rangle$ is the state of the BEC with N particles, and $|\psi(N-1)\rangle$ is the state of the BEC with $N-1$ particles. This is, of course, not a regular expectation value, since usually we use the same state for both the bra and ket sides of the expectation value. This is necessary, however, as $a(\boldsymbol{x})$ will reduce the number of bosons by one, and otherwise we will immediately obtain zero. What this means is that to explicitly work out the order parameter, we have to not only know the wavefunction of the state, but also the version of the wavefunction with one fewer particle. This is not necessarily completely trivial to obtain, which makes the definition not completely practical in every case. Nevertheless, for simple cases you can check that this gives the desired result including the phase (see Ex. 2.4.1). The benefit of defining the order parameter in this way is that this will work for a more general case, even if the many-body state is rather complicated.

An equivalent way to view this is by using the expansion of the boson operators (1.32) in terms of a complete set of states:

$$a(\boldsymbol{x}) = \psi_0(\boldsymbol{x})a_0 + \sum_{k\neq 0} \psi_k(\boldsymbol{x})a_k. \tag{3.7}$$

If we assume a BEC, then we have a large population of bosons, all occupying the same state as in (3.1). Applying (3.7) to (3.1), we obtain

$$a(\boldsymbol{x})|\text{BEC}(N)\rangle = \sqrt{N}\psi_0(\boldsymbol{x})|\text{BEC}(N-1)\rangle. \tag{3.8}$$

If the state with N and $N-1$ particles is not considerably different, and anticipating that every a_0 will give a factor of $\sqrt{N}$, we can approximately write

$$a(\boldsymbol{x}) \approx \Psi(\boldsymbol{x}) + \sum_{k\neq 0} \psi_k(\boldsymbol{x})a_k. \tag{3.9}$$

This is called the Bogoliubov approximation, and it amounts to treating the macroscopic mode a_0 classically (i.e. ignoring the commutation relations of the bosonic operators).

In the approximation (3.9), we used the same wavefunctions $\psi_k(\boldsymbol{x})$, which in the context of Chapter 1 were the eigenstates of the potential $V(\boldsymbol{x})$. But there is no rule to say that we have to use this set of states; we could equally take another set of complete states. As we will see in Section 3.3, the presence of interactions and other effects can modify the order parameter, meaning that the expansion (3.7) can be in terms of another basis that is not necessarily the same as the eigenstates of the potential $V(\boldsymbol{x})$.

Exercise 3.2.1 (a) Find $[a(\boldsymbol{x}), a_0^\dagger]$. (b) Using your result in (a), verify (3.3) by substituting (3.1) into (3.6).

3.3 The Gross–Pitaevskii Equation

In Section 3.2, we defined the order parameter of the BEC, which gives an effective single particle wavefunction of the system. While we wrote a definition in terms of the entire N-particle macroscopic wavefunction of the whole system, it would useful to derive a self-contained equation of motion for the order parameter itself. We could then completely bypass working out the N-particle wavefunction of the whole system, which can be difficult to find.

We start with the total many-particle Hamiltonian (1.42), which we repeat here for convenience:

$$H = \int dx a^\dagger(\boldsymbol{x}) \left[-\frac{\hbar^2}{2m}\nabla^2 + V(\boldsymbol{x}) \right] a(\boldsymbol{x}) + \frac{U_0}{2} \int dx n(\boldsymbol{x})(n(\boldsymbol{x}) - 1), \tag{3.10}$$

where we have used the s-wave scattering interaction (1.46) with $U_0 = \frac{4\pi\hbar^2 a_s}{m}$. The Heisenberg equation of motion for the operator $a(\boldsymbol{x})$ can be written

$$
\begin{aligned}
i\hbar \frac{\partial a(\boldsymbol{x},t)}{\partial t} &= [a(\boldsymbol{x},t), H] \\
&= \int dx \left[-\frac{\hbar^2}{2m}\nabla^2 + V(\boldsymbol{x}) + U_0 n(\boldsymbol{x},t) \right] a(\boldsymbol{x},t). \qquad (3.11)
\end{aligned}
$$

Now let us suppose that we have a state of the form (3.1), where all N bosons are occupying the same state. We write the macroscopically occupied state as

$$
a_0 = \frac{1}{\sqrt{N}} \int dx \Psi^*(\boldsymbol{x}) a(\boldsymbol{x}), \qquad (3.12)
$$

where $\Psi(\boldsymbol{x})$ is the order parameter that we are trying to find an equation of motion for. The factor of $\frac{1}{\sqrt{N}}$ is there because by convention the order parameter is normalized as $\int dx |\Psi(\boldsymbol{x})|^2 = N$. Using the prescription (3.6) by applying $|\psi(N)\rangle$ on the right and $|\psi(N-1)\rangle$ on the left for the left-hand side and the first two terms in (3.11), we can immediately obtain an expression involving the order parameter. For the interaction term, we can evaluate

$$
\langle \psi(N-1) | n(\boldsymbol{x}) a(x) | \psi(N) \rangle = |\Psi(\boldsymbol{x})|^2 \Psi(\boldsymbol{x}). \qquad (3.13)
$$

We can then obtain the Gross–Pitaevskii equation,

$$
i\hbar \frac{\partial \Psi(\boldsymbol{x},t)}{\partial t} = \left(-\frac{\hbar^2}{2m}\nabla^2 + V(\boldsymbol{x}) + U_0 |\Psi(\boldsymbol{x})|^2 \right) \Psi(\boldsymbol{x},t). \qquad (3.14)
$$

This has a similar form to the Schrodinger equation, except with the presence of a nonlinear interaction term proportional to U_0.

Several approximations have been introduced in deriving (3.14). First, we started with a BEC wavefunction of the form (3.1), which assumes that all the particles are in the ground state. This obviously neglects thermal effects, so it corresponds to a zero temperature approximation. Furthermore, (3.1) does not properly account for interactions, in the sense that it is of a form where all the bosons are in a simple product state. By assuming a wavefunction of the form (3.1), these are only included at the level of mean field theory. In addition, in (3.10) an s-wave scattering interaction was included, which is an approximation to a more realistic interatomic potential $U(\boldsymbol{x}, \boldsymbol{y})$ between particles. This amounts to the assumption that the order parameter varies slowly over the distances in the range of the interatomic potential, hence for phenomena with distances shorter than the scattering length, it is not valid to use the Gross–Pitaevskii equation.

Exercise 3.3.1 Verify (3.11) using the commutation relations for the bosonic operators.

3.4 Ground State Solutions of the Gross–Pitaevskii Equation

It is instructive to find the solution of the Gross–Pitaevskii equation for some simple examples to get a feel for how it works. Unfortunately, solving for the solution of the Gross–Pitaevskii equation is technically not quite as easy as the Schrodinger equation, due to the presence of the nonlinear term. Only very specific cases have exact solutions, which we will discuss in later sections. In many cases, we must turn to numerical methods to obtain a solution.

3.4.1 Stationary Solutions

The Gross–Pitaevskii equation possess stationary solutions much in the same way that the Schrodinger equation possess solutions that do not evolve in time up to a phase factor. Here we will be concerned with stationary solutions of the lowest energy, which correspond to the BEC state. As such, we will look for solutions of

$$\mu\Psi(\boldsymbol{x}) = \left(-\frac{\hbar^2}{2m}\nabla^2 + V(\boldsymbol{x}) + U_0|\Psi(\boldsymbol{x})|^2\right)\Psi(\boldsymbol{x}), \tag{3.15}$$

where μ is the chemical potential. Note that we use the term "chemical potential" rather than "energy" here, even though they play a rather similar role. The chemical potential is defined as

$$\mu = \frac{\partial E}{\partial N}, \tag{3.16}$$

where E is the total energy of the whole system. Thus, the chemical potential is the energy needed to add an extra particle into the system. It is roughly speaking the energy per particle, rather than the energy of the whole system together. Once the stationary solution is found, the time evolution works in exactly the same way as the Schrodinger equation

$$\Psi(\boldsymbol{x}, t) = \Psi(\boldsymbol{x})e^{-i\mu t/\hbar}. \tag{3.17}$$

3.4.2 No Confining Potential

The simplest analytical case is when there is no confining potential at all, $V(\boldsymbol{x}) = 0$. In this case, the stationary solutions can be written

$$\Psi(\boldsymbol{x}) = \sqrt{\frac{N}{V}}e^{i\boldsymbol{k}\cdot\boldsymbol{x}}, \tag{3.18}$$

where N is the number of atoms, V is the volume (assumed to extend to infinity), $\boldsymbol{k}$ is a wavenumber that can be chosen arbitrarily, and the chemical potential is

$$\mu = \frac{\hbar^2 k^2}{2m} + U_0 n_0, \tag{3.19}$$

where we define the average density to be

$$n_0 = \frac{N}{V}. \tag{3.20}$$

In this case, the solution of the Gross–Pitaevskii equation is identical to the Schrodinger equation, with the exception of the extra interaction term in the energy. In the same way as the Schrodinger equation, the solutions (3.18) are stationary solutions in the sense that the time evolution does not change the order parameter up to a global phase. This means that in principle, all the $\boldsymbol{k}$ are stable solutions in terms of the Gross–Pitaevskii dynamics. However, in the context of a BEC, usually the most relevant solution is the lowest energy one, since this is the state that is thermodynamically stable, according to the discussion in Chapter 2. Thus, we would typically say that the solution is just the $\boldsymbol{k} = 0$ case:

$$\Psi(\boldsymbol{x}) = \sqrt{n_0}. \tag{3.21}$$

This is not to say that the $\boldsymbol{k} > 0$ solutions are impossible; they would simply be described as not being in equilibrium. Such states would have to be prepared in a special way such as to favor the formation of macroscopic occupation at nonzero momentum.

3.4.3 Infinite Potential Well

For a less trivial example, we need to include a nonuniform potential $V(\boldsymbol{x})$. The obvious similarity to the Schrodinger equation suggests that we should look at the same classic example that is studied in any introductory quantum mechanics course: the one-dimensional infinite potential well, defined as

$$V(x) = \begin{cases} 0 & 0 \leq x \leq L \\ \infty & \text{otherwise} \end{cases}, \tag{3.22}$$

where L is the width of the well. This is actually not particularly realistic for a BEC for several reasons: BECs do not strictly even exist in one dimension, and producing a sharp potential that goes from zero to infinity abruptly is virtually impossible in the lab. Nevertheless, the point of this section is to get a feel for the similarities and differences from the standard Schrodinger equation.

Numerically, the Gross–Pitaevskii equation can be solved by discretizing space on a lattice and time-evolving the order parameter step by step in time. We make the approximations

$$\begin{aligned} \Psi(x_n, t_m) &\approx \Psi_{n,m} \\ \frac{\partial^2 \Psi(x_n, t_m)}{\partial x^2} &\approx \frac{\Psi_{n-1,m} - 2\Psi_{n,m} + \Psi_{n+1,m}}{\Delta x^2} \\ \frac{\partial \Psi(x_n, t_m)}{\partial t} &\approx \frac{\Psi_{n,m+1} - \Psi_{n,m}}{\Delta t}, \end{aligned} \tag{3.23}$$

where $x_n = n\Delta x$, $t_m = m\Delta t$. Substituting this into (3.14), one obtains

$$\Psi'_n = \Psi_n - \frac{i\Delta t}{\hbar}\left[-\frac{\hbar^2}{2m\Delta x^2}(\Psi_{n-1} - 2\Psi_n + \Psi_{n+1}) + V(x_n)\Psi_n + g|\Psi_n|^2\Psi_n\right], \tag{3.24}$$

where we have written $\Psi'(x_n) = \Psi_{n,m+1}$ and $\Psi(x_n) = \Psi_{n,m}$, which makes it clear that the order parameter can be written as a recursion relation in a vector containing the spatial distribution of the order parameter at a particular time. In practice, one only needs to keep the order parameter at one time instant, and it can be discarded once the next time evolution is evaluated.

The discrete Gross–Pitaevskii equation (3.24) time-evolves the order parameter. In the same way as the Schrodinger equation does not converge to a solution by simple time evolution since it is unitary, neither does the Gross–Pitaevskii equation. Thus, if we put a random order parameter in (3.24) and time-evolve it, it would not give the solution for the lowest energy state associated with a BEC. A simple trick allows us to obtain the lowest energy solution, which works rather well in practice. Making the replacement $t \to -i\tau$ changes the unitary dynamics of Hamiltonian evolution to nonunitary dissipation:

$$e^{-iHt/\hbar} \to e^{-H\tau/\hbar}. \tag{3.25}$$

Expanding in the eigenstates of $H = \sum_n E_n|E_n\rangle\langle E_n|$, this exponentially dampens high-energy states, and the lowest energy state (i.e., the ground state) will have the largest relative amplitude. Thus, by simply removing the factor of i in (3.24), we can switch to imaginary time evolution and drive the order parameter towards the lowest energy solution!

Figure 3.1 shows the numerical evolution of the Gross–Pitaevskii equation for the two-dimensional infinite well of dimension $L \times L$. For zero interaction $U_0 = 0$, the Gross–Pitaevskii equation is exactly equivalent to the Schrodinger equation, hence the spatial part of the order parameter takes the form

$$\Psi(\boldsymbol{x}) = \sqrt{\frac{4N}{L^2}} \sin\left(\frac{\pi x}{L}\right) \sin\left(\frac{\pi y}{L}\right). \tag{3.26}$$

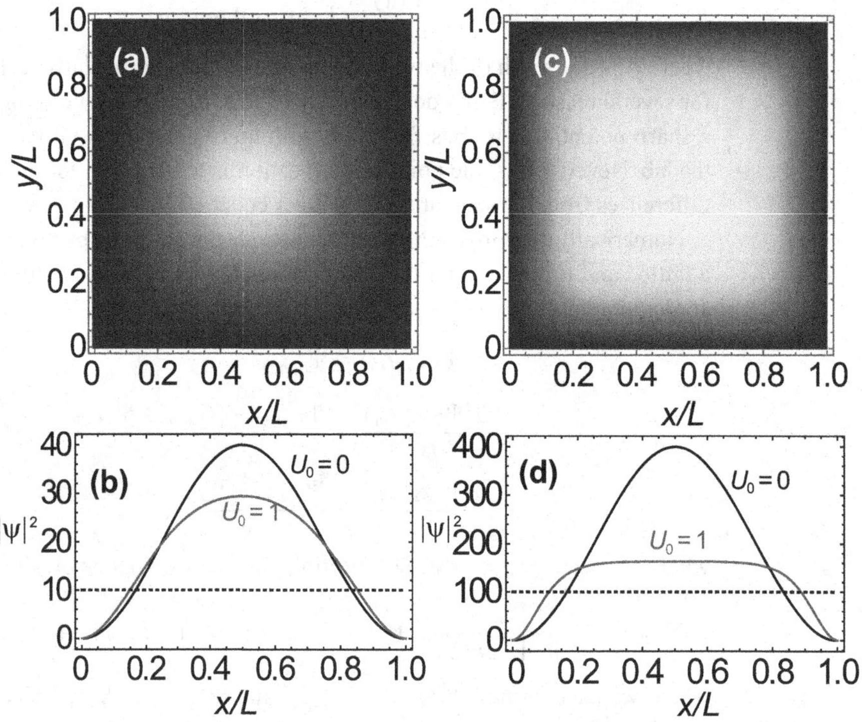

Fig. 3.1 Solution of Gross–Pitaevskii equation for a two-dimensional infinite square well with interaction U_0 as marked. Parameters used are (a) $N/L^2 = 10$ with $U_0/E_0 = 0$; (b) $N/L^2 = 10$ and interactions as marked for $y = 0$; (c) $N/L^2 = 100$ with $U_0/E_0 = 1$; and (d) $N/L^2 = 100$ and interactions as marked for $y = 0$. Solutions are numerically evolved under imaginary time for $t = 0.1\tau$. The initial conditions are (3.26) for each case. Parameters used are $L = 1$, where the units are $E_0 = \frac{\hbar^2}{2mL^2}$, $\tau = \hbar/E_0$. Dotted line shows the Thomas–Fermi approximation. Color version of this figure available at cambridge.org/quantumatomoptics.

Evolving this state under imaginary time evolution gives the solution including interactions. From Fig. 3.1(b), we see that the wavefunction becomes broadened out when interactions are included. One can interpret this to be the effect of the particles minimizing energy by spreading themselves out more than the case without interactions. As the interactions are increased, the order parameter is more evenly distributed within the allowed space. When the density is increased, the effect of the interactions becomes stronger, as can be seen in Fig. 3.1(c). This is because the higher density allows for more opportunity for particles to interact with each other.

3.4.4 Thomas–Fermi Approximation

Due to the quadratic nature of the interaction term $U_0|\Psi(\boldsymbol{x})|^2$, and remembering that the order parameter has a magnitude $\sim \sqrt{N}$, we can see that for high densities, the interaction term tends to dominate more and more. In fact, in the limit of very high density, the kinetic energy term plays a smaller and smaller role. The time-independent Gross–Pitaevskii equation (3.15) can then be approximated by

$$\mu\Psi(\boldsymbol{x}) = \left(V(\boldsymbol{x}) + U_0|\Psi(\boldsymbol{x})|^2\right)\Psi(\boldsymbol{x}), \tag{3.27}$$

which has the solution

$$\Psi(\boldsymbol{x}) = \sqrt{\frac{\mu - V(\boldsymbol{x})}{U_0}}. \tag{3.28}$$

This is called the Thomas–Fermi limit.

Figure 3.1(b) and (d) shows comparisons of the Thomas–Fermi limit versus exact stationary solutions. In Fig. 3.1(b), we see the results for the low-density case. In this case, we see that the Thomas–Fermi approximation does not work very well, since the kinetic energy term still plays an important role. As the density is increased in Fig. 3.1(d), the Thomas–Fermi approximation increasingly becomes a better approximation, with the numerically evaluated distribution flattening out. In the limit of large density or high interactions, the Thomas–Fermi approximation gives a more accurate expression for the ground state distribution.

3.4.5 Healing Length

We saw from Fig. 3.1 that as the density is increased, this has the effect of spreading the order parameter more evenly, due to the repulsive interactions. This would be true also if the interactions were increased and the density is kept constant, since the last term in the Gross–Pitaevskii equation (3.15) is the product of the interaction and the density. In Fig. 3.1(c) and (d), we see that at the edges of the well, the order parameter is zero according to the boundary conditions and approaches a relatively flat region, particularly for the higher densities. We can estimate the distance that this transition occurs, which is commonly called the *healing length*.

Let us assume that the infinite potential well has a very large area, and we consider just the region near the walls, away from any corners: Take for example the middle of the left wall in Fig. 3.1(c). We expect that as we move away from the walls, the BEC will have a similar behavior to the zero potential case $V(\boldsymbol{x}) = 0$, where the effect of the boundary is not noticeable. As we saw in (3.19), in this case for the ground state

$\boldsymbol{k} = 0$, the chemical potential is $\mu = U_0 n_0$. Substituting this into the stationary state (3.15), we obtain the equation

$$\frac{\hbar^2}{2m}\frac{d^2\Psi}{dx^2} = U_0(|\Psi(x)|^2 - n_0)\Psi(x). \tag{3.29}$$

We would like to solve this for the boundary condition that $\Psi(0) = 0$. One can easily verify that a solution of this equation with this boundary condition takes a form

$$\Psi(x) = \sqrt{n_0}\tanh\left(\frac{x}{\sqrt{2}\xi}\right), \tag{3.30}$$

where the healing length is defined as

$$\xi = \frac{\hbar}{\sqrt{2mU_0 n_0}}. \tag{3.31}$$

Physically, we can interpret this to be the length when the kinetic energy $\frac{\hbar^2}{2m\xi^2}$ is equal to to the interaction energy $U_0 n_0$. Calculating the healing length for Fig. 3.1(a) and (c), we get $\xi/L = \sqrt{\frac{E_0}{U_0 n_0}}$ to be 0.32 and 0.1, respectively, which matches the numerical calculation.

Exercise 3.4.1 Using your favorite programming language, code the recursive equation (3.24) and implement the discrete Gross–Pitaevskii equation for the potential $V(x) = x^2$. Compare the dynamics, including the factor of i (corresponding to real-time dynamics) and removing the i (imaginary time dynamics). Experiment to see the various solutions for different parameters.

Exercise 3.4.2 Check that (3.30) is actually a solution of the equation (3.29).

3.5 Hydrodynamic Equations

From our discussion in Section 2.5, we have seen that BECs can be thought of being a fluid with special properties, such as dissipationless flow. The Gross–Pitaevskii equation can be rewritten in a way that makes the analogy with fluids explicit, by writing it as a hydrodynamic equation. The first step is to parametrize the order parameter according to

$$\Psi(\boldsymbol{x}) = \sqrt{n(\boldsymbol{x})}e^{i\phi(\boldsymbol{x})}, \tag{3.32}$$

where

$$\begin{aligned} n(\boldsymbol{x}) &= |\Psi(\boldsymbol{x})|^2 \\ \phi(\boldsymbol{x}) &= \arg(\Psi(\boldsymbol{x})) \end{aligned} \tag{3.33}$$

is a spatially dependent density and phase of the order parameter, respectively. To describe the flow of the fluid, we can define the current

$$\boldsymbol{j}(\boldsymbol{x}) = -\frac{i\hbar}{2m}\left(\Psi^*(\boldsymbol{x})\boldsymbol{\nabla}\Psi(\boldsymbol{x}) - \Psi(\boldsymbol{x})\boldsymbol{\nabla}\Psi^*(\boldsymbol{x})\right) \tag{3.34}$$

$$= n(\boldsymbol{x})\frac{\hbar}{m}\boldsymbol{\nabla}\phi(\boldsymbol{x}), \tag{3.35}$$

where in the second line we used the parametrization (3.32). You may wonder why the current is written in the form (3.34). A simple way to understand this is that it can be equivalently written $\boldsymbol{j}(\boldsymbol{x}) = \frac{1}{2m}(\Psi^*(\boldsymbol{x})\boldsymbol{p}\Psi(\boldsymbol{x}) - \Psi(\boldsymbol{x})\boldsymbol{p}\Psi^*(\boldsymbol{x}))$, where $\boldsymbol{p} = -i\hbar\boldsymbol{\nabla}$ is the momentum operator. Therefore, the average current is

$$\langle \boldsymbol{j}(\boldsymbol{x})\rangle = \frac{\langle \boldsymbol{p}\rangle}{m}. \tag{3.36}$$

It is tempting to say that, classically, $\boldsymbol{p} = m\boldsymbol{v}$, and hence the right-hand side of (3.36) is a velocity. This would be correct if $\Psi(\boldsymbol{x})$ was a single particle wavefunction. However, since $\Psi(\boldsymbol{x})$ is an order parameter that is normalized to N rather than 1, $\langle \boldsymbol{j}(\boldsymbol{x})\rangle$ corresponds to a total current of all the particles combined. To obtain the local velocity, we must divide (3.34) by the local density, giving

$$\boldsymbol{v}(\boldsymbol{x}) = \frac{\hbar}{m}\boldsymbol{\nabla}\phi(\boldsymbol{x}), \tag{3.37}$$

Since for any scalar field f the curl of the gradient is zero $\boldsymbol{\nabla}\times(\boldsymbol{\nabla} f) = 0$, we then have

$$\boldsymbol{\nabla}\times\boldsymbol{v}(\boldsymbol{x}) = 0. \tag{3.38}$$

This has the physical interpretation that a superfluid is irrotational. This means that when traversing a closed loop, the particles do not experience a net rotation. For example, irrotational flow is like a carriage in a Ferris wheel, which keeps the same direction with respect to the Earth as it goes around a loop. This is in contrast to rotational flow, where a person sitting in a car on a roller coaster loop would rotate once with each revolution.

By multiplying (3.14) by Ψ^* and subtracting the complex conjugate, we obtain

$$\boldsymbol{\nabla}\cdot\boldsymbol{j}(\boldsymbol{x}) = -\frac{\partial n(\boldsymbol{x})}{\partial t}, \tag{3.39}$$

which is called the continuity equation. This is a statement of the conservation of the amount of fluid. If there is fluid flow out of a volume (the divergence of the current is positive – the left-hand side), then this must be accompanied by a decrease in the amount of fluid within the volume (the right-hand side). We can obtain another equation involving the phase by substituting (3.32) into the Gross–Pitaevskii equation (3.14):

$$\hbar\frac{\partial\phi(\boldsymbol{x})}{\partial t} + \frac{m}{2}\boldsymbol{v}^2(\boldsymbol{x}) + V(\boldsymbol{x}) + U_0 n(\boldsymbol{x}) - \frac{\hbar^2(\nabla^2\sqrt{n(\boldsymbol{x})})}{2m\sqrt{n(\boldsymbol{x})}} = 0. \tag{3.40}$$

The two equations (3.39) and (3.40) are an equivalent set of equations that has the same mathematical content as the Gross–Pitaevskii equation (3.14).

Exercise 3.5.1 Derive equations (3.39) and (3.40) using the Gross–Pitaevskii equation.

3.6 Excited State Solutions of the Gross–Pitaevskii Equation

Section 3.5 examined the ground state solutions of the Gross–Pitaevskii equation. It is also possible to examine particular excited states using the same equations. It is important to understand that the types of states that the Gross–Pitaevskii equation considers are special types of states where there is a *macroscopic occupation* of all the bosons in the system. We argued this from the point of view that Bose–Einstein condensation should occur at low enough temperatures, so that for the ground state this is a reasonable picture to have. In general, for excited states the assumption of macroscopic occupation is generally not true: A proper treatment should be described more along the lines of Chapter 1, where the full quantum many-body wavefunction is evaluated. Thus, the excited states of the Gross–Pitaevskii equation only describe specifically those states that are macroscopically populated. You may then be worried whether examining such states would have any relevance at all to states that are seen in the lab. One way to get around this is to specifically arrange the system such that the excited state of interest is the ground state of a different Hamiltonian – this is what is done for the case of vortices, as will be explained in the next section. It turns out that such states can also be prepared under nonequilibrium situations, so that the excited state solutions of the Gross–Pitaevskii equation are quite relevant in practice. We will examine two examples of such excited states: vortices and solitons.

3.6.1 Vortices

If you are given a cup of water that initially has zero flow everywhere (e.g., there are no convection currents inside, or any other kind of flow within the cup), one of the simplest ways to get it moving is to stir it. A BEC is no different, and it is possible to create states where there is a current moving circularly (i.e., vortex flow). Unlike a regular fluid, where eventually frictional forces cause the vortices to slow down and disappear, in a superfluid it is possible to have vortices that never dissipate energy. These configurations are also solutions of the Gross–Pitaevskii equation and have interesting properties that are simple examples of a topological state.

To find these solutions, let us revisit the stationary Gross–Pitaevskii equation (3.15) and examine it in polar coordinates for two dimensions. For the potential, it is natural to consider a radially symmetric case, $V(\boldsymbol{x}) = V(r)$, since we are looking for vortices. The equation then reads

$$\mu\Psi(r,\theta) = -\frac{\hbar^2}{2m}\left(\frac{\partial^2\Psi}{\partial r^2} + \frac{1}{r}\frac{\partial\Psi}{\partial r}\right) + \frac{1}{r^2}\frac{\partial^2\Psi}{\partial\theta^2} + V(r)\Psi(r,\theta) + U_0|\Psi(r,\theta)|^2\Psi(r,\theta). \tag{3.41}$$

This equation can be solved using similar techniques to that used when finding the eigenstates of the hydrogen atom or harmonic oscillator in polar coordinates. The primary difference here is that there is the nonlinear interaction term. Assuming a separable form of the order parameter $\Psi(r,\theta) = R(r)Y(\theta)$, we obtain

$$-\frac{\hbar^2}{2m}\frac{1}{R}\left(r^2\frac{\partial^2 R}{\partial r^2} + r\frac{\partial R}{\partial r}\right) + V(r) - r^2\mu + U_0 r^2|R(r)|^2|Y(\theta)|^2 = -\frac{1}{Y}\frac{\partial^2 Y}{\partial\theta^2}. \tag{3.42}$$

This would be separable if the last term on the left-hand side did not have the $|Y(\theta)|^2$ term. This looks like our separation of variables trick didn't work, but if it so happens that $|Y(\theta)|^2$ is a constant, then it might still work. Continuing with the separation of variables, we would then say that the left- and right-hand sides are equal to the same constant. Setting this to l^2, we have the equation

$$\frac{\partial^2 Y}{\partial \theta^2} = l^2 Y(\theta), \tag{3.43}$$

which has solutions

$$Y(\theta) = C_1 e^{il\theta} + C_2 e^{-il\theta}. \tag{3.44}$$

As for the hydrogen atom, we require that $Y(\theta) = Y(\theta + 2\pi)$, since this refers to the same physical spatial location. This means that l must be an integer. Furthermore, in order to have $|Y(\theta)|^2$ be a constant, we should have either $C_1 = 0$ or $C_2 = 0$. This satisfies the separability problem (3.42), which then allows the radial solution to be solved. The solution is then

$$\Psi(r, \theta) = R(r) e^{il\theta}. \tag{3.45}$$

We can check whether the solution of this form is vortex-like at this stage. Assuming that $R(r)$ is real, we can substitute it into (3.37) to obtain

$$v(r, \theta) = \frac{\hbar l}{mr} \hat{\theta}, \tag{3.46}$$

where $\hat{\theta}$ is the unit vector in the angular direction. This has a zero component in the radial direction $\hat{r}$, hence it shows that the velocities are pure rotations around the origin. This certainly looks like a vortex, as long as we have $l \neq 0$. Unlike classical rotations, the velocities are quantized and only take discrete values at a given radius.

To obtain the vortex solution (3.45), we assumed radial symmetry, but this is not necessary in general. In Fig. 3.2, we show the solutions to the same infinite potential well that we looked at in Fig. 3.1, but with a single vortex in them. To obtain these, we start with the initial condition (3.26) but multiply by the vortex phase $e^{i\theta}$ offset to the center of the trap. Starting with such a vortex-like solution, by evolving in imaginary time we obtain the stable solutions seen in Fig. 3.2. The phase distribution is virtually unchanged from the initial condition, as can be seen in Fig. 3.2(c). The most obvious difference between Figs. 3.1 and 3.2, then, is the appearance of a density dip at the location of the vortex, called a vortex core. We can see that increasing the interaction reduces the size of the vortex core, and it is always similar to the healing length ξ from the edges of the well. This is because exactly at the vortex core the density drops to zero, and a similar argument can be applied to Section 3.4.5 to get the distribution in the bulk part of the BEC. Using (3.30) gives a reasonable estimate of the distribution near the vortex. Another commonly used distribution that works rather well is found using Padé approximants,

$$R(r) = \sqrt{n_0} \frac{r/\xi}{\sqrt{(r/\xi)^2 + 2}}, \tag{3.47}$$

where n_0 is in this case the density far away from the vortex. This is plotted in Fig. 3.2(d) and (e). We see that it works quite well in reproducing the density variation near the vortex core.

Returning to (3.46), we see there is another difference to classical rotation. We know from classical rotation of a rigid object that $v = r\omega\hat{\theta}$. The proportionality to r is simply because further away from the origin, the particles have to move faster

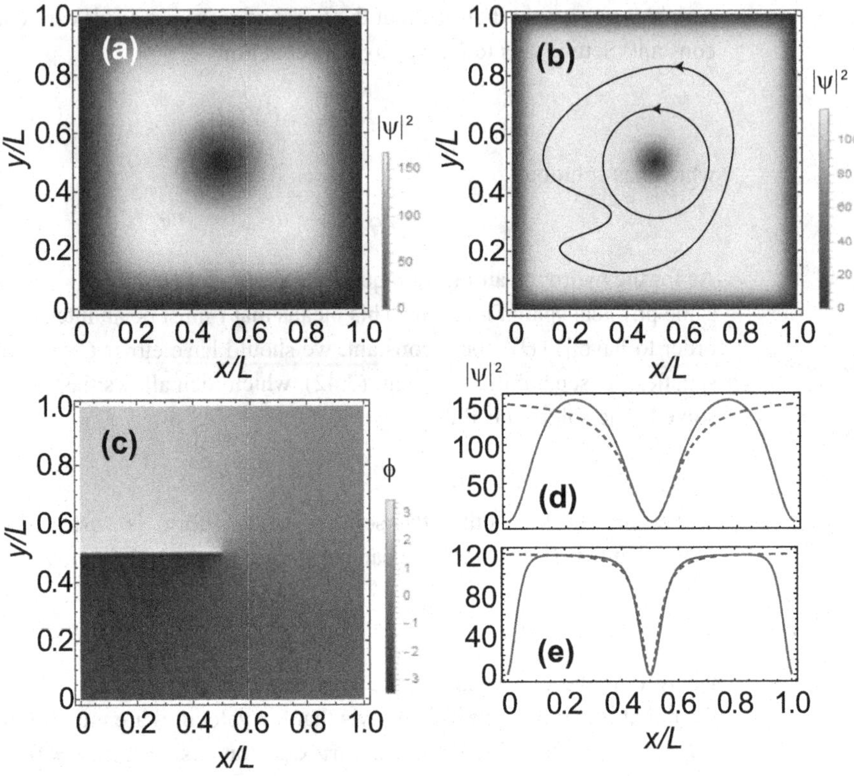

Fig. 3.2 Vortex solutions of the Gross–Pitaevskii equation for a two-dimensional infinite square well with $N/L^2 = 100$. (a and d) $U_0/E_0 = 1$; (b and e) $U_0/E_0 = 10$; and (c) the phase distribution of the solution is the same for both parameters. Solutions are numerically evolved under imaginary time for $t = 0.1\tau$. The initial condition is Eq. (3.26) multiplied by $\exp(i\Theta')$, where $\tan\Theta' = (x - L/2)/(y - L/2)$ for each case. Parameters used are $L = 1$, where the units are $E_0 = \frac{\hbar^2}{2mL^2}$, $\tau = \hbar/E_0$. Dotted line shows the Thomas–Fermi approximation. The solid line in (b) shows two examples of contours in the integral (3.48). Color version of this figure available at cambridge.org/quantumatomoptics.

to keep up with the rotation. For the BEC, we see that it is exactly the opposite of this: The velocity is *inversely* proportional to r, such that the particles move faster near the origin. This occurs because in (3.37), the velocity is proportional to the rate of change of the phase. We can see this from Fig. 3.2(c): The phase variation near the vortex core goes to a singular point. This is one of the reasons that the density at this point must drop to zero, otherwise we would have particles moving at infinite velocity, costing infinite energy!

If we calculate the line integral of the velocity around a vortex, we find that

$$\oint \boldsymbol{v} \cdot d\boldsymbol{l} = 2\pi l \frac{\hbar}{m}, \tag{3.48}$$

where for a circular contour of radius r we can take $d\boldsymbol{l} = r d\theta\hat{\theta}$. We see that the quantization occurs because of the precise cancellation of the factor of r from the line integral and velocity. This seems like a remarkable coincidence and seems to be dependent upon the somewhat arbitrary choice of a perfectly circular contour and a vortex in a radially symmetric potential. However, this is actually true no matter what choice of contour or type of potential. As long as the contour encloses the vortex core, one obtains the same result on the right-hand side of (3.48). In Fig. 3.2(b), we show two examples of contours that would give exactly the same result. Since the result of (3.48) is only dependent upon the topological properties of the contour relative to the vortex, the l is sometimes referred to as a topological charge.

But what gives rise to this remarkable property? If we look at the same contours with respect to the phase in Fig. 3.2(c), we get a better idea of where this comes from. The key aspect of a vortex is that there is a phase singularity at the vortex core, and the phase evolves by a multiple of 2π as you go around the vortex in a loop. It has to be a multiple of 2π because of the single-valuedness of $Y(\theta)$ – otherwise the wavefunction would be discontinuous. Actually, the phase is discontinuous about the vortex core, but by making the density go to zero, the discontinuity is avoided. Thus, (3.48) is really a statement that the phase of the order parameter around a vortex evolves by a factor of 2π before joining back together. This is obvious from the point of view of the definition of the velocity (3.37), which is the derivative of the phase. So if we add up all the phase changes around a vortex, then this has to add to a multiple of 2π. What is interesting is that this can be related to a physical quantity such as the velocity. Since, quantum mechanically, we have quite a different notion of velocity (i.e., the derivative of the phase) in comparison to classical physics – because we deal with waves that possess a phase – we end up with a relation that looks quite unusual.

3.6.2 Solitons

When a wave packet is evolved in time with the Schrodinger equation, we know that it starts to spread out, or disperse. Taking the potential $V(x) = 0$, and considering the one-dimensional case for simplicity, we have

$$i\hbar\frac{\partial\psi}{\partial t} = -\frac{\hbar^2}{2m}\frac{\partial^2\psi}{\partial x^2}. \tag{3.49}$$

Starting with a Gaussian wave packet moving with velocity v, the probability density evolves as

$$|\psi(x,t)|^2 = \frac{\sigma}{\sqrt{\pi(\sigma^4 + \hbar^2 t^2/m^2)}} \exp\left[-\frac{\sigma^2(x - vt)^2}{\sigma^4 + \hbar^2 t^2/m^2}\right], \tag{3.50}$$

where σ is the initial spread of the Gaussian. We can see that the wave packet spreads because the denominator in the exponential increases with time, which controls how spread out the Gaussian is. The amplitude of the Gaussian also decreases, to preserve normalization.

The origin of this behavior is that the kinetic energy term in the Hamiltonian has a diffusive effect, where the eigenstates are delocalized plane waves. In the presence of a confining potential $V(x)$, this acts as a counter to the spreading effect of the kinetic energy. But in this case, there is nothing to stop the spreading, and hence the wavefunction keeps on broadening.

Now turning to the Gross–Pitaevskii equation, we have another way to prevent the wave packets from spreading: the interaction term. If we had repulsive interactions, we can see that having a wave packet configuration is not a good idea energetically, since the particles are bunched together. What might work better is if the interactions are attractive. Considering one dimension and setting $V(x) = 0$ in (3.15), we have the solution

$$\Psi_B(x) = \sqrt{\frac{2\mu_B}{U_0}} \operatorname{sech}\left[\frac{x - vt}{\xi_B}\right] e^{-i\mu_B^0 t/\hbar + imvx/\hbar}, \tag{3.51}$$

where

$$\begin{aligned} \xi_B &= \frac{\hbar}{\sqrt{2m|\mu_B|}}, \\ \mu_B &= \mu_B^0 - \frac{1}{2}mv^2, \\ \mu_B^0 &= \frac{1}{2}U_0 n_0, \end{aligned} \tag{3.52}$$

and μ_B^0 and n_0 is the chemical potential and density at the position of the soliton for $v = 0$, respectively. Here we have $U_0 < 0$, and therefore $\mu > 0$. The special thing about this solution is that, unlike the Gaussian form that we considered initially, the shape of the wave packet doesn't change with time.

For repulsive interactions, another type of soliton can be produced that has the form of a density dip, instead of a "mound" shape that we saw above. This has the form

$$\Psi_D(x) = \sqrt{n_0}\left(\frac{iv}{c} + \sqrt{1 - \frac{v^2}{c^2}} \tanh\left[\frac{x - vt}{\xi_D}\right]\right) e^{-i\mu_D t/\hbar}, \tag{3.53}$$

where

$$\begin{aligned} \xi_D &= \frac{\hbar}{\sqrt{mU_0 n_0(1 - v^2/c^2)}}, \\ \mu_D &= U_0 n_0, \\ c &= \sqrt{\frac{U_0 n_0}{m}}. \end{aligned} \tag{3.54}$$

Here, n_0 is the density far away from the soliton, and c is the sound velocity. Because this is the form of a density dip, they are called "dark solitons," in contrast to (3.51), which are called "bright solitons."

While these solutions work perfectly well in one dimension, in higher dimensions they are unstable because they correspond to an infinitely long wave packet. Due to the attractive interactions, such long structures are not stable, and small perturbations can destabilize the solitons. In the case of attractive interactions, it is energetically more favorable to form smaller, more clumped structures. In practice, they can be produced if the trapping potential is quasi-one-dimensional, such as a long channel, where they have been observed experimentally. In Fig. 3.3b, we see an example of solitons formed in a long, one-dimensional trap. We see that the solitons are stable and can propagate for long distances.

Exercise 3.6.1 Substitute (3.51) into (3.14) and verify that it is a solution.

Exercise 3.6.2 Substitute (3.53) into (3.14) and verify that it is a solution.

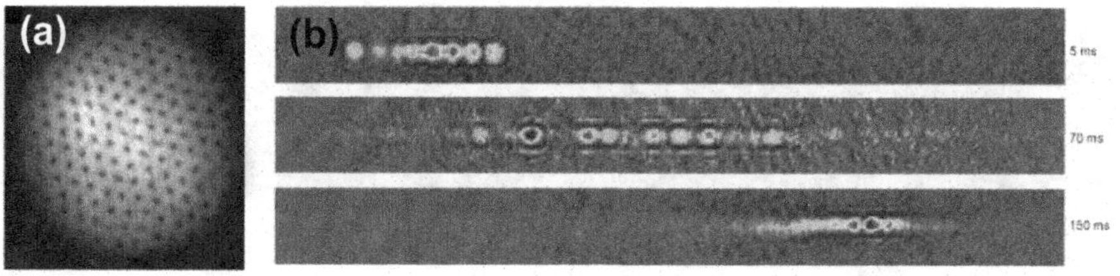

Fig. 3.3 Experimental realization of excited states in Bose–Einstein condensates. (a) Vortex array produced by stirring BECs, reproduced from [386]. (b) Bright solitons in BECs, reproduced from [445]. Color version of this figure available at cambridge.org/quantumatomoptics.

3.7 References and Further Reading

- Section 3.2: Early review on the topic of order parameters and off-diagonal long range order [499]. Textbooks detailing the concept further [375, 369].
- Section 3.3: Original references on the Gross–Pitaeveskii equation [186, 374]. Further theoretical works expanding on the approach [399, 136, 137, 267, 408].
- Section 3.4: Time-dependent solutions of the Gross–Pitaevskii equation [27, 365, 343].
- Section 3.5: For an introduction to hydrodynamic equations of the Gross–Pitaevskii equation [375, 399]. The stochastic Gross–Pitaevskii equation is derived in [162]. Theoretical analyses of the hydrodynamics are given in [267, 349]. Observation in exciton-polariton BECs is given in [12].
- Section 3.6.1: Experimental observation of vortices in BECs [323, 316, 221, 483]. Experimental observation of vortex lattices [1, 419, 135]. Experimental observation of solitons [445]. Theoretical analyses of vortices [430, 362, 403]. Review articles and books discussing vortices [143, 145, 375]. Equation (3.47) is taken from [143].
- Section 3.6.2: Experimental observation of solitons [66, 114, 13, 445, 261, 133, 320]. Theoretical analyses of solitons in BECs [366, 69, 70, 122, 357, 455]. Review articles and books discussing solitons further [181, 375].

4 Spin Dynamics of Atoms

4.1 Introduction

Up to this point, we have only considered the motional degrees of the bosons. In fact, the atoms used to form Bose–Einstein condensates (BECs) have a rich internal spin structure. This gives another degree of freedom that the many-body quantum state can occupy. In this chapter, we examine the spin dynamics that will affect the quantum state of the system. This will include processes that are naturally present in a typical setup involving trapped atoms, such as the Zeeman energy shift, collisional interactions, spontaneous emission, and loss. We also consider processes where the spin dynamics can be actively manipulated in the laboratory, using electromagnetic transitions between energy levels and Feshbach resonances.

4.2 Spin Degrees of Freedom

Many different types of atoms have been used in the context of atom trapping and cooling. In the context of BECs, the alkali atoms are a particularly popular choice, because they have a simple electronic structure and are amenable for laser cooling. In this section, we will discuss the spin structure of such atoms, taking the example of rubidium, one of the most commonly used atoms for BECs. While other atoms have different details in their atomic structure, they can be considered variations of what is described in this section.

Figure 4.1 shows the internal state structure for two isotopes of rubidium, ^{87}Rb and ^{85}Rb. The levels of atoms are typically denoted using the hyperfine spin $\boldsymbol{F}$, which consists of the total angular momentum of the electrons $\boldsymbol{J}$ and the nuclear spin $\boldsymbol{I}$:

$$\boldsymbol{F} = \boldsymbol{J} + \boldsymbol{I}. \tag{4.1}$$

The electron angular momentum is itself decomposed according to

$$\boldsymbol{J} = \boldsymbol{L} + \boldsymbol{S}, \tag{4.2}$$

where $\boldsymbol{L}$ is the orbital angular momentum and $\boldsymbol{S}$ is the electronic spin. As usual, the total angular momentum quantum numbers are eigenvalues of the total spin according to

$$\begin{aligned} \boldsymbol{F}^2 |f, m_f\rangle &= f(f+1)\hbar^2 |f, m_f\rangle \\ F_z |f, m_f\rangle &= m_f \hbar |f, m_f\rangle. \end{aligned} \tag{4.3}$$

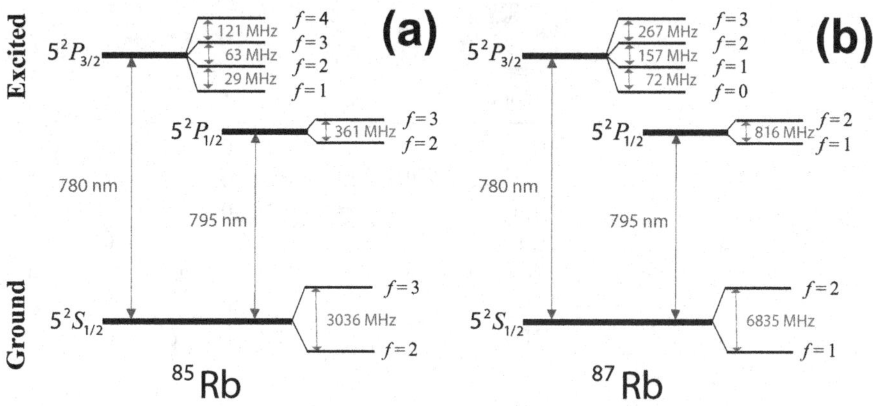

Fig. 4.1 Energy level structure for (a) ^{85}Rb and (b) ^{87}Rb. The total spins F are marked next to each level. The transition frequencies for the D_2 and D_1 transitions are given in terms of the wavelength (780 nm and 795 nm, respectively); the hyperfine splittings are given in terms of frequencies.

Since $\boldsymbol{F}^2$ and F_z commute $[\boldsymbol{F}^2, F_z] = 0$, in the above we defined the simultaneous eigenstates of $\boldsymbol{F}^2$ and F_z, where m_f is the magnetic quantum number. The magnetic quantum number takes a range $m_f \in \{-f, -f+1, \ldots, f\}$, having $2f+1$ values. For example, in Fig. 4.1(b), for ^{87}Rb, there are in fact three quantum states for $f = 1$ and five for $f = 2$.

These levels can be split with the addition of a magnetic field due to the Zeeman effect. To take this into account, we need to add an extra term to the Hamiltonian (1.2). The single-particle Hamiltonian, taking into account the first-order Zeeman effect, is

$$H_0(\boldsymbol{x}) = -\frac{\hbar^2}{2m}\nabla^2 + V(\boldsymbol{x}) + g\mu_B B m_f, \tag{4.4}$$

where $\mu_B = e\hbar/2m_e$ is the Bohr magneton, e is the elementary charge, m_e is the mass of an electron, and B is the magnetic field. The Landé g-factor is given by

$$g = \left(\frac{f(f+1) + j(j+1) - i(i+1)}{2f(f+1)}\right)\left(\frac{3}{2} + \frac{s(s+1) - l(l+1)}{2j(j+1)}\right). \tag{4.5}$$

The spin structures of several types of atoms that are used for Bose–Einstein condensates are given in Table 4.1.

In order to account for the spin quantum numbers of the atoms, we need an additional label on the bosonic operators as defined in (1.8). Including the spin, we have

$$a_{k\sigma} = \int d\boldsymbol{x}\,\psi^*_{k\sigma}(\boldsymbol{x}) a_\sigma(\boldsymbol{x}), \tag{4.6}$$

where σ is a label for the spin states that correspond to $|f, m_f\rangle$. The commutation relations then follow:

$$\begin{aligned} [a_{k\sigma}, a^\dagger_{l\sigma'}] &= \delta_{kl}\delta_{\sigma\sigma'} \\ [a_{k\sigma}, a_{l\sigma'}] &= [a^\dagger_{k\sigma}, a^\dagger_{l\sigma'}] = 0. \end{aligned} \tag{4.7}$$

Table 4.1 Spin structure of several types of atoms that are used for Bose–Einstein condensation.

Isotope	s	l	j	i	f
^{1}H	1/2	0	1/2	1/2	0, 1
^{7}Li, ^{23}Na, ^{39}K, ^{41}K, ^{87}Rb	1/2	0	1/2	3/2	1, 2
^{85}Rb	1/2	0	1/2	5/2	2, 3
^{52}Cr	3	0	3	0	3
^{133}Cs	1/2	0	1/2	7/2	3, 4
^{164}Dy	2	6	8	0	8
^{168}Dy	1	5	6	0	6

In the case of a Bose–Einstein condensate, as we discussed in Chapter 2, the bosons occupy the same spatial state. In this case, we can implicitly assume that all the spatial quantum numbers k are the same. Then, only the spin degrees of freedom are relevant, and we drop the label k and leave the spin label σ:

$$a_{k\sigma} \to a_{\sigma}. \tag{4.8}$$

Bose–Einstein condensates with such a spin degree of freedom are called *spinor Bose–Einstein condensates*, which we examine in more detail in Chapter 5. We can now write the many-body Hamiltonian including spin degrees of freedom, in the same way that we did in (1.34). This is

$$H_0 = \sum_{\sigma} (E_0 + g\mu_B B\sigma) a_{\sigma}^{\dagger} a_{\sigma}, \tag{4.9}$$

where E_0 is the ground state energy as given in (1.1), and we have assumed that all the atoms are in the ground state.

4.3 Interaction between Spins

In Section 4.2, we only wrote the single-particle Hamiltonian. For the interaction term, we extend (1.44) so that it includes spin degrees of freedom. The most general way to write this is

$$H_I = \frac{1}{2} \sum_{\sigma_1 \sigma_2 \sigma_1' \sigma_2'} g_{\sigma_1' \sigma_2' \sigma_1 \sigma_2} a_{\sigma_1'}^{\dagger} a_{\sigma_2'}^{\dagger} a_{\sigma_1} a_{\sigma_2}, \tag{4.10}$$

where the matrix elements are

$$g_{\sigma_1' \sigma_2' \sigma_1 \sigma_2} = \int d\boldsymbol{x} d\boldsymbol{y}\, \psi_{\sigma_1'}^{*}(\boldsymbol{x}) \psi_{\sigma_2'}^{*}(\boldsymbol{y}) U(\boldsymbol{x}, \boldsymbol{y}) \psi_{\sigma_1}(\boldsymbol{x}) \psi_{\sigma_2}(\boldsymbol{y}). \tag{4.11}$$

We have omitted the labels for the spatial degrees of freedom because we have assumed that all the atoms are in the ground state. Specifically, the wavefunctions are given by

$$\psi_{\sigma}(\boldsymbol{x}) \equiv \psi_{0\sigma}(\boldsymbol{x}), \tag{4.12}$$

where σ labels a particular hyperfine spin state $|f, m_f\rangle$.

In (1.46), we implied that the s-wave scattering was only dependent on the relative spatial positions of the bosons. In fact, this is not entirely true; there is a dependence on the total spin of the interacting atoms. When two atoms collide, only the outer electrons contribute to the interaction because the nuclear spin is deep within the atom and is generally unaffected to a good approximation. As such, the interaction of the spin is related to the total electron spin of the two atoms $\boldsymbol{J}_1 + \boldsymbol{J}_2$. The total spin is calculated by angular momentum addition of the interacting atoms. For example, for two ^{87}Rb atoms with $j = 1/2$, the total spin of the atoms can be either $j_{\text{tot}} = 0, 1$ (see Table 4.1). The interaction does not change the total spin, j_{tot}. We thus write the interaction in general as

$$U(\boldsymbol{x}, \boldsymbol{y}) = \frac{4\pi\hbar^2}{m}\delta(\boldsymbol{x} - \boldsymbol{y}) \sum_{j_{\text{tot}}} a_s^{(j_{\text{tot}})} P_{j_{\text{tot}}}, \tag{4.13}$$

where

$$P_{j_{\text{tot}}} = \sum_{m_{j_{\text{tot}}}=-j_{\text{tot}}}^{j_{\text{tot}}} |j_{\text{tot}}, m_{j_{\text{tot}}}\rangle\langle j_{\text{tot}}, m| \tag{4.14}$$

is the projection operator for all the states that have a total spin j_{tot}.

For alkali atoms, the orbital angular momentum is zero, $l = 0$, and only the single spin $s = 1/2$ contributes, giving $\boldsymbol{J} = \boldsymbol{S}$ in this case. In scattering theory, a particular combination of initial states that transition to a set of final states is called a *channel*. The two output total angular momenta $j_{\text{tot}} = 0, 1$ are thus called the *singlet* and *triplet channels*, respectively. The projection operators can in this case be written

$$\begin{aligned} P_0 &= \frac{1}{4} - \boldsymbol{S}_1 \cdot \boldsymbol{S}_2 \\ P_1 &= \frac{3}{4} + \boldsymbol{S}_1 \cdot \boldsymbol{S}_2. \end{aligned} \tag{4.15}$$

Thus, in this case the interaction (4.13) can be written

$$U(\boldsymbol{x}, \boldsymbol{y}) = \frac{4\pi\hbar^2}{m}\delta(\boldsymbol{x} - \boldsymbol{y}) \left[\frac{a_s^{(0)} + 3a_s^{(1)}}{4} + (a_s^{(1)} - a_s^{(0)})\boldsymbol{S}_1 \cdot \boldsymbol{S}_2\right]. \tag{4.16}$$

Substituting (4.13) into (4.11), we obtain

$$g_{\sigma_1'\sigma_2'\sigma_1\sigma_2} = \frac{4\pi\hbar^2}{m} \int d\boldsymbol{x} |\psi_0(\boldsymbol{x})|^4 \sum_{j_{\text{tot}}} a_s^{(j_{\text{tot}})} \sum_{\sigma_1\sigma_2\sigma_1'\sigma_2'} \langle\sigma_1'|\langle\sigma_2'|P_{j_{\text{tot}}}|\sigma_1\rangle|\sigma_2\rangle. \tag{4.17}$$

In the case of alkali atoms, using (4.16), we have

$$\begin{aligned} g_{\sigma_1'\sigma_2'\sigma_1\sigma_2} = {} & \frac{4\pi\hbar^2}{m} \int d\boldsymbol{x} |\psi_0(\boldsymbol{x})|^4 \\ & \times \left[\frac{a_s^{(0)} + 3a_s^{(1)}}{4}\delta_{\sigma_1\sigma_1'}\delta_{\sigma_2\sigma_2'} + (a_s^{(1)} - a_s^{(0)})\langle\sigma_1'|\langle\sigma_2'|\boldsymbol{S}_1 \cdot \boldsymbol{S}_2|\sigma_1\rangle|\sigma_2\rangle\right]. \end{aligned} \tag{4.18}$$

As was the case in Section 1.5, we see that some of the matrix elements of $g_{\sigma_1'\sigma_2'\sigma_1\sigma_2}$ are zero because of symmetries present in the interaction. These can be found by

Table 4.2 Scattering lengths of alkali atoms.

Atom	$a_s^{(0)}$	$a_s^{(1)}$
^{7}Li	34	−27.6
^{23}Li	19	65
^{41}K	85	65
^{85}Rb	2,400	−400
^{87}Rb	90	106
^{133}Cs	−208	−350

decomposing the total spins $|f, m_f\rangle$ in terms of the electronic and nuclear spins according to Clebsch–Gordan coefficients:

$$|\sigma\rangle = |f, m_f\rangle = \sum_{m_j=-j}^{j} \sum_{m_i=-i}^{i} \langle j, m_j, i, m_i | f, m \rangle | j, m_j, i, m_i \rangle. \tag{4.19}$$

Some combinations of initial and final spins are zero by virtue of being in different total spin sectors. For example, the total z-component of the spins before and after the interaction cannot change for both of the terms in (4.18):

$$m'_{f1} + m'_{f2} = m_{f1} + m_{f2}. \tag{4.20}$$

Some scattering lengths for typical atoms are shown in Table 4.2.

Exercise 4.3.1 Evaluate the s-wave interaction for ^{87}Rb for the scattering processes (a) $|f = 1, m = -1\rangle|f = 2, m = 1\rangle \to |f = 1, m = 0\rangle|f = 2, m = 0\rangle$; (b) $|f = 2, m = 2\rangle|f = 2, m = 0\rangle \to |f = 1, m = 1\rangle|f = 1, m = 1\rangle$; and (c) $|f = 1, m = 1\rangle|f = 2, m = 1\rangle \to |f = 1, m = -1\rangle|f = 2, m = -1\rangle$.

Exercise 4.3.2 (a) Verify that (4.15) are the projection operators for two spin-1/2 atoms by directly multiplying the singlet state $(|\uparrow\rangle|\downarrow\rangle - |\downarrow\rangle|\uparrow\rangle)/\sqrt{2}$ and triplet states $(|\uparrow\rangle|\downarrow\rangle + |\downarrow\rangle|\uparrow\rangle)/\sqrt{2}, |\uparrow\rangle|\uparrow\rangle, |\downarrow\rangle|\downarrow\rangle$. (b) Derive (4.15), using only the fact that the singlet and triplet states are eigenstates of the total spin operator $(\boldsymbol{S}_1 + \boldsymbol{S}_2)^2$.

4.4 Electromagnetic Transitions between Spin States

In Sections 4.2 and 4.3, we have described the Hamiltonian that is present for the spin degrees of freedom. Specifically, (4.9) described the diagonal energy of atoms including the Zeeman shift, and (4.10) described the interactions between atoms. In the absence of any applied fields, these describe the coherent spin dynamics of the BEC. However, in experiments it is often desirable and interesting to manipulate the spins by applying external electromagnetic fields to create transitions between spin levels. This can be in the form of optical, microwave, or radio frequency radiation and depends entirely upon the energy difference between the states in question. For example, in Fig. 4.1, the transitions $5^2S_{1/2} \leftrightarrow 5^2P_{3/2}, 5^2P_{1/2}$ are in the optical frequency range with a wavelength of 780 nm and 795 nm, respectively. Meanwhile, the energy differences between the hyperfine ground states are microwave frequencies with wavelength ~0.1 mm. Energy levels split by a Zeeman shift are typically in the radio frequency range.

To obtain the Hamiltonian for the interaction of the Bose–Einstein condensate with the electromagnetic field, we first look at how it interacts with a single atom. Specifically, we first look at the interaction of a single atom localized in space. To illustrate the procedure, we first consider the hydrogen atom, which only contains a single electron and gives a virtually exact way of obtaining the transition. For more complex atoms, all the electrons interact with the field in principle, but for commonly used atoms such as alkali atoms, to a good approximation, the outer electron alone determines the state of the whole atom. In this way, we can determine the interaction of an atom with the electromagnetic field by looking at a single electron.

The interaction of an electron with an electromagnetic field is given in general by

$$H_e = \frac{1}{2m_e}(-i\hbar\boldsymbol{\nabla} - e\boldsymbol{A}(\boldsymbol{x},t))^2 + eU(\boldsymbol{x},t) + V_e(\boldsymbol{x}), \tag{4.21}$$

where m_e is the electron mass and $V_e(\boldsymbol{x})$ is the binding energy for the electron within the atom. The vector and scalar potentials for the electromagnetic field are $\boldsymbol{A}(\boldsymbol{x},t)$ and $U(\boldsymbol{x},t)$, respectively. Let us start in the radiation gauge, where we take

$$U(\boldsymbol{x},t) = 0. \tag{4.22}$$

This may always be done, since scalar and vector potentials can be defined by subtracting a scalar function $\chi(\boldsymbol{x},t)$:

$$\begin{aligned} U'(\boldsymbol{x},t) &= U(\boldsymbol{x},t) - \frac{\partial\chi}{\partial t} \\ \boldsymbol{A}'(\boldsymbol{x},t) &= \boldsymbol{A}(\boldsymbol{x},t) + \boldsymbol{\nabla}\chi(\boldsymbol{x},t). \end{aligned} \tag{4.23}$$

For an electromagnetic wave, the vector potential can be taken to be

$$\boldsymbol{A}(\boldsymbol{x},t) = \boldsymbol{A}_0 e^{i(\boldsymbol{k}\cdot\boldsymbol{x}-\omega t)}, \tag{4.24}$$

where $\boldsymbol{A}_0$ is the amplitude of the vector potential, $\boldsymbol{k}$ is the wavenumber, and ω is the angular frequency. The wavelength of electromagnetic radiation is typically much larger than the size of the atom. For example, the radius of a Rb atom is 2.5×10^{-10} m, in comparison to the D_2 transition with a wavelength of 7.8×10^{-7} m. The spatial variation is therefore negligible, and the vector potential may be approximated as

$$\boldsymbol{A}(\boldsymbol{x},t) \approx \boldsymbol{A}_0 e^{i(\boldsymbol{k}\cdot\boldsymbol{x}_0-\omega t)} = \boldsymbol{A}_0(t), \tag{4.25}$$

where $\boldsymbol{x}_0$ is the location of the atom. This approximation is called the *electric dipole approximation*. The interaction Hamiltonian is then

$$H_e = \frac{1}{2m_e}(-i\hbar\boldsymbol{\nabla} - e\boldsymbol{A}_0(t))^2 + V_e(\boldsymbol{x}). \tag{4.26}$$

Since the vector potential now has no spatial dependence, we can go further and remove this from the Hamiltonian by taking

$$\chi(\boldsymbol{x},t) = -\boldsymbol{x}\cdot\boldsymbol{A}_0(t). \tag{4.27}$$

According to (4.23), this means that the scalar potential is nonzero, giving the interaction Hamiltonian

$$H_e = H_e^{(0)} + H_e^{(1)}, \tag{4.28}$$

where the unperturbed Hamiltonian is

$$H_e^{(0)} = -\frac{\hbar^2}{2m_e}\nabla^2 + V_e(\boldsymbol{x}) \tag{4.29}$$

and the electric dipole Hamiltonian is

$$H_e^{(1)} = -e\boldsymbol{x} \cdot \boldsymbol{E}_0(t). \tag{4.30}$$

Here we have used $\boldsymbol{E} = -\frac{\partial \boldsymbol{A}}{\partial t}$. The Hamiltonian (4.30) is the desired term that causes transitions between energy levels of the atoms.

Note that the gauge transformation (4.23) actually requires transformation of the wavefunction as well, according to

$$\psi'(\boldsymbol{x}, t) = e^{ie\chi(\boldsymbol{x},t)/\hbar}\psi(\boldsymbol{x}, t). \tag{4.31}$$

Thus, since the Hamiltonians (4.21), (4.26), and (4.28) are all in different gauges, the wavefunctions will differ by phase definitions according to (4.31).

Let's now take the example of the hydrogen atom to illustrate the use of (4.28). Suppose that we start with the electromagnetic field turned off, so that $\boldsymbol{E}_0 = 0$. In this case, the eigenstates of (4.28) are just the solutions of the Schrodinger equation with a potential $V_e(\boldsymbol{x}) = -\frac{e}{4\pi\epsilon_0 r}$. The ground and first excited state of the wavefunctions are

$$\begin{aligned}
\psi_{100}(r, \theta, \phi) &= \frac{1}{\sqrt{\pi a_B^3}} e^{-r/a_B} \\
\psi_{200}(r, \theta, \phi) &= \frac{1}{\sqrt{8\pi a_B^3}} \left(1 - \frac{r}{2a_B}\right) e^{-r/a},
\end{aligned} \tag{4.32}$$

where a_B is the Bohr radius. Suppose the electromagnetic field is relatively weak and is of the appropriate frequency such that the transitions only occur between these two atomic states. We will see what kinds of conditions are required for this assumption later. The electronic wavefunction can then be restricted to superpositions of the two levels

$$\psi(\boldsymbol{x}) = c_1\psi_{100}(\boldsymbol{x}) + c_2\psi_{200}(\boldsymbol{x}), \tag{4.33}$$

where c_1, c_2 are complex coefficients such that $|c_1|^2 + |c_2|^2 = 1$. In this two-dimensional space, we can write the Hamiltonian in matrix form, with matrix elements

$$H_e(n, m) = \int d^3\boldsymbol{x}\, \psi_{n00}^*(\boldsymbol{x}) H_e \psi_{m00}(\boldsymbol{x}). \tag{4.34}$$

The matrix corresponding to the unperturbed component is

$$H_e^{(0)} = \begin{pmatrix} \hbar\omega_1 & 0 \\ 0 & \hbar\omega_2 \end{pmatrix}, \tag{4.35}$$

where $\hbar\omega_n$ is the energy of the states ψ_{n00}. For the electric field, let us assume a form

$$H_e^{(1)} = -exE_0 \cos\omega_0 t, \tag{4.36}$$

where we have taken the electric field to be in the x-direction and E_0 is the amplitude. The electric dipole Hamiltonian is then

$$H_e^{(1)} = \begin{pmatrix} 0 & \hbar\Omega^* \cos\omega_0 t \\ \hbar\Omega\cos\omega_0 t & 0 \end{pmatrix}, \tag{4.37}$$

where we have defined the Rabi frequency,

$$\hbar\Omega = -eE_0 \int d^3\boldsymbol{x}\, \psi_{200}^*(\boldsymbol{x}) x \psi_{100}(\boldsymbol{x}). \tag{4.38}$$

The diagonal elements of (4.37) are zero due to the fact that

$$\int d^3\boldsymbol{x} |\psi_{n00}(\boldsymbol{x})|^2 x = 0. \tag{4.39}$$

This follows from the fact that due to the spherical symmetry of the atom, the wavefunctions are all odd or even functions, $\psi_{nlm}(-\boldsymbol{x}) = (-1)^l \psi_{nlm}(\boldsymbol{x})$. Since (4.30) is an odd function, this ensures that all diagonal components evaluate to zero.

We can remove the time dependence of the Hamiltonian by working in the interaction picture. To do this, we separate the Hamiltonian into two parts according to

$$\begin{aligned} H_e &= H_e^{(0)\prime} + H_e^{(1)\prime} \\ H_e^{(0)\prime} &= \hbar\omega_1 |1\rangle\langle 1| + \hbar(\omega_1 + \omega_0)|2\rangle\langle 2| \\ H_e^{(1)\prime} &= \hbar\Delta|2\rangle\langle 2| + \hbar\Omega\cos\omega_0 t|2\rangle\langle 1| + \hbar\Omega^*\cos\omega_0 t|1\rangle\langle 2|, \end{aligned} \tag{4.40}$$

where we have switched to bra-ket notation, $\Delta = \omega_2 - \omega_1 - \omega_0$, and the state $|n\rangle$ corresponds to the wavefunction ψ_{n00}. An operator O in the interaction picture is related to that in the Schrodinger picture according to

$$[O]_{\rm I} = U_e^{(0)\dagger} O U_e^{(0)}, \tag{4.41}$$

where $U_e^{(0)}$ is the time evolution operator for $H_e^{(0)\prime}$. We denote interaction picture operators by $[\dots]_{\rm I}$ and leave Schrodinger picture operators unlabeled. The time evolution operator can be evaluated to give

$$U_e^{(0)}(t) = e^{-iH_e^{(0)\prime}t/\hbar} = e^{-i\omega_1 t}|1\rangle\langle 1| + e^{-i(\omega_1+\omega_0)t}|2\rangle\langle 2|. \tag{4.42}$$

We see that each of the states picks up a time-evolving phase related to its energy. The states then evolve according to only the interaction Hamiltonian,

$$\begin{aligned} [H_e^{(1)\prime}]_{\rm I} = U_e^{(0)\dagger} H_e^{(1)\prime} U_e^{(0)} &= \frac{(e^{2i\omega_0 t} + 1)\hbar\Omega}{2}|2\rangle\langle 1| \\ &+ \frac{(e^{-2i\omega_0 t} + 1)\hbar\Omega^*}{2}|1\rangle\langle 2| + \hbar\Delta|2\rangle\langle 2|. \end{aligned} \tag{4.43}$$

We see that in the interaction picture there are two terms that contribute to each off-diagonal matrix element. The term that has no time dependence is an energy-conserving term, since the energy difference between the atomic levels $\hbar(\omega_2 - \omega_1)$ is approximately matched by the energy of the photon $\hbar\omega_0$. Physically, this corresponds to the absorption of a photon by the atom and a transition to a higher energy state, or emission of a photon by the atom and a transition to a lower energy state. The other term does not conserve energy, as it corresponds to the absorption of a photon *and* a transition to a lower energy state, or the reverse. A common step at this point is to neglect such energy-nonconserving terms, which amounts to the *rotating wave-approximation*. The name comes from neglecting terms in (4.43) that involve phases $e^{i\omega t}$, where the ωt is a large number within the relevant timescale such that the average integrates to zero. Under this approximation, we finally have

$$[H_e^{(1)}]_{\text{I}} = \hbar \begin{pmatrix} 0 & \Omega^*/2 \\ \Omega/2 & \Delta \end{pmatrix}. \tag{4.44}$$

In the above example, we considered two simple states of the hydrogen atom for the sake of a simple example. More generally, the transitions that the electric dipole Hamiltonian allows depends upon selection rules, which depend on the matrix element as was calculated in (4.38). There are generally two considerations that determine whether a given transition is zero or nonzero. The first is the parity effect, as we discussed above. Since $\psi_{nlm}(-\boldsymbol{x}) = (-1)^l \psi_{nlm}(\boldsymbol{x})$, this means that if the initial and final states are both the same parity, then the transition matrix element evaluates to zero since (4.30) is odd. The other consideration is due to angular momentum conservation. To see this, first write

$$\boldsymbol{x} \cdot \boldsymbol{E} = \sqrt{\frac{4\pi}{3}} \left(E_z Y_{10} + \frac{-E_x + iE_y}{\sqrt{2}} Y_{11} + \frac{E_x + iE_y}{\sqrt{2}} Y_{1-1} \right), \tag{4.45}$$

where $Y_{10} = \sqrt{\frac{3}{4\pi}} \cos\theta$, $Y_{1\pm1} = \mp\sqrt{\frac{3}{8\pi}} \sin\theta e^{\pm i\phi}$ are the spherical harmonics. Then by the rules of angular momentum addition, the angular momentum of the final and initial states must obey $l' = l-1, l, l+1$, respectively. The purely spatial form of the electric dipole moment means that the spin is unaffected, hence $s' = s$. Between the parity and angular momentum addition rules, we thus only have transitions such that $l' = l \pm 1$. In terms of the total angular momentum quantum numbers, for hydrogen we have the selection rules

$$\begin{aligned} \Delta j &= 0, \pm 1 \\ \Delta m &= 0, \pm 1. \end{aligned} \tag{4.46}$$

For more complex atoms involving more than one electron, similar rules apply. The parity selection rule is not an exact symmetry for more complex atoms, but the angular momentum addition rules still apply, hence we again obtain (4.46). In this case $\Delta l = 0, \pm 1$ is allowed, except for the case $l' = l = 0$, but $s' = s$ again.

In the above discussion, we were only concerned with a single atom. We can straightforwardly write the Hamiltonian for the many-atom case, such as in a BEC. The transitions simply occur for each atom individually, and we have

$$H_\Omega = \frac{\hbar\Omega e^{i\boldsymbol{k}\cdot\boldsymbol{x}}}{2} a_2^\dagger(\boldsymbol{x}) a_1(\boldsymbol{x}) + \frac{\hbar\Omega^* e^{-i\boldsymbol{k}\cdot\boldsymbol{x}}}{2} a_1^\dagger(\boldsymbol{x}) a_2(\boldsymbol{x}) + \hbar\Delta a_2^\dagger(\boldsymbol{x}) a_2(\boldsymbol{x}). \tag{4.47}$$

One difference from the single atom case is that the transition matrix now takes a spatial dependence due to the phase of the electromagnetic field. In a BEC, atoms can be delocalized over distances that are larger than optical wavelengths, hence it becomes important to take into account the spatial dependence. As a result of the phase dependence, the atoms experience a momentum shift equal to $\boldsymbol{k}$ according to momentum conservation when a photon is absorbed or emitted. We will show an example of this in Section 9.4.4.

Exercise 4.4.1 Show that (4.42) is true by expanding the exponential operator, and using the fact that the identity operator is $I = |1\rangle\langle 1| + |2\rangle\langle 2|$.

Exercise 4.4.2 Verify (4.43).

4.5 The ac Stark Shift

In Section 4.4, we saw that light will in general cause a transition to an excited state of the atom, according to (4.44) for a single atom and (4.47) for a multi-atom case. We now consider the case where the light field has a large detuning with respect to the transition $\Delta \gg |\Omega|$. Diagonalizing (4.44), the eigenstates and energies are

$$\begin{aligned} |\pm\rangle &= \frac{1}{N_\pm}\left[(\Delta \pm \sqrt{\Delta^2 + |\Omega|^2})|1\rangle - \Omega|2\rangle\right], \\ E_\pm &= \frac{1}{2}\left(\Delta \mp \sqrt{\Delta^2 + |\Omega|^2}\right), \end{aligned} \tag{4.48}$$

where $|1\rangle$ is a ground state of the atom, $|2\rangle$ is an excited state, and $N_\pm$ is a suitable normalization factor. The frequency of the laser is chosen so that it is detuned from the excited state, such that $|\Omega| \ll \Delta$. We can thus expand the square roots to give

$$|+\rangle \approx |1\rangle - \frac{\Omega}{2\Delta}|2\rangle \tag{4.49}$$

$$|-\rangle \approx |2\rangle + \frac{\Omega^*}{2\Delta}|2\rangle \tag{4.50}$$

$$E_+ \approx -\frac{|\Omega|^2}{4\Delta} \tag{4.51}$$

$$E_- \approx \Delta + \frac{|\Omega|^2}{4\Delta}. \tag{4.52}$$

The energetically lower state $|+\rangle$ is to a good approximation the same as the state $|1\rangle$, except that it is shifted lower by an energy $-\frac{|\Omega|^2}{4\Delta}$. The energy shift is called the *ac Stark shift*.

We can then write an effective Hamiltonian that describes the ac Stark shift for the many-atom case straightforwardly. Working in the diagonal basis (4.48), the Hamiltonian (4.47) reads

$$\begin{aligned} H_\Omega &= \hbar\left[E_+ b_+^\dagger b_+ + E_- b_-^\dagger b_-\right] \\ &\approx \hbar\left[-\frac{|\Omega|^2}{4\Delta} b_+^\dagger b_+ + \left(\Delta + \frac{|\Omega|^2}{4\Delta}\right) b_-^\dagger b_-\right], \end{aligned} \tag{4.53}$$

where $b_\pm$ are associated with the states $|\pm\rangle$. Since the energy level of the diagonalized states has corrections to the original state, it is often interpreted as a second-order effect in perturbation theory, where the off-diagonal terms in (4.44) are the perturbative terms, which modify the energy of the state $|1\rangle$.

The ac Stark shift is a very common effect that is exploited in many situations in atomic physics. For instance, in optical lattices a light field at a frequency detuned with the atomic resonance is prepared with a spatially varying intensity. Since the light field intensity is $\propto |\Omega|^2$, an energy potential (4.51) proportional to the intensity of the light at that point in space is produced. It is also the basis for atom trapping methods such as that using optical dipole traps. In the treatment of Section 4.4, the light was considered to be a classical field, which is valid as long as the intensities are high. However, in some applications, such as nondestructive measurements, it is also important to take into account the quantum nature of light. The quantum version of the ac Stark shift is used as the basis of techniques such as optical imaging of atoms and entanglement generation.

4.6 Feshbach Resonances

We now discuss another effect that originates from a second-order transition, which affects the interactions between the atoms. We consider a scattering picture, such that there is a particular initial state of two atoms that are initially far away from each other, that undergo some interaction at close range, and then scatter to leave the atoms at distant locations. Depending upon the initial and final states, there are various scattering processes, or channels, which have different forms of interatomic potential. Two such channels are shown in Fig. 4.2(a). An open channel refers to a type of interaction that is allowed by energy conservation, such that at large interatomic distances, the energy matches that of the free atoms. Meanwhile, a closed channel refers to a type of interaction where there is an energy barrier between the initial and final states.

Although the closed channel may be highly off-resonant when the atoms are highly separated, at close distances the potential may possess some bound molecular states that have a similar energy. The atoms in the open channel cannot directly scatter to the closed channel, because by definition there are no continuum states in closed channels. This means that the first-order correction in perturbation theory is zero. However, the atoms in the open channel can be affected by a second-order process in a similar way to that seen in Section 4.5. This affects the interatomic potential of the open channel, which in turn affects the scattering length. The scattering length is modified according to the form

$$\frac{4\pi\hbar^2}{m}a_s = \frac{4\pi\hbar^2}{m}\tilde{a}_s + \sum_n \frac{|\langle\psi_n|V|\psi_0\rangle|^2}{E - E_n}, \tag{4.54}$$

where $a_s, \tilde{a}_s$ are the modified and nonresonant scattering lengths, respectively, $|\psi_0\rangle$ are the incoming and outgoing spherical waves, $|\psi_n\rangle$ is the closed channel bound state with energies E_n, V is the Hamiltonian, which causes the transitions between the open and closed channel states, and E is the energy of the particles in the open channel.

In order to adjust the energy of the bound states to match the initial states of the atoms, a magnetic field is typically used to tune the energy via the Zeeman effect. A measurement of the scattering length across a Feshbach resonance is shown in

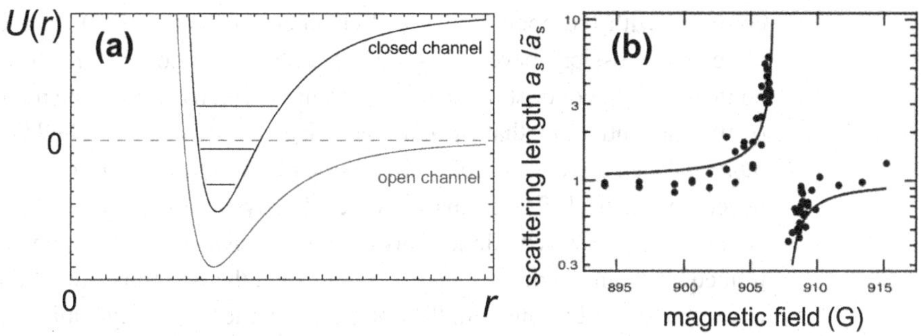

Fig. 4.2 (a) Schematic of the process involved in a Feshbach resonance. (b) Experimental data showing the scattering length variation with magnetic field in a Feshbach resonance in a gas of sodium atoms from [237]. Color version of this figure available at cambridge.org/quantumatomoptics.

Fig. 4.2(b). We see the typical form of the scattering length, which follows the form of (4.54). Depending upon the sign of $E - E_n$, the additional term to the background $\tilde{a}_s$ can be positive or negative. This means that it is possible to tune the atomic interactions such that they are either attractive or repulsive, by applying a magnetic field. Another possible way of achieving the coupling between the closed and open channels is by using optical methods, which also produces an effective tuning of the interactions.

4.7 Spontaneous Emission

The spin processes described up to now are all coherent processes. Coherent processes are describable by a Hamiltonian, which implies that they are reversible, since they are described by the Schrodinger equation. By changing the Hamiltonian from $H \to -H$, one can always reverse the dynamics and "undo" the time evolution $e^{-iHt/\hbar}$. Not all processes are reversible, however. The process we examine in this section is a prime example: spontaneous emission. This is the process where an atom relaxes from a high to a low energy state. The reverse process of starting in a low energy state and ending up in a high energy state usually does not happen unless the atom is made to do so, perhaps by illuminating it with laser light, as we saw in Section 4.4. Such irreversible processes are called incoherent processes, and they cannot be described by a Hamiltonian evolution.

But first, what is the physical origin of this difference? Isn't the universe ultimately described by a huge wavefunction that evolves according to a Hamiltonian? The reason for the irreversibility can be seen in the spontaneous emission case, because the relaxation from high to low energy state is tied to the emission of a photon. But unless the photon is reflected back in some way, it usually flies away and doesn't come back. The reverse process, on the other hand, requires a photon for the atom to go from a low to a high energy state. Therefore, if we don't specifically arrange for a photon to be near the atom, it cannot undergo the reverse process.

The key difference between the irreversible and reversible cases comes down to whether we consider the system as being an open or closed system (not to be confused with the open and closed channels of Section 4.6!). In a closed system, we cannot possibly have any situation where the photon "flies away and doesn't come back." In contrast, in an open system we can have the system coupled to many (usually a continuum, that is, an infinity) degrees of freedom, called the *bath* or *environment*. Then there is the possibility of the photon escaping into the environment and never coming back. To obtain the irreversible dynamics, one then obtains an effective equation for the system alone. This is called the *master equation* and is a time evolution equation in the density matrix, rather than the wavefunction of the system. This is capable of describing both the coherent and incoherent dynamics of the system.

In this book, we shall not go through the explicit derivation of the master equation. Often the derivation does not give much insight into the dynamics anyway; one usually starts at the master equation and uses that as a starting point to examine the dynamics. Fortunately, there is a simple prescription for most incoherent processes that enables writing the master equation without explicitly working through the derivation, as we explain below. We will show just the starting and ending points of the master equation derivation and examine a few simple consequences.

For spontaneous emission, the model to describe the system (S) and environment (E) is

$$H_{\text{spon}} = H_{\text{S}} + H_{\text{E}} + H_{\text{I}} \tag{4.55}$$

$$\begin{aligned} H_{\text{S}} &= \frac{\hbar(\omega_2 - \omega_1)}{2}\sigma^z + \frac{\hbar(\omega_2 + \omega_1)}{2} I \\ H_{\text{E}} &= \hbar \sum_k \nu_k b_k^\dagger b_k \\ H_{\text{I}} &= \hbar \sum_k \left(g_k \sigma^+ b_k + g_k^* \sigma^- b_k^\dagger \right), \end{aligned} \tag{4.56}$$

where we have used the definitions

$$\begin{aligned} \sigma^z &= |2\rangle\langle 2| - |1\rangle\langle 1| \\ \sigma^- &= |1\rangle\langle 2| = (\sigma^+)^\dagger \\ I &= |2\rangle\langle 2| + |1\rangle\langle 1| \end{aligned} \tag{4.57}$$

for the two atomic levels $|1\rangle, |2\rangle$ with energies $\hbar\omega_1, \hbar\omega_2$, respectively, and a continuum of electromagnetic modes b_k with energy $\hbar\nu_k$. The interaction (I) between the system and environment has a coupling g_k. Schematically, the coupling takes a form that is shown in Fig. 4.3.

At this point, everything is described by a Hamiltonian, so there are no irreversible dynamics in the system. The irreversibility comes about because one would like to obtain an effective equation that only involves the system in question, which is the atom. Needless to say, treating an infinity of modes is rather inconvenient if one would like to perform simple calculations to capture the dynamics of the atom. After a standard derivation under the Markovian assumption and the initial state of the environment being a vacuum state for all the modes, the final result that is obtained is

$$\frac{d\rho}{dt} = \Gamma L(\sigma^-, \rho), \tag{4.58}$$

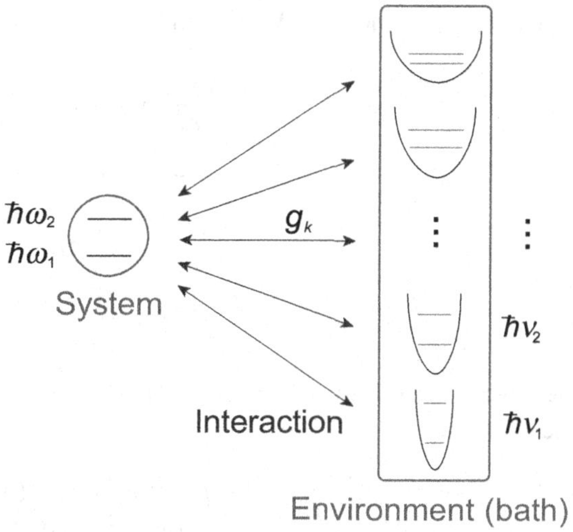

Fig. 4.3 Schematic of model to describe the spontaneous emission given in the Hamiltonian (4.55). The system comprises two atomic levels that are coupled to a large number of electromagnetic modes, which form the environment. Color version of this figure available at cambridge.org/quantumatomoptics.

where the Lindblad operator for an arbitrary operator X is defined as

$$L(X, \rho) \equiv X\rho X^\dagger - \frac{1}{2}(X^\dagger X\rho + \rho X^\dagger X). \tag{4.59}$$

The constant Γ is a decay rate,

$$\Gamma = \frac{(\omega_2 - \omega_1)^3 |\langle 1|e\boldsymbol{x}|2\rangle|^2}{3\pi\epsilon_0\hbar c^3}, \tag{4.60}$$

where ϵ_0 is the permittivity of free space and c is the speed of light. The equation (4.58) describes the irreversible dynamics due to spontaneous emission.

In (4.58), the only effect that we have taken account of is spontaneous emission – there are no other dynamics for the system alone or for other types of baths causing different types of dynamics. In the presence of other dynamics, one can simply include them by adding the corresponding terms to the master equation. Thus, the effects of spontaneous emission can be included by simply adding the Lindblad operator (4.59) to the master equation with a suitable decay coefficient that determines its rate. As a rule of thumb, the operator X is the same operator that appears in the interaction Hamiltonian H_I. We will see below that simply by changing the operator X, all types of other incoherent processes can be easily included.

Let us solve (4.58) to show explicitly that it indeed captures the spontaneous emission process. The way to solve a matrix equation is to look at each matrix element and see how it evolves with time. Since we are looking at a two-level atom in this example, the density matrix is 2×2, and there are four matrix elements that we must find the evolution of. We can do this by multiplying (4.58) by $\langle\sigma|$ from the left and $|\sigma'\rangle$ from the right. Writing this out, we obtain the four coupled equations

$$\frac{d\rho_{11}}{dt} = \Gamma\rho_{22}$$
$$\frac{d\rho_{22}}{dt} = -\Gamma\rho_{11}$$
$$\frac{d\rho_{12}}{dt} = -\frac{\Gamma}{2}\rho_{12}$$
$$\frac{d\rho_{21}}{dt} = -\frac{\Gamma}{2}\rho_{21}, \tag{4.61}$$

where we have defined $\rho_{\sigma\sigma'} = \langle\sigma|\rho|\sigma'\rangle$. These can be solved explicitly to give the time evolution

$$\rho(t) = \begin{pmatrix} 1 - \rho_{22}(0)e^{-\Gamma t} & \rho_{12}(0)e^{-\Gamma t/2} \\ \rho_{21}(0)e^{-\Gamma t/2} & \rho_{22}(0)e^{-\Gamma t} \end{pmatrix}, \tag{4.62}$$

where $\rho_{\sigma\sigma'}(0)$ are the initial matrix elements of the density matrix.

We see that (4.62) has exactly the expected behavior. The probability of the excited state ρ_{22} decays exponentially from its initial value. The loss of ρ_{22} is the gain of ρ_{11}, hence it increases in probability accordingly. The off-diagonal elements of the density matrix are associated with coherence between the levels $|1\rangle$ and $|2\rangle$. For example, an equal superposition state between the two states $(|1\rangle + |2\rangle)/\sqrt{2}$ has a density matrix with elements $\rho_{\sigma\sigma'} = 1/2$ everywhere. We see that the off-diagonal terms also are damped, so that spontaneous emission also reduces coherence between the states, albeit at half the rate of the diagonal terms.

Generalizing to the many-atom case is again straightforward. If we consider the annihilation operator for the two levels $|1, 2\rangle$ as $a_{1,2}$, respectively, then all that is required in this case is to change the operator in the Lindblad operator to $X = a_1^\dagger a_2$ such that the master equation reads

$$\begin{aligned}\frac{d\rho}{dt} &= \Gamma L(a_1^\dagger a_2, \rho) \\ &= \frac{\Gamma}{2}\left(2a_1^\dagger a_2 \rho a_1 a_2^\dagger - n_2(n_1+1)\rho - \rho n_2(n_1+1)\right),\end{aligned} \tag{4.63}$$

where $n = a^\dagger a$ is the number operator for the two levels as labeled.

Let us consider a simple example of the dynamics for the many-atom case. Suppose that we start in a state with N atoms occupying the level denoted by a_2. We can derive an equation that describes the number of atoms under the presence of spontaneous emission from the master equation above. Multiplying (4.63) by $n_2 = a_2^\dagger a_2$ and taking the trace, we obtain the equation

$$\frac{d\langle n_2\rangle}{dt} = -\Gamma(N+1)\langle n_2\rangle + \Gamma\langle (n_2)^2\rangle. \tag{4.64}$$

To exactly solve for the dynamics, we would at this point need the evolution equation for $\langle (n_2)^2\rangle$, which can be obtained by multiplying (4.63) by n_2^2 and taking the trace. One approach is to make the approximation that $\langle (n_2)^2\rangle \approx (\langle n_2\rangle)^2$, which amounts to saying that the variance of n_2 is rather small, which is usually obeyed if $N \gg 1$.

Another way to view this is that it is a mean-field approximation,

$$\begin{aligned} \langle XY \rangle &= \langle ((\langle X \rangle + \delta X)((\langle Y \rangle + \delta Y)) \rangle \\ &= \langle X \rangle \langle Y \rangle + \langle X \rangle \langle \delta Y \rangle + \langle \delta X \rangle \langle Y \rangle + \langle \delta X \delta Y \rangle \qquad (4.65) \\ &\approx \langle X \rangle \langle Y \rangle, \qquad (4.66) \end{aligned}$$

where $\delta X = X - \langle X \rangle$ are operators that offset by the mean, and we have used $\langle \delta X \rangle = 0$ and discarded the last term in (4.65) since it is a product of two small quantities. We can then approximate

$$\frac{d\langle n_2 \rangle}{dt} = -\Gamma(N+1)\langle n_2 \rangle + \Gamma \langle n_2 \rangle^2, \qquad (4.67)$$

which can be solved analytically to obtain the dynamics

$$\langle n_2 \rangle = \frac{N(N+1)}{N + e^{(N+1)\Gamma t}}. \qquad (4.68)$$

We can see that the timescale of the exponential decay is $\sim \frac{1}{(N+1)\Gamma}$. This means that the more particles that occupy the system, the faster the decay rate. This is an example of the collective bosonic enhancement effect and an example of superradiance. The factor of $N+1$ often arises and originates from the way bosonic creation and annihilation operators work. Adding an extra boson to a system that already has N bosons gives a factor of $\sqrt{N+1}$:

$$a^\dagger |N\rangle = a^\dagger \frac{(a^\dagger)^N}{\sqrt{N!}} |0\rangle = \sqrt{N+1}|N+1\rangle. \qquad (4.69)$$

Thus, matrix elements have this bosonic factor, and probabilities then have factors that are increased by $N+1$.

Exercise 4.7.1 Solve (4.58) and verify (4.62). Keep in mind that for a density matrix, the diagonal terms always add to 1, so that $\rho_{11} + \rho_{22} = 1$ at any time.

Exercise 4.7.2 Work out the matrix elements of (4.63) for a Fock state

$$|N-k, k\rangle = \frac{(a_1^\dagger)^{N-k}(a_2^\dagger)^k}{\sqrt{(N-k)!\,k!}}, \qquad (4.70)$$

where the total number of atoms is N. Assuming that the system starts in the state $|1, N-1\rangle$, find the dynamics of the system thereafter. Compare the dynamics to the single-atom case and (4.68). Does it decay faster, the same, or slower due to the presence of many atoms in the lower energy state?

4.8 Atom Loss

Another source of decoherence that acts on many-atom systems is loss. The atoms in a trapped BEC are not perfectly trapped, and generally they will be lost through various processes. Various mechanisms exist for atoms to be lost from the trap. The simplest is one-body loss, where an atom is lost from the trap because it exceeds the energy barrier imposed by the trap. This can occur due to the background interaction with untrapped atoms or absorption of stray photons. This is described by a master equation

$$\begin{aligned}\frac{d\rho}{dt} &= \gamma_1 L(a,\rho)\\ &= \frac{\gamma_1}{2}\left(2a\rho a^\dagger - n\rho - \rho n\right),\end{aligned} \tag{4.71}$$

where the rate of the one-body loss is γ_1 and the bosonic operators for a particular atomic species are given by a. For the one-body loss term alone, it is easy to show that the atom number exponentially decays, by multiplying (4.71) by n and taking the trace. This gives

$$\frac{d\langle n_l\rangle}{dt} = \gamma_1\langle n_l\rangle, \tag{4.72}$$

which has the solution

$$\langle n\rangle = Ne^{-\gamma_1 t}, \tag{4.73}$$

where N is the initial number of atoms.

Atoms can also be lost from the BEC by the interactions between the atoms. In a two-body loss process, two atoms scatter and change the internal state of the atoms. The dominant process is spin-exchange, and it occurs in a similar way to that described in Section 4.3. Two-body loss is described according to the master equation

$$\begin{aligned}\frac{d\rho}{dt} &= \gamma_2 L(a^2,\rho)\\ &= \frac{\gamma_2}{2}\left(2a^2\rho(a^\dagger)^2 - (a^\dagger)^2a^2\rho - \rho(a^\dagger)^2a^2\right),\end{aligned} \tag{4.74}$$

where γ_2 is the two-body loss rate.

For three-body loss, the relevant process is molecule formation. Two atoms can be unstable towards the formation of a molecule; for example, in Rb it is energetically favorable to form a Rb_2 molecule. The binding energy of the molecule must be carried away by a third atom in order to conserve energy and momentum. The energy of the molecule and the third atom have larger energies than the trapping potential, which results in the loss of three atoms. Three-body loss is described according to the master equation

$$\begin{aligned}\frac{d\rho}{dt} &= \gamma_3 L(a^3,\rho)\\ &= \frac{\gamma_3}{2}\sum_j\left(2a^3\rho(a^\dagger)^3 - (a^\dagger)^3a^3\rho - \rho(a^\dagger)^3a^3\right),\end{aligned} \tag{4.75}$$

where γ_3 is the three-body loss rate.

4.9 Quantum Jump Method

In Sections 4.7 and 4.8, we introduced methods for simulating the time dynamics of several incoherent processes. These are generally described by master equations that take the general form

$$\frac{d\rho}{dt} = -\frac{i}{\hbar}[H_0, \rho] + \sum_j \Gamma_j L(X_j, \rho), \tag{4.76}$$

where X_j is the Lindblad operator for the particular process, Γ_j is its rate, and H_0 is a Hamiltonian describing the coherent processes. For processes involving several incoherent processes, the Lindblad terms simply sum together. A straightforward way to solve the master equations is to take matrix elements of both sides of (4.76), such that it is a set of coupled first-order differential equations

$$\frac{d\rho_{nm}}{dt} = -\frac{i}{\hbar}H_{nm} + \sum_j \Gamma_j L_{nm}(X_j, \rho), \tag{4.77}$$

where

$$\rho_{nm} = \langle n|\rho|m\rangle \tag{4.78}$$

$$H_{nm} = \langle n|[\rho, H_0]|m\rangle = \sum_{n'} \left[\langle n|H_0|n'\rangle\rho_{n'm} - \langle n'|H_0|m\rangle\rho_{nn'}\right] \tag{4.79}$$

$$\begin{aligned} L_{nm}(X_j, \rho) &= \langle n|L(X_j, \rho)|m\rangle \\ &= \langle n| \left[X\rho X^\dagger - \frac{1}{2}(X^\dagger X\rho + \rho X^\dagger X)\right] |m\rangle, \end{aligned} \tag{4.80}$$

and $|n\rangle, |m\rangle$ are an orthogonal set of states. For the coherent Hamiltonian evolution H_0, we introduced a complete set of states $I = \sum_{n'} |n'\rangle\langle n'|$. The set of equations (4.77) involves the square of the dimension of the Hilbert space and is quite a numerically intensive task. An alternative method, called the *quantum jump method*, gives equivalent results that can be more efficient in terms of calculation in many circumstances, since it can be calculated using only states that have the dimension of the Hilbert space.

The procedure for the quantum jump method is as follows. Suppose at time t the state is $|\psi(t)\rangle$. One then evolves the state at some small time δt later, following the algorithm:

1. Calculate the quantity

$$\delta p_j = \langle \psi(t)|X_j^\dagger X_j|\psi(t)\rangle \Gamma_j \delta t, \tag{4.81}$$

   and define $\delta p = \sum_j \delta p_j$.
2. Draw a random number ϵ from the range $[0, 1]$.
3. If $\epsilon > \delta p$, then evolve the state according to

$$|\psi(t + \delta t)\rangle = \frac{1}{\sqrt{1 - \delta p}} \left(1 - \frac{i}{\hbar} H_{\text{eff}} \delta t\right) |\psi(t)\rangle, \tag{4.82}$$

   where

$$H_{\text{eff}} = H_0 - \frac{i\hbar\Gamma_j}{2} \sum_j X_j^\dagger X_j. \tag{4.83}$$

4. If $\epsilon \leq \delta p$, randomly choose one of the jumps labeled by j according to the probability $\delta p_j/\delta p$ and perform the time evolution

$$|\psi(t+\delta t)\rangle = \sqrt{\frac{\Gamma_j \delta t}{\delta p_j}} X_j |\psi(t)\rangle. \tag{4.84}$$

5. Go to step (1) and repeat until the desired time evolution is achieved.

The above procedure obviously has a stochastic nature and will give a different time evolution for each run. Let us label the trajectory for a single run as $|\psi^{(k)}(t)\rangle$. In running the time evolution M times, the statistical mixture of these random runs approaches the true time-evolved density matrix:

$$\frac{1}{M}\sum_{k=1}^{M} |\psi^{(k)}(t)\rangle\langle\psi^{(k)}(t)| \to \rho(t). \tag{4.85}$$

As long as M is sufficiently large, the the time evolution of any pure state can be found. For an initially mixed state, one first writes this as a mixture of pure states:

$$\rho(0) = \sum_l P_l |\psi_l(0)\rangle\langle\psi_l(0)|. \tag{4.86}$$

Randomly choosing the initial state with probability P_l, one performs the steps as before to find the ensemble average.

Exercise 4.9.1 Solve (4.58) using the quantum jump method and compare it to the exact solution (4.62). Do you obtain the same results?

4.10 References and Further Reading

- Section 4.2: For a more in-depth introduction on the spin degrees of freedom of atoms, see the textbooks [148, 56]. For more details on the atomic structure of ^{87}Rb, [438].
- Section 4.3: Introductions discussing atomic interactions can be found in [369, 272, 56, 102].
- Section 4.4: Textbooks discussing electromagnetic transitions between spin states, [148, 151, 439, 421, 477, 333].
- Section 4.5: The original reference introducing the ac Stark effect, [24]. For further textbook-level introductions, see [369, 333, 151]. Experimental demonstration of atom trapping, [382]. Reviews relating to atom trapping are given in [373, 93, 331, 2, 21, 372, 489, 182, 22, 332]. Treating the ac Stark shift quantum mechanically (the optical fields are treated mechanically) gives the related quantum nondemolition Hamiltonian, [201]. Theoretical works examining this are given in [278, 340, 281, 234].
- Section 4.6: First experimental works observing Feshbach resonance, [237, 98]. Further experiments using Feshbach resonances, [97, 117, 509, 248, 130]. Reviews and books further discussing Feshbach resonances, [90, 369, 266, 460]. For the full derivation of (4.54), see [369].

- Section 4.7: Early theories of spontaneous emission, [116, 485]. For further textbook-level introductions on spontaneous emission, see [168, 421, 414, 439, 369, 151, 345, 333].
- Section 4.8: For textbooks and reviews discussing atom loss, see [421, 477, 168, 182, 148, 369, 372, 269]. Specific discussions of the role of atom loss in [465, 73, 433].
- Section 4.9: For reviews and books discussing the quantum jump method, see [377, 493, 80, 269]. Experimental observation of quantum jumps are reported in [409, 36, 35, 172]. Theoretical works developing the topic are performed in [103, 128, 208, 341].

5 Spinor Bose–Einstein Condensates

5.1 Introduction

We have seen in Chapter 4 that the atoms constituting a Bose–Einstein condensate (BEC) typically have a rich spin structure. For the case of ^{87}Rb, there are eight spin states (the $F = 1, 2$ states), which have energies that are relatively close. The energies of these states are close enough such that they can be considered to be approximately degenerate; these are collectively called the "ground states." By external manipulation of the spin states, various spin states of the BEC can be produced. In this chapter, we will describe these states and methods of how to characterize them. We will examine several archetypal states, such as spin coherent states and squeezed states. Many of the techniques in this chapter are borrowed from quantum optics, where useful methods to visualize states such as the *P*-, *Q*-, and Wigner functions were first developed. These can also be defined for the spin case, which serves as a powerful visualization of the quantum state. We also show a gallery of typical states for a spinor BEC and how they appear for various quasiprobability distributions.

5.2 Spin Coherent States

The simplest spin system that we can consider is that involving two levels. The annihilation operators for the two levels will be labeled by a, b, and we assume that there are a total of N atoms. The general wavefunction of a state can then be written in terms of the Fock states, which we define as

$$|k\rangle = \frac{(a^\dagger)^k (b^\dagger)^{N-k}}{\sqrt{k!\,(N-k)!}}|0\rangle, \tag{5.1}$$

where we only label the number of atoms in the state a in the ket because we always assume a total number of atoms as being N. Recall that the meaning of (5.1) is that the spatial degrees of freedom are all taken to be in the same state for both the levels a, b (see Section 4.2). This means that the only degrees of freedom that are left are the spin, and thus the different states arise from different combinations of occupations of the two levels. The total number operator,

$$N = a^\dagger a + b^\dagger b, \tag{5.2}$$

satisfies

$$N|k\rangle = N|k\rangle. \tag{5.3}$$

The general wavefunction of a state for N particles is then

$$|\psi\rangle = \sum_{k=0}^{N} \psi_k |k\rangle. \tag{5.4}$$

This state is an eigenstate of the total number operator

$$N|\psi\rangle = N|\psi\rangle. \tag{5.5}$$

In a Bose–Einstein condensate, we typically start with all the atoms occupying the same quantum state. For example, this might be the energetically lowest hyperfine state such as that given in Fig. 4.1. Taking this to be the state a, the state would be written

$$|k = N\rangle = \frac{(a^\dagger)^N}{\sqrt{N!}}|0\rangle. \tag{5.6}$$

If all the bosons were in the other state, then the state would be

$$|k = 0\rangle = \frac{(b^\dagger)^N}{\sqrt{N!}}|0\rangle. \tag{5.7}$$

It is then natural to consider the general class of states where the a and b states are in an arbitrary superposition,

$$|\alpha, \beta\rangle\rangle = \frac{1}{\sqrt{N!}}(\alpha a^\dagger + \beta b^\dagger)^N |0\rangle, \tag{5.8}$$

where α, β are coefficients such that $|\alpha|^2 + |\beta|^2 = 1$. This class of states is called a *spin coherent state* and plays a central role in manipulations of BECs. Spin coherent states are aptly named in the sense that they have analogous properties to coherent states of light. We will see how such states arise naturally from a physical process in the next section, but for now let us take a mathematical point of view and explore its properties.

First, we can write the spin coherent state in the Fock representation (5.4) by expanding the brackets in (5.8) with a binomial series,

$$\begin{aligned} |\alpha, \beta\rangle\rangle &= \frac{1}{\sqrt{N!}} \sum_{k=0}^{N} \binom{N}{k} (\alpha a^\dagger)^k (\beta b^\dagger)^{N-k} |0\rangle \\ &= \sum_{k=0}^{N} \sqrt{\binom{N}{k}} \alpha^k \beta^{N-k} |k\rangle, \end{aligned} \tag{5.9}$$

where we used (5.1). We see that the spin coherent state wavefunction takes the form of (5.4), so that indeed it is a state of fixed atom number N. Equation (5.9) has an analogous role to the expansion of an optical coherent state in terms of Fock states. Recall that

$$\begin{aligned} |\alpha\rangle &= e^{-|\alpha|^2/2} e^{\alpha a^\dagger} |0\rangle \\ &= e^{-|\alpha|^2/2} \sum_{n=0}^{\infty} \frac{\alpha^n}{\sqrt{n!}} |n\rangle, \end{aligned} \tag{5.10}$$

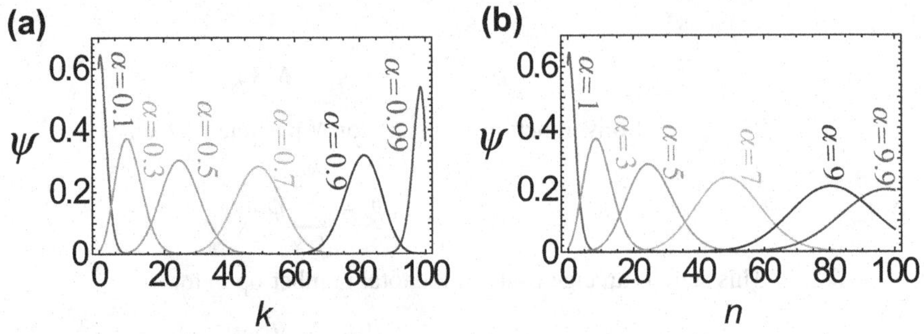

Fig. 5.1 Amplitude of Fock states for (a) spin coherent states and (b) optical coherent states. For the spin coherent state, we take a value of $\beta = \sqrt{1-\alpha^2}$ and $N = 100$. Color version of this figure available at cambridge .org/quantumatomoptics.

where for this equation α is any complex number and the optical Fock states are $|n\rangle = \frac{(a^\dagger)^n}{\sqrt{n!}}|0\rangle$. The functional forms of (5.9) and (5.10) have rather similar forms, as can be seen in Fig. 5.1. Since $|\alpha|^2 + |\beta|^2 = 1$, we see almost an identical distribution of the coefficients for $|\alpha| \ll 1$ and $n \ll N$. The reason for this can be seen by observing that one can approximate the amplitudes of the spin coherent states as

$$\psi_k = \sqrt{\frac{N!}{k!\,(N-k)!}}\alpha^k \beta^{N-k} \approx \frac{(\sqrt{N}\alpha)^k}{\sqrt{k!}} e^{-N\frac{|\alpha|^2}{2}}, \tag{5.11}$$

since $\exp[(N-k)\ln\beta] = \exp[(N-k)\ln\sqrt{1-|\alpha|^2}] \approx \exp[-N|\alpha|^2/2]$ and $N!/(N-k)! \approx N^k$. Thus, by rescaling the α by a factor of $\sqrt{N}$, one can approximate the same coefficients as in (5.10). From Fig. 5.1, we see that the approximation starts to fail when $\alpha \approx \beta$. Optical coherent states have a distribution where the average photon number and variance are equal to $|\alpha|^2$. The spin coherent state distribution is symmetrical such that when $\alpha \to 1$, it again narrows. This is due to the finite Hilbert space of the spins, whereas the photonic Hilbert space is infinite.

Exercise 5.2.1 Verify (5.11).

5.3 The Schwinger Boson Representation

We now examine some observables of the spin coherent state (5.8). For optical coherent states, the natural observables are the position and momentum, defined in terms of bosonic operators as

$$\begin{aligned} X &= \sqrt{\frac{\hbar}{2\omega}}(a + a^\dagger), \\ P &= -i\sqrt{\frac{\hbar\omega}{2}}(a - a^\dagger), \end{aligned} \tag{5.12}$$

which have commutation relations as $[X, P] = i\hbar$ and $\hbar\omega$ is the photon energy. Typically, it is more convenient to work in dimensionless units, defining instead

$$x = \frac{1}{\sqrt{2}}(a + a^\dagger),$$
$$p = -i\frac{1}{\sqrt{2}}(a - a^\dagger), \tag{5.13}$$

which have the commutation relation $[x, p] = i$.

In our case, due to the two-level nature we are working with, it is more natural to define operators that have a form that closely resembles Pauli spin operators. We define the dimensionless total spin operators as

$$S_x = a^\dagger b + b^\dagger a,$$
$$S_y = -ia^\dagger b + ib^\dagger a,$$
$$S_z = a^\dagger a - b^\dagger b. \tag{5.14}$$

These operators are also called the *Schwinger boson operators*. These operators have many of the same properties as Pauli spin operators, such as

$$[S_j, S_k] = 2i\epsilon_{jkl} S_l, \tag{5.15}$$

where $j, k, l \in \{x, y, z\}$ and ϵ_{jkl} is the Levi–Civita antisymmetric tensor. Although the *commutation* relations of the operators are identical to Pauli operators, they do not have identical properties to them. For example, $(S_j)^2 \neq I$, where I is the identity operator, and the anticommutator is $\{S_i, S_j\} \neq 0$, unlike Pauli operators. We note that when $N = 1$, the spin operators reduce to Pauli operators:

$$\sigma_x = |a\rangle\langle b| + |b\rangle\langle a|,$$
$$\sigma_y = -i|a\rangle\langle b| + i|b\rangle\langle a|,$$
$$\sigma_z = |a\rangle\langle a| - |b\rangle\langle b|. \tag{5.16}$$

Raising and lowering operators are defined according to

$$S_+ = \frac{S_x + iS_y}{2} = a^\dagger b,$$
$$S_- = \frac{S_x - iS_y}{2} = b^\dagger a. \tag{5.17}$$

These operators have the effect of changing the Fock state by one unit:

$$S_+|k\rangle = \sqrt{(k+1)(N-k)}|k+1\rangle,$$
$$S_-|k\rangle = \sqrt{k(N-k+1)}|k-1\rangle. \tag{5.18}$$

The raising and lowering operators can be related to the S_x, S_y spin operators by

$$S_x = S_+ + S_-,$$
$$S_y = -iS_+ + iS_-. \tag{5.19}$$

Operating the spin operators on the Fock states gives

$$\begin{aligned} S_x|k\rangle &= \sqrt{(k+1)(N-k)}|k+1\rangle + \sqrt{k(N-k+1)}|k-1\rangle, \\ S_y|k\rangle &= -i\sqrt{(k+1)(N-k)}|k+1\rangle + i\sqrt{k(N-k+1)}|k-1\rangle, \\ S_z|k\rangle &= (2k-N)|k\rangle. \end{aligned} \tag{5.20}$$

The Fock states are eigenstates of the S_z operator.

What are the eigenstates of the S_x and S_y operators? These are also Fock states, but the bosons are defined in a different basis. If we define bosonic operators

$$\begin{aligned} c &= \frac{a+b}{\sqrt{2}}, \\ d &= \frac{a-b}{\sqrt{2}}, \end{aligned} \tag{5.21}$$

the eigenstates of the S_x operator take the form

$$\begin{aligned} |k\rangle_x &= \frac{(c^\dagger)^k (d^\dagger)^{N-k}}{\sqrt{k!\,(N-k)!}}|0\rangle \\ &= \frac{(a^\dagger + b^\dagger)^k (a^\dagger - b^\dagger)^{N-k}}{\sqrt{2^N k!\,(N-k)!}}|0\rangle \end{aligned} \tag{5.22}$$

and the eigenstates of the S_y operator are

$$|k\rangle_y = \frac{(a^\dagger + ib^\dagger)^k (ia^\dagger + b^\dagger)^{N-k}}{\sqrt{2^N k!\,(N-k)!}}|0\rangle. \tag{5.23}$$

Then, the eigenvalue equations are written

$$\begin{aligned} S_x|k\rangle_x &= (2k-N)|k\rangle_x \\ S_y|k\rangle_y &= (2k-N)|k\rangle_y. \end{aligned} \tag{5.24}$$

The Schwinger boson operators conserve the total number of particles. The particle number is thus a conserved quantity that arises because

$$[N, S_x] = [N, S_y] = [N, S_z] = 0. \tag{5.25}$$

The eigenstates of the spin operators and total number operators thus have simultaneous eigenstates. This is apparent from (5.5), which shows that states of this form are eigenstates of the total number operator and that the eigenstates (5.22) and (5.23) can be expanded in Fock states (5.1).

Exercise 5.3.1 Verify (5.15). Why isn't $(S_j)^2 \neq I$ and $\{S_i, S_j\} \neq 0$? If $N = 1$, do any of your conclusions change?

Exercise 5.3.2 Show that (5.22) and (5.23) satisfy the eigenvalue equations (5.24).

Exercise 5.3.3 Expand (5.22) and (5.23), and show that it can be written in the form of (5.5).

5.4 Spin Coherent State Expectation Values

Let us now evaluate the expectation values of the Schwinger boson operators (5.14) with respect to the spin coherent states (5.8). First, turning to the expectation value of S_x, we require evaluation of the expression

$$\begin{aligned}\langle S_x \rangle &= \langle\langle \alpha, \beta | S_x | \alpha, \beta \rangle\rangle \\ &= \frac{1}{N!} \langle 0 | (\alpha^* a + \beta^* b)^N S_x (\alpha a^\dagger + \beta b^\dagger)^N | 0 \rangle. \end{aligned} \tag{5.26}$$

To evaluate this, our strategy will be to commute the S_x operator one by one through the $(\alpha a^\dagger + \beta b^\dagger)$ product to the far right-hand side until it immediately precedes $|0\rangle$. Then, using the fact that $S_x|0\rangle = 0$, we can remove the operator from the expression. The commutator that we will need is

$$[S_x, \alpha a^\dagger + \beta b^\dagger] = \beta a^\dagger + \alpha b^\dagger. \tag{5.27}$$

Every time the S_x is passed through, we obtain the above factor; hence, we obtain

$$\langle S_x \rangle = \frac{1}{(N-1)!} \langle 0 | (\alpha^* a + \beta^* b)^N (\beta a^\dagger + \alpha b^\dagger)(\alpha a^\dagger + \beta b^\dagger)^{N-1} | 0 \rangle. \tag{5.28}$$

Next, we will commute the $(\beta a^\dagger + \alpha b^\dagger)$ factor one by one through all the terms on the left so that it is immediately after the $\langle 0|$. Then, using the fact that $\langle 0|(\beta a^\dagger + \alpha b^\dagger) = 0$, we can eliminate this factor. The commutation relation that we will need is

$$[\alpha^* a + \beta^* b, \beta a^\dagger + \alpha b^\dagger] = \alpha^* \beta + \beta^* \alpha, \tag{5.29}$$

which gives

$$\langle S_x \rangle = N(\alpha^* \beta + \beta^* \alpha) \frac{1}{(N-1)!} \langle 0 | (\alpha^* a + \beta^* b)^{N-1} (\alpha a^\dagger + \beta b^\dagger)^{N-1} | 0 \rangle. \tag{5.30}$$

Noting that the remaining expectation value is simply the normalization condition for a spin coherent state with $N - 1$ particles $\langle\langle \alpha, \beta | \alpha, \beta \rangle\rangle_{N-1} = 1$, we have

$$\langle S_x \rangle = N(\alpha^* \beta + \beta^* \alpha). \tag{5.31}$$

We can work out the expectation values of the other operators in a similar way, which yields

$$\begin{aligned}\langle S_y \rangle &= N(-i\alpha^* \beta + i\beta^* \alpha), \\ \langle S_z \rangle &= N(|\alpha|^2 - |\beta|^2). \end{aligned} \tag{5.32}$$

Comparing the expectation values of (5.31) and (5.32), we see that they are identical to the expectation values of Pauli operators (5.16) for the state

$$\alpha|a\rangle + \beta|b\rangle, \tag{5.33}$$

up to a multiplicative factor of N. For standard qubits, the above quantum state can be visualized on the Bloch sphere using the parametrization

$$\begin{aligned}\alpha &= \cos\frac{\theta}{2} e^{-i\phi/2}, \\ \beta &= \sin\frac{\theta}{2} e^{i\phi/2}, \end{aligned} \tag{5.34}$$

where the angles θ, ϕ give a position on the unit sphere. Using this parametrization, the expectation values take a form

$$\begin{aligned} \langle S_x \rangle &= N \sin\theta \cos\phi, \\ \langle S_y \rangle &= N \sin\theta \sin\phi, \\ \langle S_z \rangle &= N \cos\theta. \end{aligned} \tag{5.35}$$

It is easy to see that the expectation values for the spin coherent state exactly coincide with the usual x, y, z coordinates on a sphere of radius N:

$$\langle S_x \rangle^2 + \langle S_y \rangle^2 + \langle S_z \rangle^2 = N^2. \tag{5.36}$$

Figure 5.2 shows the Bloch sphere representation of a spin coherent state with $N = 10$. The set of all spin coherent states (5.8) form a sphere in the expectation values $(\langle S_x \rangle, \langle S_y \rangle, \langle S_z \rangle)$.

Equation (5.36), which is only true for spin coherent states, should not be confused with the total spin operator, which in the language of group theory is a Casimir operator for SU(2). This is

$$\begin{aligned} S^2 &= S_x^2 + S_y^2 + S_z^2 \\ &= N(N+2), \end{aligned} \tag{5.37}$$

where we have used (5.2). Since this can be written in terms of the number operator N, this means that any state with fixed number as given in (5.4) is an eigenstate with eigenvalue $N(N+2)$. This is the same as the usual expression for the eigenvalue of total spin operator $s(s+1)$, if we note that our dimensionless spin operators have an extra factor of 2 throughout.

Exercise 5.4.1 Evaluate (5.32) using the same method as (5.31). You will need to first evaluate $[S_{y,z}, \alpha a^\dagger + \beta b^\dagger]$ and the similar relations to (5.29).

Exercise 5.4.2 Check that the expectation value of the Pauli operators for the state (5.33) gives the same values as (5.31) and (5.32) up to a factor of N.

Exercise 5.4.3 Substitute the expressions for the spin operators into (5.37) and show that all the off-diagonal terms cancel to give the expression in terms of the number operator.

5.5 Preparation of a Spin Coherent State

Previously, we defined the spin coherent state without describing how it can be prepared physically. How would one prepare a state such as (5.8)? Suppose we then apply some electromagnetic radiation of a suitable frequency such that transitions now take place between the levels a and b, such as that described in (4.47). Taking the detuning to be zero, we have the Hamiltonian

$$H_x = \frac{\hbar\Omega}{2}\left(a^\dagger b + b^\dagger a\right). \tag{5.38}$$

In terms of the Schwinger boson operators, the transition Hamiltonian takes a form

$$H_x = \frac{\hbar\Omega}{2} S_x. \tag{5.39}$$

It is reasonable to assume that (5.6) can be created through the process of Bose–Einstein condensation, since it is the lowest energy state. If the transition Hamiltonian is applied for a time t, the new state is

$$|\psi(t)\rangle = e^{-iH_x t/\hbar}|k = N\rangle = e^{-iS_x \Omega t/2}|k = N\rangle. \tag{5.40}$$

For Pauli operators, evaluating the time evolution operator is straightforward, because we can take advantage of the relation

$$e^{-i\sigma_x \Omega t/2} = \cos(\Omega t/2) - i\sigma_x \sin(\Omega t/2), \tag{5.41}$$

which was possible because $\sigma_x^2 = I$. Unfortunately, we cannot do the same here, so we must use a different method. The way to evaluate the time evolution operator is to diagonalize the S_x operator by performing the transformation

$$\begin{aligned} a &= \frac{c+d}{\sqrt{2}}, \\ b &= \frac{c-d}{\sqrt{2}}. \end{aligned} \tag{5.42}$$

The spin operators (5.14) then take the form

$$\begin{aligned} S_x &= c^\dagger c - d^\dagger d, \\ S_y &= ic^\dagger d - id^\dagger c, \\ S_z &= c^\dagger d + d^\dagger c. \end{aligned} \tag{5.43}$$

We see that in this representation, S_x looks like how the S_z operator appeared, in diagonal form. The state then can be written

$$|\psi(t)\rangle = \frac{1}{\sqrt{2^N N!}} e^{-in_c\Omega t/2} e^{in_d \Omega t/2} (c^\dagger + d^\dagger)^N |0\rangle, \tag{5.44}$$

where $n_c = c^\dagger c$ and $n_d = d^\dagger d$. This can be evaluated, noting that for an arbitrary function f,

$$\begin{aligned} f(n_c)c^\dagger &= \left(f(0) + f'(0)n_c + \frac{f''(0)}{2}(n_c)^2 + \cdots \right) c^\dagger \\ &= c^\dagger f(n_c + 1), \end{aligned} \tag{5.45}$$

where the function was expanded as a Taylor series and we used

$$n_c c^\dagger = c^\dagger (n_c + 1), \tag{5.46}$$

and similarly for n_d. Our strategy in evaluating (5.44) will be to commute the exponentials to the right since they can be removed once they act on the vacuum according to $n_{c,d}|0\rangle = 0$. Moving the exponentials through one factor of $(c^\dagger + d^\dagger)$, we find

$$e^{-in_c\Omega t/2} e^{in_d\Omega t/2} (c^\dagger + d^\dagger) = (e^{-i\Omega t/2} c^\dagger + e^{i\Omega t/2} d^\dagger) e^{-in_c \Omega t/2} e^{in_d \Omega t/2}. \tag{5.47}$$

After commuting exponentials through all N factors of $(c^\dagger + d^\dagger)$, we obtain

$$|\psi(t)\rangle = \frac{1}{\sqrt{2^N N!}}(e^{-i\Omega t/2}c^\dagger + e^{i\Omega t/2}d^\dagger)^N|0\rangle, \tag{5.48}$$

Reverting back to the original bosonic variables by inverting (5.42), we have

$$\begin{aligned}|\psi(t)\rangle &= \frac{1}{\sqrt{N!}}(\cos(\Omega t/2)a^\dagger - i\sin(\Omega t/2)b^\dagger)^N|0\rangle \\ &= |\cos(\Omega t/2), -i\sin(\Omega t/2)\rangle\rangle.\end{aligned} \tag{5.49}$$

This is the form of a spin coherent state (5.8), where $\alpha = \cos(\Omega t/2)$ and $\beta = -i\sin(\Omega t/2)$. By choosing the time t appropriately, the magnitudes of α, β can be chosen arbitrarily. From (5.31), (5.32), and Fig. 5.2, we see that the electromagnetic radiation can produce states that are along the circle

$$\langle S_y\rangle^2 + \langle S_z\rangle^2 = N^2, \tag{5.50}$$

which is the great circle with axis along the S_x-direction.

But what about the other points on the Bloch sphere? To obtain spin coherent states on an arbitrary location of the Bloch sphere, we must perform an additional operation with a different axis of rotation. This time, suppose we apply a magnetic field such that the levels experience a Zeeman effect. Then according to (1.34), we will have the Hamiltonian

$$\begin{aligned}H_z &= (E_0 + g\mu_B\sigma_a)a^\dagger a + (E_0 + g\mu_B\sigma_b)b^\dagger b \\ &= E_0' N + \frac{\hbar\Delta}{2}S_z,\end{aligned} \tag{5.51}$$

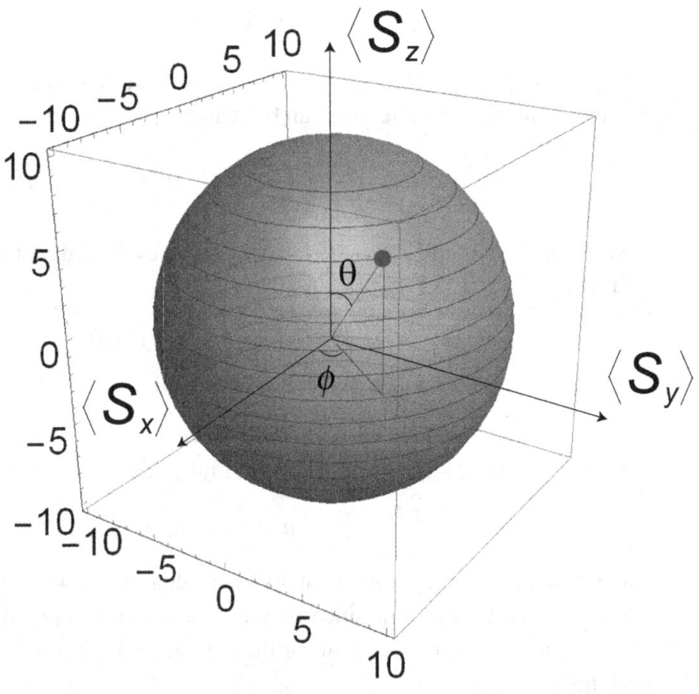

Fig. 5.2 The Bloch sphere representation of a spin coherent state for $N = 10$. Color version of this figure available at cambridge.org/quantumatomoptics.

where we have defined

$$\begin{aligned} E_0' &= E_0 + g\mu_B \frac{\sigma_a + \sigma_b}{2}, \\ \hbar\Delta &= g\mu_B(\sigma_a - \sigma_b), \end{aligned} \tag{5.52}$$

and used (5.2) and (5.14). As we have seen from (5.5), the spin coherent state is an eigenstate of the total number operator. Thus, the application of N only creates a global phase on the state that is physically unobservable. The only important part of the Hamiltonian is then the S_z component, which as we will evaluate below creates a rotation around the S_z axis.

Applying (5.51) to (5.49), we then would like to evaluate

$$\begin{aligned} |\psi(t,\tau)\rangle &= e^{-iH_z\tau/\hbar}|\psi(t)\rangle \\ &= e^{-iE_0'N\tau/\hbar} e^{-i\frac{\Delta}{2}S_z\tau} \sum_{k=0}^{N} \sqrt{\binom{N}{k}} \cos^k(\Omega t/2) i^{N-k} \sin^{N-k}(\Omega t/2)|k\rangle, \end{aligned} \tag{5.53}$$

where we used (5.9) to expand the spin coherent state, and the number operator N was evaluated to its eigenvalue N. The exponential can be evaluated, noting that the Fock states are eigenstates of the S_z operator (5.20). The exponential of S_z can then be evaluated by expanding it as a Taylor series:

$$\begin{aligned} e^{-i\frac{\Delta}{2}S_z\tau}|k\rangle &= \left(1 - i\frac{\Delta}{2}S_z\tau + \left(i\frac{\Delta}{2}S_z\tau\right)^2 \Big/ 2 + \cdots\right)|k\rangle \\ &= \left(1 - i\frac{\Delta}{2}(2k-N)\tau + \left(i\frac{\Delta}{2}(2k-N)\tau\right)^2 \Big/ 2 + \cdots\right)|k\rangle \\ &= e^{-i\frac{\Delta}{2}(2k-N)\tau}|k\rangle. \end{aligned} \tag{5.54}$$

We thus obtain

$$\begin{aligned} |\psi(t,\tau)\rangle &= e^{-iE_0'N\tau/\hbar} e^{-i\frac{\Delta}{2}S_z\tau} \sum_{k=0}^{N} \sqrt{\binom{N}{k}} [e^{-i\frac{\Delta}{2}\tau}\cos(\Omega t/2)]^k [ie^{i\frac{\Delta}{2}\tau}\sin(\Omega t/2)]^{N-k}|k\rangle \\ &= e^{-iE_0'N\tau/\hbar} |e^{-i\frac{\Delta}{2}\tau}\cos(\Omega t/2), ie^{i\frac{\Delta}{2}\tau}\sin(\Omega t/2)\rangle\rangle. \end{aligned} \tag{5.55}$$

Comparing this to (5.34), we see that an arbitrary position on the Bloch sphere can be attained if we associate

$$\begin{aligned} \theta &= \Omega t, \\ \phi &= \Delta\tau + \frac{\pi}{2} \end{aligned} \tag{5.56}$$

up to a physically irrelevant global phase factor. In summary, we have seen that an arbitrary state on the Bloch sphere can be attained by two successive rotations, first around the S_x axis and then around the S_z axis:

$$e^{-i\phi S_z/2} e^{-i\theta S_x/2}|1,0\rangle\rangle = |e^{-i\phi/2}\cos(\theta/2), -ie^{i\phi/2}\sin(\theta/2)\rangle\rangle, \tag{5.57}$$

where we used the fact that

$$\begin{aligned} |k=N\rangle &= |1,0\rangle\rangle, \\ |k=0\rangle &= |0,1\rangle\rangle. \end{aligned} \tag{5.58}$$

It is not a coincidence that the axis of the great circle for the first rotation is along the same operator that appeared in (5.39), and the axis for the second rotation was the same as that in (5.51). In general, for a Hamiltonian containing a sum of S_x, S_y, S_z operators, the axis of rotation will be along the unit vector $\boldsymbol{n} = (n_x, n_y, n_z)$ given by the coefficients of these operators. The general rotation is

$$e^{-i\theta \boldsymbol{n}\cdot\boldsymbol{S}}|\alpha,\beta\rangle\rangle = |\alpha',\beta'\rangle\rangle, \tag{5.59}$$

where

$$\begin{pmatrix} \alpha' \\ \beta' \end{pmatrix} = \begin{pmatrix} \cos\theta - in_z \sin\theta & -(n_y + in_x)\sin\theta \\ (n_y - in_x)\sin\theta & \cos\theta + in_z\sin\theta \end{pmatrix} \begin{pmatrix} \alpha \\ \beta \end{pmatrix} \tag{5.60}$$

and

$$\boldsymbol{S} = (S_x, S_y, S_z). \tag{5.61}$$

Thus, by a suitable combination of spin operators, any point can be rotated to any other point on the Bloch sphere via a great circle rotation. Special cases of this when the spins are aligned along the x, y, z axes are given by

$$\begin{aligned} e^{-i\theta S_x/2}|\alpha,\beta\rangle\rangle &= |\alpha\cos(\theta/2) - i\beta\sin(\theta/2), \beta\cos(\theta/2) - i\alpha\sin(\theta/2)\rangle\rangle, \\ e^{-i\theta S_y/2}|\alpha,\beta\rangle\rangle &= |\alpha\cos(\theta/2) - \beta\sin(\theta/2), \beta\cos(\theta/2) + \alpha\sin(\theta/2)\rangle\rangle, \\ e^{-i\theta S_z/2}|\alpha,\beta\rangle\rangle &= |e^{-i\theta/2}\alpha, e^{i\theta/2}\beta\rangle\rangle. \end{aligned} \tag{5.62}$$

These correspond to a rotation around the S_x, S_y, and S_z axes by an angle θ, respectively.

Exercise 5.5.1 Evaluate (5.49) from (5.44) in a different way. First, expand the factor $(c^\dagger + d^\dagger)^N$ using a binomial expansion and obtain a superposition in terms of Fock states

$$|k\rangle' = \frac{(c^\dagger)^{N-k}(d^\dagger)^k}{\sqrt{(N-k)!\,k!}}|0\rangle. \tag{5.63}$$

Then, using the fact that $e^{i\theta n_c}|k\rangle' = e^{i\theta(N-k)}|k\rangle'$ and $e^{i\theta n_d}|k\rangle' = e^{i\theta k}|k\rangle'$, evaluate the exponentials and then refactorize the expression to obtain (5.48).

Exercise 5.5.2 Verify that (5.59) with (5.60) produces the correct spin coherent state. Hint: make a linear transformation of the a, b operators such that unit vector $\boldsymbol{n}$ is rotated to $(0, 0, 1)$ in the new basis. Then you can use the same method as that used in (5.55) to evaluate the S_z rotation. Reverting back to the original basis then gives the final result.

Exercise 5.5.3 Substitute the parametrization (5.34) into (5.62) and show that original spin coherent state is rotated by an angle θ.

5.6 Uncertainty Relations

One of the central properties of quantum mechanics is the Heisenberg uncertainty relation, which in the most familiar form states that

$$\sigma_X \sigma_P \geq \frac{\hbar}{2}, \tag{5.64}$$

where σ_X, σ_P are the standard deviations of the position and momentum operators (5.12), respectively. For a coherent state, we can evaluate the standard deviations to be

$$\sigma_X = \sqrt{\langle X^2 \rangle - \langle X \rangle^2} = \sqrt{\frac{\hbar}{2\omega}},$$
$$\sigma_P = \sqrt{\langle P^2 \rangle - \langle P \rangle^2} = \sqrt{\frac{\hbar\omega}{2}}, \tag{5.65}$$

which satisfies the relation (5.64).

The uncertainty relation as written in (5.64) is in fact a special case of an inequality that can be derived for two arbitrary operators. As we have seen in the previous sections, the main operators that we will be dealing with are spin operators that have a different commutation relation to the position and momentum operators. For two arbitrary operators A, B, the *Schrodinger uncertainty relation* holds:

$$\sigma_A^2 \sigma_B^2 \geq \left| \frac{1}{2} \langle \{A, B\} \rangle - \langle A \rangle \langle B \rangle \right|^2 + \left| \frac{1}{2i} \langle [A, B] \rangle \right|^2. \tag{5.66}$$

Dropping the first term on the right-hand side of the inequality then coincides with the *Robertson uncertainty relation*. Evaluating the Schrodinger uncertainty relation for a coherent state again achieves a minimum uncertainty state where the left- and right-hand sides are equal.

The Schrodinger Uncertainty Relation

To prove the Schrodinger uncertainty relation, define the auxiliary states

$$|f_A\rangle = (A - \langle A \rangle)|\Psi\rangle,$$
$$|f_B\rangle = (B - \langle B \rangle)|\Psi\rangle, \tag{5.67}$$

which then can be used to write the variances as

$$\sigma_A^2 = \langle \Psi | (A - \langle A \rangle)^2 | \Psi \rangle = \langle f_A | f_A \rangle,$$
$$\sigma_B^2 = \langle \Psi | (B - \langle B \rangle)^2 | \Psi \rangle = \langle f_B | f_B \rangle. \tag{5.68}$$

Applying the Cauchy–Schwarz inequality, we have

$$\langle f_A | f_A \rangle \langle f_B | f_B \rangle \geq |\langle f_A | f_B \rangle|^2. \tag{5.69}$$

Using the fact that the modulus squared is the square of the real and imaginary parts of a complex number, we can write this in an equivalent form,

$$\sigma_A^2 \sigma_B^2 \geq \left(\frac{\langle f_A | f_B \rangle + \langle f_B | f_A \rangle}{2} \right)^2 + \left(\frac{\langle f_A | f_B \rangle - \langle f_B | f_A \rangle}{2i} \right)^2, \tag{5.70}$$

where we used (5.68). Then, using

$$\langle f_A | f_B \rangle = \langle AB \rangle - \langle A \rangle \langle B \rangle, \tag{5.71}$$

we obtain (5.66).

Since optical coherent states are minimum uncertainty states, it is reasonable to expect that spin coherent states are also minimum uncertainty states. Let us now directly evaluate the Schrodinger uncertainty relation to see whether this is so. For convenience, we take $A = S_z$ and $B = S_x$ and evaluate all the terms in (5.66) for the spin coherent state (5.8).

On the left-hand side, we have the variances of the spin operators, which can be evaluated to be

$$\begin{aligned} \sigma^2_{S_z} &= \langle S_z^2 \rangle - \langle S_z \rangle^2 \\ &= 4N|\alpha\beta|^2 = N \sin^2\theta, \end{aligned} \tag{5.72}$$

where we used the parametrization (5.34). For the other spin operators, the variances are

$$\begin{aligned} \sigma^2_{S_x} &= \langle S_x^2 \rangle - \langle S_x \rangle^2 \\ &= N(1 - (\alpha\beta^* + \beta\alpha^*)^2) = N(1 - \cos^2\phi \sin^2\theta), \\ \sigma^2_{S_y} &= \langle S_y^2 \rangle - \langle S_y \rangle^2 \\ &= N(1 - (i\alpha\beta^* - i\beta\alpha^*)^2) = N(1 - \sin^2\phi \sin^2\theta). \end{aligned} \tag{5.73}$$

The above clearly shows that the sum of the variances evaluates to

$$\sigma^2_{S_x} + \sigma^2_{S_y} + \sigma^2_{S_z} = 2N, \tag{5.74}$$

which can equally be seen by combining (5.36) and (5.37).

On the right-hand side, the anticommutator can be written

$$\langle \{S_z, S_x\} \rangle = 2\langle S_x S_z \rangle + 2i\langle S_y \rangle. \tag{5.75}$$

Evaluating the first correlation we obtain (see Ex. 5.6.3 for a workthrough)

$$\langle S_x S_z \rangle = -i\langle S_y \rangle + \frac{N-1}{N}\langle S_x \rangle \langle S_z \rangle. \tag{5.76}$$

For the commutator, we can evaluate

$$\langle [S_x, S_z] \rangle = 2i\langle S_y \rangle. \tag{5.77}$$

Putting all this together, we find that the left and right sides are equal as expected, yielding

$$\begin{aligned} \sigma^2_{S_z}\sigma^2_{S_x} &= \left| \frac{1}{2}\langle \{S_z, S_x\} \rangle - \langle S_z \rangle \langle S_x \rangle \right|^2 + \left| \frac{1}{2i}\langle [S_z, S_x] \rangle \right|^2 \\ &= N^2 \sin^2\theta (1 - \cos^2\phi \sin^2\theta). \end{aligned} \tag{5.78}$$

Thus, all spin coherent states are minimum uncertainty states in the same way as optical coherent states.

Exercise 5.6.1 Evaluate the Schrodinger uncertainty relation for an optical coherent state (5.10) and the position and momentum operators (5.12).

Exercise 5.6.2 (a) Find the variance of S_z giving the result (5.72). (b) Find the variance of S_x giving the result (5.73). Hint: The simplest way to do this is by reusing the result that you found in (a). Perform a basis transformation on the spin coherent state to make S_x take the form of S_z using

$$a_x = \frac{1}{\sqrt{2}}(a + b),$$
$$b_x = \frac{1}{\sqrt{2}}(a - b). \tag{5.79}$$

Put this in the definition of the spin coherent state (5.8) to transform it, and then use the new coefficients in the formula (5.72).

Exercise 5.6.3 Verify (5.76). Hint: First evaluate the commutator relations

$$[S_z, \alpha a^\dagger + \beta b^\dagger]$$
$$[S_x, \alpha a^\dagger + \beta b^\dagger]. \tag{5.80}$$

Use this result to commute first S_z to the right side of the expectation value, then do the same for S_x. Then simplify the expression using the fact that

$$[\alpha^* a + \beta^* b, \alpha a^\dagger + \beta b^\dagger] = 1. \tag{5.81}$$

Finally, use the expressions (5.31) and (5.32) to write the correlator in the form (5.76).

5.7 Squeezed States

5.7.1 Squeezed Optical States

In the previous sections, we saw that spin coherent states are minimum uncertainty states and play an analogous role to optical coherent states. Optical coherent states have an uncertainty that is equal in both position and momentum operators. In terms of dimensionless variables, we have

$$\sigma_x = \sigma_p = \frac{1}{\sqrt{2}}, \tag{5.82}$$

where we used (5.65) and (5.13). Another type of state that is important in the context of quantum optics is the squeezed state, where uncertainties in one direction are suppressed at the expense of the other. The optical squeezed vacuum state is

$$|re^{i\Theta}\rangle = \exp\left[-\frac{re^{-i\Theta}}{2}a^2 + \frac{re^{i\Theta}}{2}(a^\dagger)^2\right]|0\rangle, \tag{5.83}$$

where r is the squeezing parameter and Θ specifies the direction of the squeezing. This has uncertainties for $\Theta = 0$,

$$\sigma_x = \frac{e^r}{\sqrt{2}},$$
$$\sigma_p = \frac{e^{-r}}{\sqrt{2}}, \tag{5.84}$$

which is a squeezed state for the momentum. This is also a minimum uncertainty state, as the Heisenberg uncertainty relation is satisfied as an equality, $\sigma_x\sigma_p = 1/2$.

A different choice of Θ would produce squeezing in a different direction. In this section, we will explore two types of squeezed states for spin ensembles.

5.7.2 One-Axis Twisting Squeezed States

A similar type of squeezed state for a spin coherent state can be produced by the *one-axis twisting* Hamiltonian

$$H_{1A1S} = \hbar\kappa(S_z)^2, \tag{5.85}$$

where κ is the interaction constant. Here, 1A1S stands for "one-axis one-spin," in anticipation of two-spin generalizations that we will encounter in Chapter 10. This type of Hamiltonian arises from a diagonal interaction between the bosons. This corresponds to a particular form of the interaction Hamiltonian (4.10). For example, an interaction of the form

$$H_I = \frac{g_{aa}}{2}n_a^2 + g_{ab}n_a n_b + \frac{g_{bb}}{2}n_b^2 \tag{5.86}$$

contains the interaction (5.85), where g_{aa}, g_{bb} is the intraspecies interaction and g_{ab} is the interspecies interaction. To see this explicitly, we may use the relations

$$\begin{aligned} n_a &= a^\dagger a = \frac{N+S_z}{2} \\ n_b &= b^\dagger b = \frac{N-S_z}{2} \end{aligned} \tag{5.87}$$

to obtain

$$H_I = \frac{N^2}{8}(g_{aa}+g_{bb}+2g_{ab}) + \frac{N}{4}(g_{aa}-g_{bb})S_z + \frac{1}{8}(g_{aa}-2g_{ab}+g_{bb})(S_z)^2. \tag{5.88}$$

The first term contributes to a physically irrelevant global phase, the second term produces a S_z rotation if $g_{aa} \neq g_{bb}$, and the third term produces squeezing.

Let us now see what effect of the squeezing is on a spin coherent state. From (5.72) and (5.73), we can see that the variance depends upon the location of the state on the Bloch sphere. Let us choose an initial spin coherent state with parameters $\theta = \pi/2, \phi = 0$, which corresponds to spins fully polarized in the S_x direction. Applying the squeezing operation, we obtain

$$e^{-i(S_z)^2\tau}|\frac{1}{\sqrt{2}}, \frac{1}{\sqrt{2}}\rangle\rangle = \frac{1}{\sqrt{2^N}}\sum_{k=0}^{N}\sqrt{\binom{N}{k}}e^{-i(2k-N)^2\tau}|k\rangle, \tag{5.89}$$

where $\tau = \kappa t$ is the dimensionless squeezing parameter. To see the effect of the squeezing, we can calculate the variance of the spin in a generalized direction,

$$\sigma_{\boldsymbol{n}}^2 \equiv \langle(\boldsymbol{n}\cdot\boldsymbol{S})^2\rangle - \langle\boldsymbol{n}\cdot\boldsymbol{S}\rangle^2, \tag{5.90}$$

where

$$\boldsymbol{n} = (\sin\theta\cos\phi, \sin\theta\sin\phi, \cos\theta). \tag{5.91}$$

It will turn out that the largest variation in the variance occurs when $\phi = \pi/2$, which gives a spin in the direction

$$\boldsymbol{n}\cdot\boldsymbol{S} = \sin\theta S_y + \cos\theta S_z. \tag{5.92}$$

The reason for this is that the spin coherent state is originally polarized in the S_x direction, hence for small τ it remains an approximate eigenstate with small variance. For spins in the S_y–S_z plane, the variance is

$$\sigma_\theta^2 = \sin^2\theta\sigma_{S_y}^2 + \cos^2\theta\sigma_{S_z}^2 + \sin\theta\cos\theta(\langle S_y S_z + S_z S_y\rangle - 2\langle S_y\rangle\langle S_z\rangle). \tag{5.93}$$

The various quantities can be evaluated as

$$\begin{aligned}
\langle S_y^2\rangle &= \frac{N(N+1)}{2} - \frac{N(N-1)}{2}\cos^{N-2} 8\tau\cos 16\tau,\\
\langle S_y\rangle &= N\cos^{N-1} 4\tau\sin 4\tau,\\
\langle S_z^2\rangle &= N,\\
\langle S_z\rangle &= 0,\\
\langle S_y S_z + S_z S_y\rangle &= N(N+2)\cos^{N-1} 4\tau\sin 8\tau.
\end{aligned} \tag{5.94}$$

Putting these together, we obtain a dependence that is plotted in Fig. 5.3(a). For $\tau = 0$ (i.e. no squeezing), there is no dependence on the angle and we have

$$\sigma_\theta^2 = \sigma_{S_y}^2 = \sigma_{S_z}^2 = N. \tag{5.95}$$

This is again analogous to an unsqueezed optical coherent state where the variance is the same in all directions. As the squeezing is increased, $\tau > 0$, the variance depends upon the spin direction and can become smaller than (5.95). The angle at which the variance is minimized can be found by solving for $\frac{d\sigma_\theta}{d\theta} = 0$. To obtain an approximate expression for the minimum variance, we can first expand $\frac{d\sigma_\theta}{d\theta}$ to second order in τ, which yields

$$\theta_{\text{opt}} \approx -\frac{1}{2}\tan^{-1}\left(\frac{1}{2N\tau}\right), \tag{5.96}$$

where we assumed $N \gg 1$. Fig. 5.3(b) compares the exact minimum angle with the above formula. We see that the formula agrees extremely well for $N \gg 1$. For very small $\tau \ll 1$, the angle starts at $\theta_{\text{opt}} = \pi/4$ and approaches $\theta_{\text{opt}} = 0$ for large τ.

In Fig. 5.3(c), we show the variance (5.93) at the minimum angle θ_{opt}. For example, for $N = 1{,}000$ atoms, the minimum value of squeezing achieved in this case is $\sigma_\theta^2 \approx 7$ – much smaller than the unsqueezed value. The optimal squeezing time can be found for the curve Fig. 5.3(c), and the results are plotted in Fig. 5.3(d) for various N. The log-log plot reveals a power law behavior to the optimal time, which depends as

$$\tau_{\text{opt}} \approx \frac{0.3}{N^{2/3}}. \tag{5.97}$$

Kitagawa and Ueda (1993) found that the optimal squeezing that can be attained scales as

$$\frac{\sigma_\theta}{N} \propto \frac{1}{N^{2/3}}. \qquad \text{(one-axis twisting)} \tag{5.98}$$

We can now compare the variances with and without squeezing. For the original unsqueezed state, we can see from (5.95) that the standard deviation scales as

$$\frac{\sigma_\theta}{N} \propto \frac{1}{\sqrt{N}}, \qquad \text{(standard quantum limit)} \tag{5.99}$$

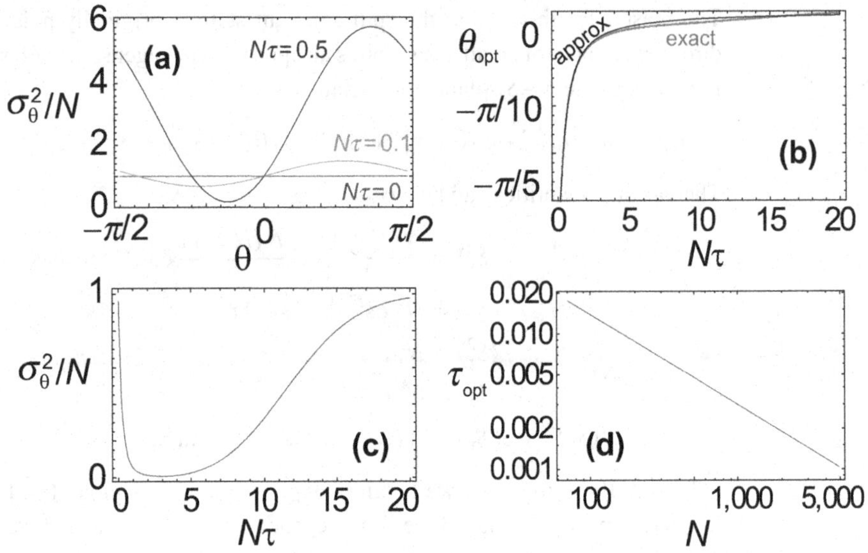

Fig. 5.3 (a) The variance of the one-axis squeezed state (5.89) given by the expression (5.93). (b) The angle for which the minimum uncertainty occurs. The points are an exact minimization of (5.93) and the line is the approximate expression (5.96). (c) Log-log plot of variance of the squeezed state (5.89) for the optimum angle θ_{opt} that gives the minimum variance. (d) The time τ_{opt} that gives the minimum squeezing for a given N plotted on a log-log scale. $N = 1{,}000$ is used for all calculations. Color version of this figure available at cambridge.org/quantumatomoptics.

which is a typical result that is characteristic of systems where the noise is dominated by quantum mechanics (as opposed to technical noise in the experiment) and is called the *standard quantum limit*. Since (5.98) decreases with N faster than (5.99), this is an indication of the squeezing effect of the one-axis twisting Hamiltonian.

Exercise 5.7.1 Evaluate the expression (5.93) including the expectation values (5.94). Hint: For these expectation values, try evaluating them by using the expanded form (5.89). For the S_z expectation values, you will see that the time dependence immediately disappears, hence you can use the results of the previous section. For the expectation values involving S_y, you should still have the time dependence. Use the fact that

$$S_y|k\rangle = -i\sqrt{(k+1)(N-k)}|k+1\rangle + i\sqrt{k(N-k+1)}|k-1\rangle \tag{5.100}$$

and

$$\sqrt{\binom{N}{k}} = \sqrt{\frac{N!}{k!\,(N-k)!}} \tag{5.101}$$

to simplify the square root factors.

5.7.3 Two-Axis Countertwisting Squeezed States

We have seen that the noise fluctuations in the one-axis twisting Hamiltonian have a scaling with N that is less than the standard quantum limit; hence it is a demonstration of squeezing. There are, however, other types of Hamiltonians that can produce a higher level of squeezing. In many systems, it has been observed that the limit of squeezing that can be obtained scales as $\sigma_\theta/N \propto 1/N$ and is commonly called the *Heisenberg limit*. This is the typical best scaling of a quantum mechanical system in a single collective state such that noise fluctuations are further suppressed. In this section, we introduce the *two-axis countertwisting* squeezed state, which attains this type of scaling.

The two-axis countertwisting Hamiltonian is given by

$$H_{2A1S} = \frac{\hbar\kappa}{2}[(S_x)^2 - (S_y)^2] = \hbar\kappa[(S_+)^2 + (S_-)^2], \tag{5.102}$$

where κ is the interaction constant. Here, 2A1S stands for "two-axis one-spin" in anticipation of the two-spin generalizations we will encounter in Chapter 10. Realizing the two-axis countertwisting Hamiltonian experimentally is not quite as straightforward as the one-axis version. One way that it can be produced is by taking advantage of the Suzuki–Trotter expansion:

$$e^{(A+B)} = \lim_{n\to\infty}\left(e^{\frac{A}{n}}e^{\frac{B}{n}}\right)^n. \tag{5.103}$$

If one wishes to perform a time evolution with the Hamiltonian (5.102), then we can perform a decomposition

$$e^{-iH_{2A1S}t/\hbar} \approx \left(e^{-i(S_x)^2\tau/2n}e^{i(S_y)^2\tau/2n}\right)^n, \tag{5.104}$$

where $\tau = \kappa t$ again. The above terms can be realized using a combination of one-axis squeezing terms and rotations according to [310]

$$\begin{aligned} e^{-i(S_x)^2\tau} &= e^{-iS_y\pi/4}e^{-i(S_z)^2\tau}e^{iS_y\pi/4} \\ e^{i(S_y)^2\tau} &= e^{-iS_x\pi/4}e^{-i(S_z)^2\tau}e^{iS_x\pi/4}. \end{aligned} \tag{5.105}$$

Let us now see the effect of the two-axis countertwisting Hamiltonian on a spin coherent state, which we choose to be the state at $\theta = 0$ this time. We define the state

$$|\psi_{2A1S}(\tau)\rangle = e^{-i[(S_x)^2-(S_y)^2]\tau}|1,0\rangle\rangle. \tag{5.106}$$

Unlike the one-axis twisting Hamiltonian, for which we can easily write down the exact wavefunction at any time, the time evolution for the two-axis countertwisting must in general be evaluated numerically. One nice feature of the two-axis countertwisting state is that the direction of the squeezing and antisqueezing occurs for the variables

$$\begin{aligned} \tilde{S}^x &= \frac{S^x + S^y}{\sqrt{2}} \\ \tilde{S}^y &= \frac{S^y - S^x}{\sqrt{2}}, \end{aligned} \tag{5.107}$$

respectively, hence no optimization needs to be performed as was required for the one-axis case.

Figure 5.4(a) shows the variance of the operator $\tilde{S}^x$. Comparing to the one-axis twisting squeezed states (Fig. 5.3(c)), we see that the standard deviation reaches

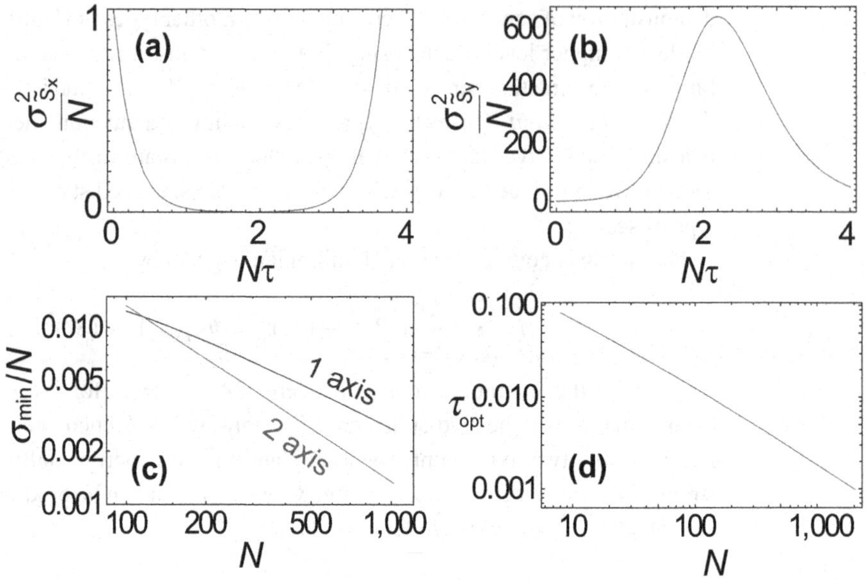

Fig. 5.4 The variance of the two-axis countertwisting squeezed state (5.106) for the spins (a) $\tilde{S}^x$ and (b) $\tilde{S}^y$ as defined in (5.107) for $N = 1{,}000$. (c) The minimum standard deviations for the one-axis and two-axis squeezed states. (d) The time τ_{opt} that gives the minimum squeezing for a given N plotted on a log-log scale. Color version of this figure available at cambridge.org/quantumatomoptics.

a smaller value for the same $N = 1{,}000$, giving a value $\sigma_{\tilde{S}^x} \approx 2$. Figure 5.4(a) shows the variance of the operator $\tilde{S}^y$ in the orthogonal direction, which displays antisqueezing. In Fig. 5.4(c), we show the scaling of standard deviations for the one-axis and two-axis squeezed states. The two-axis squeezed states indeed have a scaling that is close to the Heisenberg scaling of $\sigma_{\tilde{S}^x}/N \propto 1/N$. We finally show the optimal times to reach the minimum squeezing in 5.4(d) for the two-axis countertwisting Hamiltonian, which scales as $\tau_{\text{opt}} \propto 1/N$.

5.8 Entanglement in Spin Ensembles

We saw in the previous section that it is possible to reduce the variance of spin operators such that they are below the value in a spin coherent state. The reason for this can be understood as follows. Initially, the state is prepared in a spin coherent state (5.8), which can be viewed as an ensemble of noninteracting spins. When the squeezing operation is introduced, the spins interact, and a collective quantum state is formed. Since the state is now a single, unseparable quantum object, suitable observables can take a reduced value since it is a single quantum system. This means that due to the interactions between the bosons, we might expect that there is entanglement between the bosons. Several ways of detecting this have been proposed for spin ensemble states. In this section, we list two entanglement witnesses that serve this purpose.

5.8.1 Spin Squeezing Criterion

To quantify the degree of squeezing, it is useful to introduce a parameter

$$\xi^2 = \frac{N\sigma^2_{\theta_{\min}}}{\langle S_x\rangle^2}, \tag{5.108}$$

where $\sigma^2_{\theta_{\min}}$ is the variance as defined in (5.93) along the minimal angular direction. For zero squeezing $\tau = 0$, the variance is $\sigma^2_{\theta_{\min}} = N$ as given in (5.95), and the initial state is polarized in the S_x direction such that $\langle S_x\rangle = N$. The unsqueezed initial state thus takes a value $\xi = 1$.

Sørensen, Duan, Cirac, and Zoller found that this squeezing parameter can be used to signal entanglement. The criterion takes the simple form

$$\xi^2 < 1. \qquad \text{(for entangled states)} \tag{5.109}$$

Thus, any squeezed state also signals the presence of entanglement. If the criterion (5.109) is satisfied, it guarantees that the state is entangled. However, there are some entangled states that have $\xi^2 \geq 1$, hence this is a *sufficient* but not *necessary* criterion for entanglement.

5.8.2 Optimal Spin Squeezing Inequalities

A more sensitive entanglement criterion than the spin squeezing criterion was developed by Toth, Knapp, Gühne, and Briegel, only involving the first and second moments of the spin operators. Violation of any of the following inequalities implies entanglement:

$$\langle S_x^2\rangle + \langle S_y^2\rangle + \langle S_z^2\rangle \leq N(N+2), \qquad \text{(for separable states)} \tag{5.110}$$

$$\sigma_x^2 + \sigma_y^2 + \sigma_z^2 \geq 2N, \qquad \text{(for separable states)} \tag{5.111}$$

$$\langle S_i^2\rangle + \langle S_j^2\rangle - 2N \leq (N-1)\sigma_k^2, \qquad \text{(for separable states)} \tag{5.112}$$

$$(N-1)(\sigma_i^2 + \sigma_j^2) \geq \langle S_k^2\rangle + N(N-2), \qquad \text{(for separable states)} \tag{5.113}$$

where i, j, k take all the possible permutations of x, y, z. These criteria can detect a large range of entangled states, such as Fock states, and ground states of an antiferromagnetic Heisenberg chain.

Exercise 5.8.1 Evaluate the criteria (5.110)–(5.113) for a Fock state (5.1) with $k = N/2$. Is there any difference if $k = 0$ or $k = N$? Why?

5.9 The Holstein–Primakoff Transformation

The spin operators $S_{x,y,z}$ involve two bosonic operators a, b in the Schwinger boson representation. As we saw in (5.25), the spin operators $S_{x,y,z}$ conserve the total boson number. Suppose that we restrict the total number of bosons to N, such that we consider the space of states

$$N|\psi\rangle = N|\psi\rangle \tag{5.114}$$

where the wavefunction takes the form (5.4). In this case, we may write

$$b^\dagger b = N - a^\dagger a. \tag{5.115}$$

This suggests that we may try to eliminate the b operators and only deal with the a operators alone.

For the S_z operator, this is straightforwardly done by substituting (5.115) into the definition (5.14)

$$S_z = 2a^\dagger a - N. \tag{5.116}$$

Defining the Fock states with only the a operators,

$$|k\rangle = \frac{(a^\dagger)^k}{\sqrt{k!}}|0\rangle, \tag{5.117}$$

we observe that we obtain the same relation as (5.20).

To find a relation for the remaining off-diagonal operators, it is easiest to look at the raising and lowering operators as defined in (5.17). Looking at the action of these on the Fock states (5.18), we can see that $S_\pm$ has the effect of shifting the number of a bosons, hence it should be similar to a and $a^\dagger$, respectively. Applying these operators to (5.117) has similar bosonic factors of $\sqrt{k}$ and $\sqrt{k+1}$, except that there is another bosonic factor arising from the b operator. To account for these, we can add appropriate number operators,

$$\begin{aligned} S_+ &= a^\dagger \sqrt{N - a^\dagger a}, \\ S_- &= \sqrt{N - a^\dagger a}\, a, \end{aligned} \tag{5.118}$$

which gives the same relations as (5.18) with the use of the Fock states (5.117). The relations for the other spin operators can be obtained simply by substitution into (5.19).

The relations (5.118) and (5.116) are together taken to be the *Holstein–Primakoff transformation*, which relates spin operators to bosonic operators. The price to be paid for making the transformation is the nonlinear square root factor, which gives an infinite series in terms of its expansion as a Taylor series. Keeping just the leading order gives the approximate relation

$$\begin{aligned} S_+ &\approx \sqrt{N} a^\dagger \\ S_- &\approx \sqrt{N} a. \end{aligned} \tag{5.119}$$

This approximation is often used in the context of large spins $N \gg 1$, where the leading order is often enough to capture the basic physics.

Exercise 5.9.1 Verify that the Holstein–Primakoff transformations (5.118) acting on the Fock states (5.117) give the same relations as (5.18).

5.10 Equivalence between Bosons and Spin Ensembles

So far, we have only considered BECs consisting of indistinguishable bosons that populate several levels. In fact, under certain conditions it is possible to have ensembles of spins that are not indistinguishable to give the same types of states. For example, instead of a BEC where all the atoms occupy the same spatial quantum state, a hot gas of atoms where condensation has not occurred can be put into a spin coherent state (5.8) or a squeezed state (5.89).

How is this possible? After all, the dimension of the Hilbert space for indistinguishable and distinguishable particles do not even match. Recall from Fig. 2.1 that for N two-level indistinguishable bosons, the dimension of the Hilbert space is $N + 1$, but for N two-level distinguishable particles, the dimension is 2^N. But as we discussed in Section 2.2, there is the set of states for the spin ensemble that has analogous properties to the indistinguishable bosons. Specifically, the states (2.8), which we repeat here,

$$|k\rangle^{(\text{ens})} = \sum_{d_k=1}^{\binom{N}{k}} |k, d_k\rangle, \tag{5.120}$$

are the analogue of the Fock states for the indistinguishable bosons (5.1). These states are symmetric under particle interchange, which is a general property of any state of a BEC. Then the analogue of the general state (5.4) is

$$|\psi\rangle^{(\text{ens})} = \sum_{k=0}^{N} \psi_k |k\rangle^{(\text{ens})}. \tag{5.121}$$

The remaining states (i.e., $2^N - (N + 1)$ of them) are then unpopulated.

How can we ensure that the remaining unsymmetric states are unpopulated? This can be guaranteed if the initial state and the Hamiltonian used to prepare the given state are both symmetric. Suppose we are following the procedure to prepare the spin coherent state as in Section 5.5. There, we first prepared the state in $|k = N\rangle$, which for a spin ensemble corresponds to

$$|k = N\rangle^{(\text{ens})} = \prod_{j=1}^{N} |a\rangle_j = |aa\ldots a\rangle. \tag{5.122}$$

We have taken the state $|1\rangle \to |a\rangle$ and $|0\rangle \to |b\rangle$ in the language of Section 2.2. This is obviously symmetric under particle interchange. Next, we apply symmetric total spin operators for the ensembles

$$\begin{aligned} S_x^{(\text{ens})} &= \sum_{j=1}^{N} \sigma_x^{(j)} \\ S_y^{(\text{ens})} &= \sum_{j=1}^{N} \sigma_y^{(j)} \\ S_z^{(\text{ens})} &= \sum_{j=1}^{N} \sigma_z^{(j)}, \end{aligned} \tag{5.123}$$

where the spin operator for the jth spin is defined in Eq. (5.16). Applying the $S_x^{(\text{ens})}$ operation in this case is simple, because all the Pauli operators in (5.123) commute. The resulting state is

$$
\begin{aligned}
e^{-iS_x^{(\text{ens})}\Omega t/2}|k=N\rangle^{(\text{ens})} &= \prod_{j=1}^{N} e^{-i\sigma_x^{(j)}\Omega t/2}|a\rangle_j \\
&= \prod_{j=1}^{N}\left[\cos(\Omega t/2)|a\rangle_j - i\sin(\Omega t/2)|b\rangle_j\right],
\end{aligned}
\tag{5.124}
$$

where we used (5.41). This state is symmetric under particle interchange, since all the spins are in the same state. The general spin coherent state for ensembles is thus written

$$
|\alpha,\beta\rangle\rangle^{(\text{ens})} = \prod_{j=1}^{N}\left[\alpha|a\rangle_j + \beta|b\rangle_j\right]. \tag{5.125}
$$

The equivalence between the total spin operators (5.123) and the Schwinger boson operators (5.14) is the result of the *Jordan map*, which historically was used by Schwinger to derive the theory of quantum angular momentum in a simpler way.

The example above can be generalized to any operation involving symmetric Hamiltonians. As long as the initial state is symmetric (typically a state such as (5.122)) and the applied Hamiltonians are symmetric, such as those involving (5.123), the final state is also symmetric and can be written in the form (5.4).

We note that the symmetric Fock states (5.120) are eigenstates of the total spin operator,

$$
(\boldsymbol{S}^{(\text{ens})})^2|k\rangle^{(\text{ens})} = N(N+2)|k\rangle^{(\text{ens})}, \tag{5.126}
$$

where

$$
(\boldsymbol{S}^{(\text{ens})})^2 = (S_x^{(\text{ens})})^2 + (S_y^{(\text{ens})})^2 + (S_z^{(\text{ens})})^2. \tag{5.127}
$$

This is the equivalent result of (5.37) for the spin ensemble. This has the interpretation that the symmetric Fock states are maximum spin eigenstates for N spins. The remaining $2^N - (N+1)$ states will have a spin magnitude that is smaller than this value.

Exercise 5.10.1 Show that the state (5.125) can be expanded into the form (5.4), where the Fock states only involve the symmetric Fock states (5.120).

Exercise 5.10.2 Verify (5.126) for a symmetric Fock state (5.120).

5.11 Quasiprobability Distributions

The order parameter and Gross–Pitaevskii equation gives an effective way to understand the spatial degrees of freedom of a Bose–Einstein condensate. The usefulness of this approach is that it makes an effective single particle representation that is easily visualized. In fact, the Bose–Einstein condensate has an extremely complicated many-body wavefunction, as we explored in Chapter 1, so the order parameter is a great simplification to the full wavefunction. However, due to the weakly interacting nature of a typical BEC, it often contains much of the important information of interest. In the same way, the spin degrees of freedom alone can give a complex wavefunction, which is not always easily graspable. Methods to visualize the wavefunction are useful to understand the nature of the quantum state in a simple way.

In quantum optics, phase space representations are an indispensable tool to visualize quantum states of light. The reason is that the quantum state of a single mode of light has an infinite Hilbert space, and hence potentially an infinite number of parameters to specify it. Quasiprobability distributions such as the Q- and Wigner functions allow for a simple plot that can be interpreted in terms of the amplitude and phase of the light. For spins, we have a similar situation. Typically the number of spins N is large, and the full quantum state is then specified within a large Hilbert space. What is desirable is to again convert an arbitrary quantum state to a distribution specified by phase space. For the spin case, the natural phase space is the Bloch sphere, as already discussed in Fig. 5.2. Thus, in contrast to optical quantum states that have a distribution on a two-dimensional plane, spin states are plotted on a sphere. We can then define the spin versions of the Q- and Wigner functions.

In all the quasiprobability distributions that we discuss here, we only discuss the visualization of a general bosonic state (5.4) and the equivalent general state with spin ensembles (5.121). For the bosonic states, the state (5.4) is completely general, but (5.121) is a subclass of a more general state within the full 2^N Hilbert space. We also note that we do not discuss the P-function because it is more difficult than the analogous optical case to give a simple and consistent way of calculating it. For this reason, the Q- and Wigner functions are more commonly used, and for most applications these are sufficient to visualize the state.

5.11.1 Q-Function

The Q-function is defined in terms of the overlap of the state in question with a spin coherent state,

$$Q(\theta, \phi) = \frac{N+1}{4\pi} \langle\langle \cos\frac{\theta}{2} e^{-i\phi/2}, \sin\frac{\theta}{2} e^{i\phi/2} | \rho | \cos\frac{\theta}{2} e^{-i\phi/2}, \sin\frac{\theta}{2} e^{i\phi/2} \rangle\rangle, \quad (5.128)$$

where the factor of $\frac{N+1}{4\pi}$ is needed for the normalization. We note that the same normalization factor appears in the completeness relation for spin coherent states,

$$\frac{N+1}{4\pi} \int \sin\theta d\theta d\phi | \cos\frac{\theta}{2} e^{-i\phi/2}, \sin\frac{\theta}{2} e^{i\phi/2} \rangle\rangle\langle\langle \cos\frac{\theta}{2} e^{-i\phi/2}, \sin\frac{\theta}{2} e^{i\phi/2} | = I, \quad (5.129)$$

where I is the identity operator. We then have for any state,

$$\int Q(\theta, \phi) \sin\theta d\theta d\phi = 1. \quad (5.130)$$

The Q-function can also be interpreted as the probability of making a measurement in the spin coherent state basis, where the outcome is the spin coherent state with parameters θ, ϕ.

5.11.2 Wigner Function

The Wigner function is defined as

$$W(\theta, \phi) = \sum_{l=0}^{2j} \sum_{m=-l}^{l} \rho_{lm} Y_{lm}(\theta, \phi), \tag{5.131}$$

where $Y_{lm}(\theta, \phi)$ are the spherical harmonic functions. Here, ρ_{lm} is defined as

$$\rho_{lm} = \sum_{m_1=-j}^{j} \sum_{m_2=-j}^{j} (-1)^{j-m_1-m} \langle j m_1; j - m_2 | l m \rangle \langle j m_1 | \rho | j m_2 \rangle, \tag{5.132}$$

where $\langle j_1 m_1; j_2 m_2 | JM \rangle$ is a Clebsch–Gordan coefficient for combining two angular momentum eigenstates, $|j_1 m_1\rangle$ and $|j_2 m_2\rangle$ to $|JM\rangle$. The angular momentum eigenstates are equivalent to the Fock states (5.1), which are also eigenstates of the S_z operator. We temporarily use a different notation for these states, since this is more natural in terms of the above formula,

$$|jm\rangle = |k = j + m\rangle = \frac{(a^\dagger)^{j+m} (b^\dagger)^{j-m}}{\sqrt{(j+m)!\,(j-m)!}} |0\rangle, \tag{5.133}$$

which are also called Dicke states. In the above,

$$j = \frac{N}{2} \tag{5.134}$$

throughout. Since the Clebsch–Gordan coefficients are zero unless $m = m_1 - m_2$, one of the summations may be removed to give the formula

$$W(\theta, \phi) = \sum_{l=0}^{2j} \sum_{m_1=-j}^{j} \sum_{m_2=-j}^{j} (-1)^{j-2m_1+m_2} \langle j m_1; j - m_2 | l m_1 - m_2 \rangle Y_{l m_1 - m_2}(\theta, \phi) \times \langle j m_1 | \rho | j m_2 \rangle. \tag{5.135}$$

5.11.3 Examples

We now show several examples of quasiprobability distributions of spinor BEC states. Both the Q- and Wigner functions have a distribution that is defined in terms of angular variables θ, ϕ, and the natural way to plot this is on the surface of a sphere, much as we represented the spin coherent state on the Bloch sphere in Fig. 5.2.

5.11.3.1 Spin Coherent States

Figure 5.5 shows the Q-distribution for two examples of a spin coherent state. In both cases, the distributions are of a Gaussian form, centered at different locations on the sphere. The Gaussian nature of the Q-functions can be easily deduced using

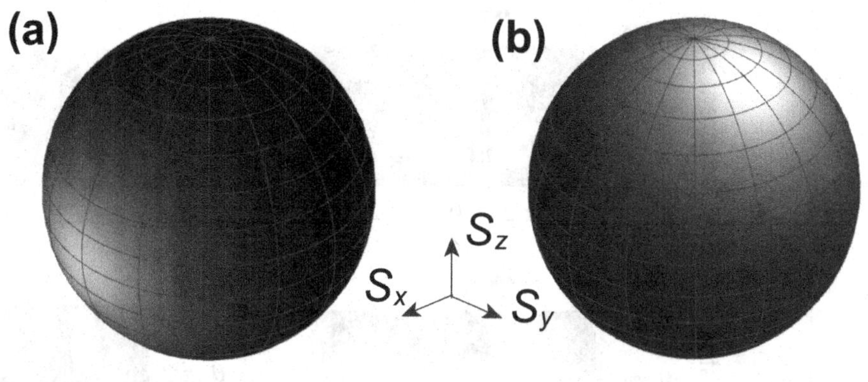

Fig. 5.5 Q-distributions for spin coherent states (5.8) of two-component Bose–Einstein condensates with $N = 10$ atoms. (a) $\theta = \pi/2, \phi = 0$, (b) $\theta = \pi/8, \phi = \pi/2$ for the parametrization (5.34). Color version of this figure available at cambridge.org/quantumatomoptics.

(5.153), which shows that the overlap of two spin coherent states are of this form. For example, for the spin coherent state centered at θ_0, ϕ_0, the Q-distribution is

$$\begin{aligned} Q(\theta,\phi) &= \frac{N+1}{4\pi}\left|\langle\langle \cos\frac{\theta_0}{2}e^{-i\phi_0/2}, \sin\frac{\theta_0}{2}e^{i\phi_0/2} | \cos\frac{\theta}{2}e^{-i\phi/2}, \sin\frac{\theta}{2}e^{i\phi/2}\rangle\rangle\right|^2 \\ &= \frac{N+1}{4\pi}\cos^{2N}\left(\frac{\Delta\Theta}{2}\right) \\ &\approx \frac{N+1}{4\pi}e^{-N(\Delta\Theta)^2/4}, \end{aligned} \tag{5.136}$$

where

$$\cos\Delta\Theta = \cos\theta\cos\theta_0 + \sin\theta\sin\theta_0\cos(\phi-\phi_0). \tag{5.137}$$

From (5.136), it is clear that for larger N, the distributions become narrower.

Although plotting the quasiprobability distribution on sphere as shown in Fig. 5.5 is the most natural, it is not convenient for obvious reasons – you cannot see the other side of the sphere without rotating it. A simple solution to this is to flatten the sphere as you would with a world map, which inevitably distorts the distribution. Fig. 5.6(a) and (b) shows the same spin coherent states as shown in Fig. 5.5 using the Mercator projection. Near the center of the projection (Fig. 5.6(a)) the distribution appears Gaussian, but near the poles the distribution is heavily distorted (Fig. 5.6(b)). We must thus be careful to interpret the distribution in the correct way when using such projected maps.

Figure 5.7(a) and (b) shows the Wigner distributions for the same spin coherent states. We again see a similar Gaussian-shaped distribution as given by the Q-functions. A simplified expression for the Wigner function of a general spin coherent state can obtained by noticing that it must be the same as that of a displaced spin coherent state starting from the poles. Recall from Section 5.12.2 that there are two special spin coherent states that are also Fock states. For these cases, the evaluation of the Wigner function simplifies, and in displacing the distribution, one obtains

$$W(\theta,\phi) = N!\sum_{l=0}^{N}\sqrt{\frac{2l+1}{(N-l)!\,(l+1+N)!}}Y_{l0}(\Delta\Theta,0), \tag{5.138}$$

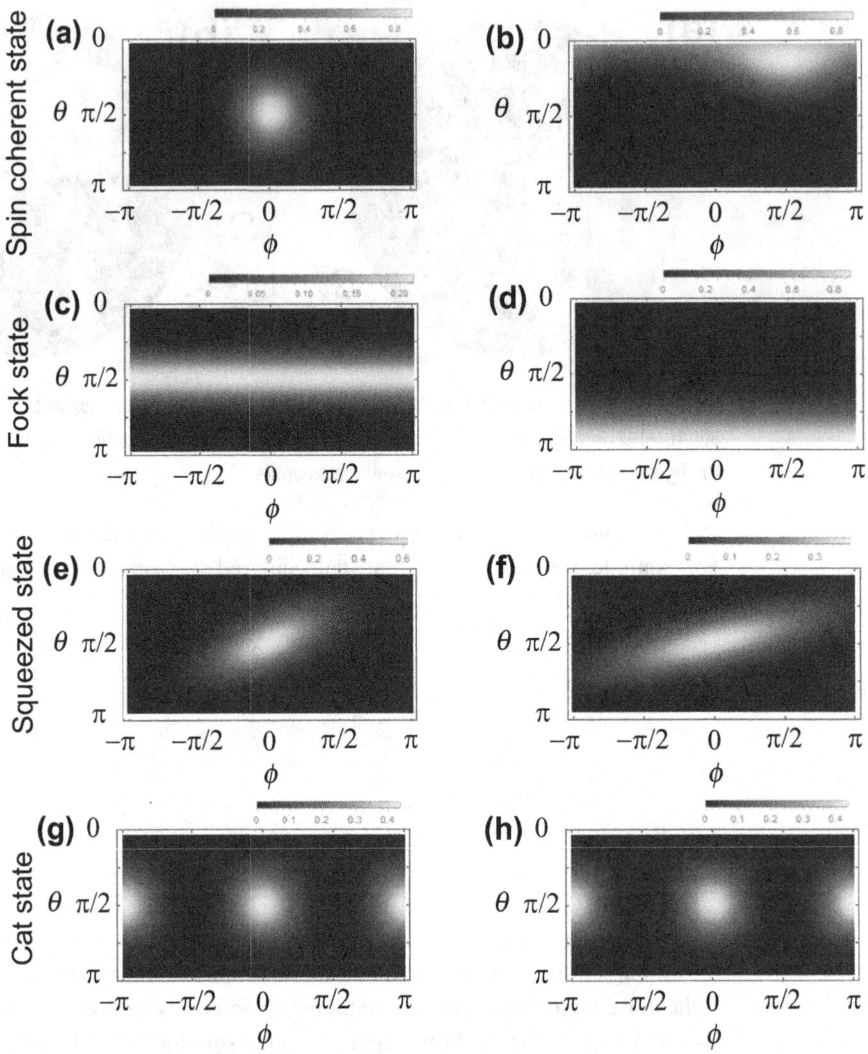

Fig. 5.6 Q-distributions for various states of a two-component Bose–Einstein condensate with $N = 10$ atoms. States shown are: spin coherent state (5.8) (a) $\theta = \pi/2, \phi = 0$, (b) $\theta = \pi/8, \phi = \pi/2$ for the parametrization (5.34); Fock state (5.1) (c) $k = N/2$, (d) $k = 0$; one-axis twisting squeezed state (5.89) (e) $\tau = 1/2N$, (f) $\tau = 1/N$; Schrodinger cat state (5.146) (g) $\sigma = +1$, (h) $\sigma = -1$. Color version of this figure available at cambridge.org/quantumatomoptics.

where the spherical harmonic function in this case is given by

$$Y_{l0}(\theta, \phi) = \sqrt{\frac{2l+1}{4\pi}} P_l(\cos\theta) \tag{5.139}$$

and $P_l(x)$ is the Legendre polynomial. We see generally the same type of distribution, but the Wigner functions have a narrower distribution. This is also true for optical Q- and Wigner functions, which share this same characteristic.

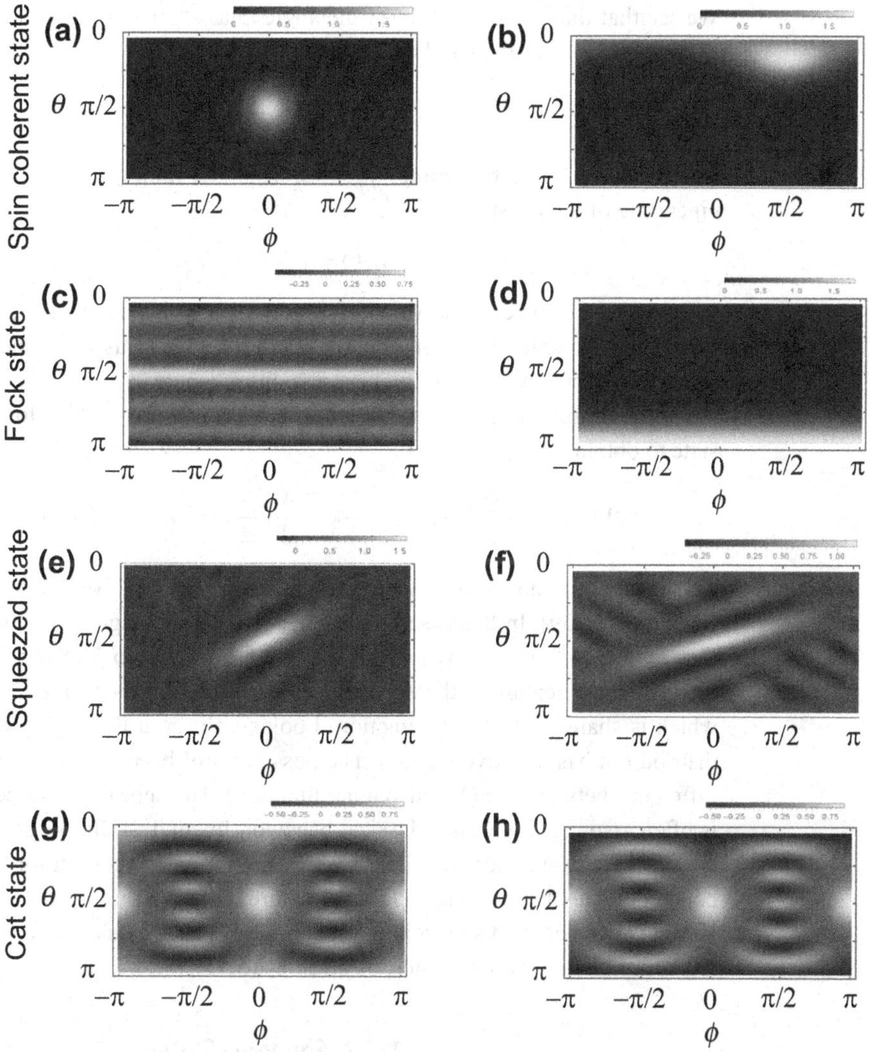

Fig. 5.7 Wigner distributions for various states of a two-component Bose–Einstein condensate with $N = 10$ atoms. States shown are: spin coherent state (5.8) (a) $\theta = \pi/2, \phi = 0$, (b) $\theta = \pi/8, \phi = \pi/2$ for the parametrization (5.34); Fock state (5.1) (c) $k = N/2$, (d) $k = 0$; one-axis twisting squeezed state (5.89) (e) $\tau = 1/2N$, (f) $\tau = 1/N$; Schrodinger cat state (5.146) (g) $\sigma = +1$, (h) $\sigma = -1$. Color version of this figure available at cambridge.org/quantumatomoptics.

5.11.3.2 Fock States

The Q-function of a Fock state $|k\rangle$ is

$$\begin{aligned} Q(\theta, \phi) &= \frac{N+1}{4\pi} |\langle k| \cos\frac{\theta}{2} e^{-i\phi/2}, \sin\frac{\theta}{2} e^{i\phi/2} \rangle\rangle|^2 \\ &= \frac{N+1}{4\pi} \binom{N}{k} \cos^{2k}\frac{\theta}{2} \sin^{2N-2k}\frac{\theta}{2}. \end{aligned} \tag{5.140}$$

We see that this is independent of the angle ϕ as shown in Fig. 5.6(c) and (d). The Q-distribution is peaked at the location

$$\cos\theta = \frac{2k - N}{N}, \tag{5.141}$$

which can be found by setting $\frac{dQ}{d\theta} = 0$. This is a rather intuitive result, since the eigenstate of a Fock state is

$$S_z|k\rangle = (2k - N)|k\rangle. \tag{5.142}$$

Meanwhile, the expectation value of S_z for a spin coherent state is given by (5.35). Since the Q-function is nothing but the overlap with various spin coherent states, the maximum occurs when these two expressions coincide.

For the Wigner function, we can evaluate the expression (5.135) for a general Fock state to obtain

$$W(\theta, \phi) = \sum_{l=0}^{N} (-1)^{N-k} \langle \frac{N}{2} k - \frac{N}{2}; \frac{N}{2}\frac{N}{2} - k|l0\rangle Y_{l0}(\theta, \phi). \tag{5.143}$$

This again has no ϕ dependence because it only involves spherical harmonic functions (5.139). In this case, the Wigner function has significant differences from that of the Q-function, as can be observed from Fig. 5.7(c) and (d). The main distinguishing feature is the appearance of several ripples around the largest peak, which is shared with the Q-function. Looking closer at the scale, we note that the distribution has negative regions. The possibility of becoming negative is one large difference between the Q- and Wigner functions. The appearance of negative regions is often associated with nonclassical behavior. In contrast, the Wigner functions for the spin coherent states were positive. A common interpretation for this is that spin coherent states have a quasi-classical nature, particularly when N becomes large. However, we note that spin coherent states always have a quantum nature in that they are minimum uncertainty states, as given by (5.95).

5.11.3.3 Squeezed States

The Q-function for the squeezed state (5.89) can be evaluated to be

$$Q(\theta, \phi) = \frac{N+1}{2^{N+2}\pi} \left| \sum_{k=0}^{N} \binom{N}{k} e^{i(N-2k)(\phi/2+(N-2k)\tau)} \cos^k \frac{\theta}{2} \sin^{N-k} \frac{\theta}{2} \right|^2, \tag{5.144}$$

which is plotted in Fig. 5.6 for two squeezing times. We see that the Gaussian distribution becomes stretched in a diagonal direction. For longer squeezing times, the distribution gets stretched to a greater degree, and the angle tends towards a smaller gradient in the θ–ϕ plane.

An intuitive explanation for why the distribution becomes elongated in a diagonal direction can be obtained by rewriting the form of the squeezing interaction as

$$e^{-i(S_z)^2\tau} = \sum_{k=0}^{N} e^{-i(2k-N)\tau S_z} |k\rangle\langle k|. \tag{5.145}$$

Recalling from (5.62) that application of an operator $e^{-i\theta S_z}$ produces a rotation around the S_z axis, we can interpret the squeezing term to be a rotation of the state around the S_z axis by an angle $2(2k-N)\tau$. The factor $(2k-N)$ can be either positive or negative, depending on the Fock state in question. Since the Fock states are represented by a position (5.141) on the Bloch sphere, the lower hemisphere rotates in a clockwise direction, while the upper hemisphere rotates in an anticlockwise direction.

Comparing to the discussion of Section 5.7, we see a clearer picture of the effects of the squeezing. Previously, we analyzed the variances in various spin directions and found that there was a particular direction where it was at a minimum, as seen in Fig. 5.3(a) and (5.96). We can attribute this to the elongation of the distribution, which minimizes the variance in a direction perpendicular to the stretching. According to (5.96), the angle of elongation starts in the 45° direction and rotates to 0° in the θ–ϕ plane.

The Wigner functions for the squeezed states are shown in Fig. 5.7(e) and (f). The main diagonal feature is similar to the Q-function, except that it has a narrower distribution, as was observed for the spin coherent and Fock states. The major difference is that the remaining parts of the distribution are less than even the Q-function case. These arise primarily because of the relatively small number of atoms $N = 10$ that are used in these plots. Such small N Wigner functions tend to always have fluctuations that arise due to the sum of a relatively small number of spherical harmonic functions. An interesting feature of Fig. 5.7(e) and (f) is the presence of negative regions of the Wigner functions, which are associated with nonclassical behavior. The negative regions tend to increase with τ, showing that the squeezing operation produces increasingly nonclassical states.

5.11.3.4 Schrodinger Cat States

The final state we examine are Schrodinger cat states. These are defined as

$$|\text{cat}^{\pm}\rangle = \frac{1}{\sqrt{2}}\left(|\cos\frac{\theta_0}{2}e^{-i\phi_0/2}, \sin\frac{\theta_0}{2}e^{i\phi_0/2}\rangle\rangle \pm |-\sin\frac{\theta_0}{2}e^{i\phi_0/2}, \cos\frac{\theta_0}{2}e^{-i\phi_0/2}\rangle\rangle \right), \tag{5.146}$$

where the two states in the superposition are two spin coherent states at antipodal points and are orthogonal (5.155). For the special case of $\alpha = \beta = 1/\sqrt{2}$, the superposition causes either the even or odd Fock states to completely cancel,

$$\left|\text{cat}^{\pm}\left(\alpha = \beta = \frac{1}{\sqrt{2}}\right)\right\rangle = \frac{1}{\sqrt{2^{N-1}}} \sum_{k \in \text{even, odd}} \sqrt{\binom{N}{k}}|k\rangle, \tag{5.147}$$

where the even terms are for the + superposition and odd for the − superposition.

The Q-functions for both types of Schrodinger cat states are given by

$$\begin{aligned} Q(\theta,\phi) &= \frac{N+1}{8\pi}\left[\cos^{2N}\left(\frac{\Theta}{2}\right) + \cos^{2N}\left(\frac{\Theta'}{2}\right) \pm 2\cos^{N}\left(\frac{\Theta}{2}\right)\cos^{N}\left(\frac{\Theta'}{2}\right)\right] \\ &\approx \frac{N+1}{8\pi}\left[\cos^{2N}\left(\frac{\Theta}{2}\right) + \cos^{2N}\left(\frac{\Theta'}{2}\right)\right], \end{aligned} \tag{5.148}$$

where

$$\cos \Delta\Theta = \cos\theta \cos\theta_0 + \sin\theta \sin\theta_0 \cos(\phi - \phi_0)$$
$$\cos \Delta\Theta' = -\cos\theta \cos\theta_0 - \sin\theta \sin\theta_0 \cos(\phi + \phi_0). \tag{5.149}$$

Here, $\Delta\Theta, \Delta\Theta'$ is the great circle angle from (θ, ϕ) to $(\theta_0, \phi_0), (\theta_0 + \pi, -\phi_0)$, respectively. In the second line of (5.148), we have dropped the last term, since this is small for $N \gg 1$. The reason this term is small is that the cross term is a product of two overlaps between the point (θ, ϕ) and two antipodal points. This term is thus exponentially suppressed according to (5.153), regardless of the choice of (θ, ϕ).

A numerical plot of the Q-function is shown in Fig. 5.6(g) and (h). We see that both distributions for the even and odd cat states are identical, according to the approximation given in (5.148). The dominant structure is simply a combination of two spin coherent state distributions, as predicted by (5.148). The Wigner function as shown in Fig. 5.7(g) and (h) shows a very different distribution. In addition to the peaks associated with the spin coherent states, strong fringes appear connecting the two peaks. The fringes show strong negative regions, indicating the highly nonclassical nature of the state. The two types of cat states also show a complementary structure, where the negative regions of one are positive in the other. For optical Wigner functions, superpositions of coherent states also exhibit a similar structure, with fringes appearing connecting the coherent states. In this case, the Wigner function provides more information about the structure of the state, distinguishing between the types of quantum state.

Exercise 5.11.1 Verify (5.140) and show that the peak of the distribution occurs at the angle (5.141).

Exercise 5.11.2 Verify (5.138). First evaluate (5.135) for the case

$$\rho = |k = N\rangle\langle k = N| = |1, 0\rangle\rangle\langle\langle 1, 0|. \tag{5.150}$$

Then use the fact that the great circle angle between (θ, ϕ) and (θ_0, ϕ_0) is (5.137).

5.12 Other Properties of Spin Coherent States and Fock States

Here we list several other important relations between spin coherent states and Fock states.

5.12.1 Overlap of Two Spin Coherent States

Like optical coherent states, the spin coherent states are in general not orthogonal. The overlap of two spin coherent states can be evaluated to be

$$\left\langle\left\langle \cos\frac{\theta'}{2} e^{-i\phi'/2}, \sin\frac{\theta'}{2} e^{i\phi'/2} \right| \left. \cos\frac{\theta}{2} e^{-i\phi/2}, \sin\frac{\theta}{2} e^{i\phi/2} \right\rangle\right\rangle = \cos^N\left(\frac{\Delta\Theta}{2}\right), \tag{5.151}$$

where

$$\cos\Delta\Theta = \cos\theta\cos\theta' + \sin\theta\sin\theta'\cos(\phi - \phi') \tag{5.152}$$

is the central angle (i.e. the angle at the center of the sphere for the great-circle distance) between the two spin coherent states. For $N \gg 1$, the powers of cosine can be well approximated by a Gaussian; hence, we have

$$\left\langle\!\left\langle \cos\frac{\theta'}{2}e^{-i\phi'/2}, \sin\frac{\theta'}{2}e^{i\phi'/2} \middle| \cos\frac{\theta}{2}e^{-i\phi/2}, \sin\frac{\theta}{2}e^{i\phi/2} \right\rangle\!\right\rangle \approx e^{-N(\Delta\Theta)^2/8}. \tag{5.153}$$

For a given spin coherent state, there is always another spin coherent state that is exactly orthogonal to it. From (5.151), one can easily deduce that this is when $\Delta\Theta = \pi$, which is the state that is at the antipodal point to it. That is, if

$$\begin{aligned} \theta' &= \theta + \pi \\ \phi' &= -\phi, \end{aligned} \tag{5.154}$$

then

$$\langle\langle -\beta^*, \alpha^* | \alpha, \beta \rangle\rangle = 0. \tag{5.155}$$

5.12.2 Equivalence of Fock States and Spin Coherent States

In general, the Fock states (5.1) are a completely different class of state to spin coherent states (5.8). However, there are two states where the two classes coincide. These are the extremal Fock states:

$$\begin{aligned} |k=0\rangle &= \frac{(b^\dagger)^N}{\sqrt{N!}} = |0,1\rangle\rangle \\ |k=N\rangle &= \frac{(a^\dagger)^N}{\sqrt{N!}} = |1,0\rangle\rangle. \end{aligned} \tag{5.156}$$

In both these cases, all the bosons are in the same state, which is essentially the definition of a spin coherent state.

5.12.3 Transformation between Fock States in Different Bases

The Fock states (5.1) are eigenstates of the S_z operator. For a more general spin operator,

$$\boldsymbol{n} \cdot \boldsymbol{S} = \sin\theta\cos\phi S_x + \sin\theta\sin\phi S_y + \cos\theta S_z, \tag{5.157}$$

where $\boldsymbol{n} = (\sin\theta\cos\phi, \sin\theta\sin\phi, \cos\theta)$ is a unit vector, the Fock states are defined as

$$|k\rangle_{\boldsymbol{n}} = \frac{(c^\dagger)^k (d^\dagger)^{N-k}}{\sqrt{k!\,(N-k)!}}|0\rangle, \tag{5.158}$$

where

$$\begin{aligned} c &= \cos\frac{\theta}{2}e^{-i\phi/2}a + \sin\frac{\theta}{2}e^{i\phi/2}b \\ d &= \sin\frac{\theta}{2}e^{-i\phi/2}a - \cos\frac{\theta}{2}e^{i\phi/2}b. \end{aligned} \tag{5.159}$$

The Fock states satisfy

$$\boldsymbol{n} \cdot \boldsymbol{S}|k\rangle_{\boldsymbol{n}} = (2k - N)|k\rangle_{\boldsymbol{n}} \tag{5.160}$$

in the same way as we have seen already for the special cases of $\boldsymbol{n} = (1, 0, 0)$ and $\boldsymbol{n} = (0, 1, 0)$ in (5.24).

In order to convert the Fock state $|k\rangle_{\boldsymbol{n}}$ to the S_z basis operators $|k\rangle$, one may use the result

$$\langle k'|e^{-iS_y\theta/2}|k\rangle = \sqrt{k!\,(N-k)!\,k'!\,(N-k')!}$$
$$\times \sum_{n=\max(k'-k,0)}^{\min(k',N-k)} \frac{(-1)^n \cos^{k'-k+N-2n}(\theta/2)\sin^{2n+k-k'}(\theta/2)}{(k'-n)!\,(N-k-n)!\,n!\,(k-k'+n)!}. \quad (5.161)$$

We may then write

$$\begin{aligned} |k\rangle_{\boldsymbol{n}} &= e^{-iS_z\phi/2}e^{-iS_y\theta/2}|k\rangle \\ &= e^{-iS_z\phi/2}\left(\sum_{k'}|k'\rangle\langle k'|\right)e^{-iS_y\theta/2}|k\rangle \\ &= \sum_{k'} e^{-i(2k'-N)\phi/2}\langle k'|e^{-iS_y\theta/2}|k\rangle|k'\rangle, \end{aligned} \quad (5.162)$$

where the terms on the expansion are given by (5.161). For example, for eigenstates of the S_x operator where $\theta = \pi/2, \phi = 0$, the Fock states are

$$|k\rangle_x = \frac{1}{\sqrt{2^N}}\sum_{k'=0}^{N}\sqrt{k!\,(N-k)!\,k'!\,(N-k')!}$$
$$\times \sum_{n=\max(k'-k,0)}^{\min(k',N-k)} \frac{(-1)^n}{(k'-n)!\,(N-k-n)!\,n!\,(k-k'+n)!}|k'\rangle. \quad (5.163)$$

For the eigenstates of the S_y operator, we use parameters $\theta = \pi/2, \phi = \pi/2$ and obtain a similar expression to (5.163) but with an extra factor of $e^{i(N-2k')\pi/4}$ in the sum. We note that for $k \ll N/2$ or $k \gg N/2$, the overlap $|\langle k'|k\rangle_x|^2$ is well approximated by the modulus squared of the quantum harmonic oscillator eigenstates.

Exercise 5.12.1 Set $\phi = \phi'$ and confirm that the formula (5.151) gives the correct result.

Exercise 5.12.2 Expand the first three terms of (5.151) as a Taylor series and confirm that it agrees with the Taylor expansion of (5.153).

5.13 Summary

In this chapter, we have seen that there is a close analogy between optical states and spin states. We summarize the correspondences in Table 5.1. The spin coherent states play an analogous role to coherent states and are minimum uncertainty states. These form a set of nonorthogonal states with a Gaussian overlap with respect to the great circle distance between two spin coherent states. Due to three operators, rather than two for the optical case, the phase space is a two-dimensional space with

Table 5.1 Equivalencies between optics and spins.

Quantity	Optics	Spin
Coherent state	$\lvert\alpha\rangle = e^{-\lvert\alpha\rvert^2/2} e^{\alpha a^\dagger} \lvert 0\rangle$	$\lvert\alpha,\beta\rangle\rangle = \frac{1}{\sqrt{N!}}(\alpha a^\dagger + \beta b^\dagger)^N \lvert 0\rangle$
Fock state	$\lvert n\rangle = \frac{1}{\sqrt{n!}}(a^\dagger)^n \lvert 0\rangle$	$\lvert k\rangle = \frac{1}{\sqrt{k!(N-k)!}}(a^\dagger)^k (b^\dagger)^{N-k} \lvert 0\rangle$
Overlap of coherent states	$\lvert\langle\alpha'\lvert\alpha\rangle\rvert^2 = e^{-\lvert\alpha-\alpha'\rvert^2}$	$\lvert\langle\langle\alpha',\beta'\lvert\alpha,\beta\rangle\rangle\rvert^2 = \cos^{2N}(\Delta\Theta/2)$ $\approx e^{-N(\Delta\Theta)^2/4}$
Observables	$X = \frac{a+a^\dagger}{\sqrt{2}}$ $P = \frac{a+a^\dagger}{\sqrt{2}}$	$S_x = a^\dagger b + b^\dagger a$ $S_y = -ia^\dagger b + ib^\dagger a$ $S_z = a^\dagger a - b^\dagger b$
Phase space	2-dimensional plane (X, P)	3-dimensional sphere (S_x, S_y, S_z)
Squeezing Hamiltonian	$e^{-i\Theta}a^2 + e^{i\Theta}(a^\dagger)^2$	$(S_z)^2$
Displacement Hamiltonian	$e^{-i\Theta}a + e^{i\Theta}a^\dagger$	$\boldsymbol{n} \cdot \boldsymbol{S}$
Quasiprobability distributions	P-, Q-, Wigner functions	P-, Q-, Wigner functions

a spherical topology. Analogous quasiprobability distributions such as the Q- and Wigner functions can be defined for the spin case equally, in the same way as the optical case.

5.14 References and Further Reading

- Section 5.2: Theoretical works discussing the mathematics of spin coherent states [384, 19, 225]. Edited volume on coherent states [48]. Book on coherent states taking a group theoretical perspective [363]. Effect of interactions in spinor Rb BECs [265].
- Section 5.3: Original work introducing the Schwinger boson representation [250]. Edited volume by Schwinger on angular momentum [420]. Textbooks further discussing Schwinger boson representation [407, 23].
- Section 5.4: Original work discussing spin coherent state expectation values [384]. Further discussion can be found in the textbook [23].
- Section 5.5: Experimental preparation of a spin coherent state has been performed in [202, 467, 50]. The original work for Ramsey interference [388]. Textbook-level discussion of spin transitions and interference [345].
- Section 5.6: Original work on the Robertson uncertainty relation [397]. Heisenberg's original paper on the uncertainty relation [209]. Textbooks further discussing the topic [421, 477, 168]. Original works showing that atomic spin coherent states have minimal uncertainty [19, 25]. Review article focusing on atomic spin squeezing [183].
- Section 5.7.1: Experimental observation of an optical squeezed state [431]. Using cold atoms to squeeze light [284]. Early review article on optical squeezed states [476]. Textbooks and review articles discussing squeezed optical states [59, 421, 477, 168].

- Sections 5.7.2 and 5.7.3: Original theoretical works introducing spin squeezing [262, 492, 491]. Experimental demonstrations of spin squeezing [192, 279, 355, 18, 451, 138, 394, 297]. Further theoretical approaches to spin squeezing [280, 434, 457, 52]. Review articles on spin squeezing [315, 183]. Theoretical scheme to turn one-axis squeezing to two-axis squeezing [310].
- Section 5.8: Theoretical works relating entanglement in spin ensembles to squeezing [434, 435, 491, 479, 268, 463, 464, 462, 106]. Other theoretical works investigating entanglement in BECs [380, 210, 123, 334, 217, 206]. Review article on entanglement detection [188]. Experimental observation of entanglement [192, 138, 159, 276, 287, 141].
- Section 5.9: Original paper introducing the Holstein–Primakoff transformation [224]. Textbooks discussing the transformation [83, 263]. Review article for atomic ensemble systems that use the transformation [199].
- Section 5.10: Original works discussing the equivalence between bosons and spin ensembles [250, 420]. Textbook-level discussion of the topic [407].
- Section 5.11: Original reference introducing the spin Q-function [293] and the spin Wigner function [121]. Other definitions of quasiprobability distributions for spins [422, 459].
- Section 5.12.1: Original references discussing spin coherent states [384, 19].
- Section 5.12.3: Equation (5.161) can be found in chapter 6 of [456]. Other textbooks that discuss the transformation [61, 407].

6 Diffraction of Atoms Using Standing Wave Light

6.1 Introduction

Diffraction of waves – like sound waves, water waves, and light waves – is characterized by the spreading or "flaring" of waves after passing through an aperture. For diffraction to occur, the size of the aperture must be comparable to the wavelength of the incident wave. A classic diffraction experiment was performed by Thomas Young in 1801, where a monochromatic beam of light is coherently split into two distinct beams after passing through two slits. In the region beyond the aperture, the split beams diffract and produce interference fringes on a screen placed a large distance away in comparison with the distance between the slits. As with other waves mentioned above, matter waves like electrons and neutrons also diffract after passing through an aperture. For instance, in electron beam diffraction, highly energetic electrons that are incident on a periodic crystalline structure diffract in a predictable manner prescribed by Bragg's law.

Beyond particles such as photons, electrons, and neutrons, Young's double slit experiment has been performed with atoms. In 1991, a highly excited helium atom was interfered after passing through two slits in gold foil. In contrast to the gold foil technique, most modern experiments use apertures made from light crystals to diffract atoms. This will be the focus of this chapter. We begin our discussion in Section 6.2 by reviewing the diffraction of light and matter waves. Next, we describe the interaction with detuned off-resonant light and solve the resulting dynamics under Bragg conditions and Raman transitions. The results of this chapter are used in the subsequent two chapters discussing metrology. For experiments that use the results we refer the reader to references found at the end of this chapter.

6.2 Theory of Diffraction

Diffraction gratings consist of many slits that are spaced at a constant interval of distance d between the slits. Gratings are often used to study different light sources. When light passes through a slit, it diffracts in accordance with Huygens's wave theory, where each slit acts as a source of light waves that produces a secondary wavefront. The wavefronts spread as they advance and give rise to diffraction. Thus, it is possible for a light wave from one slit to interfere with a light wave from another slit. The final interference pattern depends on the directional angle measured from the horizontal at which the rays are examined. To see this, consider a grating illuminated by a light wave of wavelength λ as shown in Fig. 6.1. Each slit on the grating is a

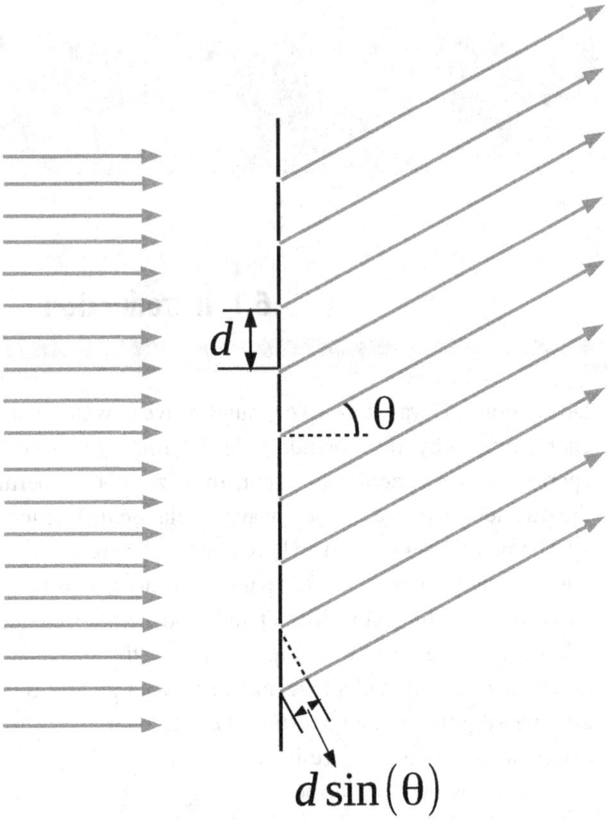

Fig. 6.1 The diffraction of light by a grating. Each slit on the grating is a source of light waves, and the slits are at equal distances d from one another. The wave from each slit is nearly parallel to the others, and the path difference of the light waves from any adjacent slits is $d \sin \theta$.

source of a light wave that emits rays of light that are mutually parallel to each other. They are also in phase as they leave the slit. As the rays propagate further toward a point on a screen, it is observed that the path difference between the rays depends on the angle a ray makes with the horizontal. For any two adjacent rays, the path difference between them is $d \sin \theta$. On the screen placed a large distance from the slit, bright bands are observed when all the waves arriving at the point on the screen are in phase. This is realized if the path difference is equal to a wavelength or integral multiples of a wavelength. Therefore, the maximum in a bright band is observed if the path difference is

$$d \sin \theta = m\lambda, \tag{6.1}$$

where $m = 0, \pm 1, \pm 2, \pm 3, \ldots$ is the diffraction order. To observe diffraction from the grating, the separation distance d between the slits should be large in comparison with the wavelength λ of light. Otherwise, the slits would appear to be the same source and no diffraction pattern would be observed. Because of this requirement, a diffraction grating made for one wave of a particular wavelength may not be suitable in the study of another wave of a very different wavelength.

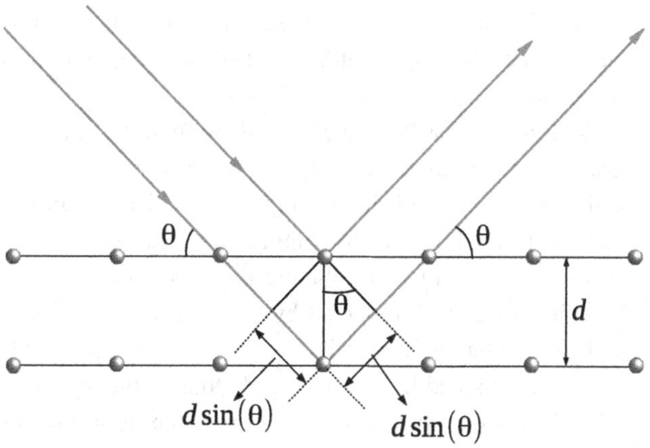

Fig. 6.2 The diffraction of a matter wave by a crystal. Each plane consists of atoms arranged in a regular fashion, and the planes are separated by a distance d. For waves incident on the crystal at angle θ to the plane, waves from the bottom plane travel an extra distance of $2d \sin \theta$ relative to those from the upper plane. Color version of this figure available at cambridge.org/quantumatomoptics.

Perhaps the simplest way to realize matter wave interference is to observe the diffraction of electrons in a crystal lattice. The slits in the crystal – NaCl, for instance – are made of atoms, which are arranged in a specific order. When matter waves are incident on a crystal, the atoms forming the crystal reflect the matter wave at regular intervals. The intensity of the emerging matter wave is very strong in a particular direction, which corresponds to constructive interference. For example, suppose that the incident matter wave makes an angle θ with the planes of the crystal, as shown in Fig. 6.2. Each plane is at a constant spacing d, and we consider two planes as shown in Fig. 6.2. Matter waves reflected from the upper plane travel a smaller distance compared with those reflected from the bottom plane. Waves from the two planes would reinforce at a point on the screen if the path difference $2d \sin \theta$ traveled by the waves is an integer multiple of the wavelength λ of the matter waves. Thus, a maximum exists within a constructive interference if

$$2d \sin \theta = m\lambda, \tag{6.2}$$

where $m = 1, 2, 3, \ldots$.

6.3 Ultra-cold Atom Interaction with Standing Light Wave

Material gratings, such as the crystals described in Section 6.2, work well to perform diffraction of particles such as electrons and neutrons but are less suited to diffraction of atoms. This is partly because atoms have a large mass compared to that of electrons and neutrons, which results in a much smaller de Broglie wavelength for a given velocity. Also, due to the charge neutrality of atoms, producing an energetic atom with high velocity was almost impossible until high-energy helium was produced via high-impact electron excitation. But even at that, the challenge still remained on how to diffract a stationary atom or an atom with negligible velocity, as found in ultra-cold atoms. The answer came with advances in techniques for atomic manipulation and better understanding of atom-light interactions that are used to impart momentum on an atom.

Photons possess both energy and momentum. A photon in a light wave with wave vector κ_l has a momentum $\hbar\kappa_l$. When an atom absorbs a photon, it absorbs its energy and momentum, and the excited atom will move in the direction of the absorbed photon. To see how the momentum transfer is used to provide directed motion for the atoms, consider a two-level atom that has a mirror to the right of it. Light traveling from left to right is reflected by the mirror. The reflected light traveling to the left and the incident light traveling to the right superpose to form a standing wave. Thus, the atom is bathed by streams of photons in the light wave traveling to the right and left. If the atom absorbs a photon from the light traveling to the right, it receives a momentum kick to the right, and the excited atom would start moving to the right with momentum $\hbar\kappa_l$. The excited atom can interact with either of the light waves, which will result in a stimulated emission. If the excited atom interacts with the light wave traveling to the right, it emits a photon with momentum $\hbar\kappa_l$ and receives a momentum kick to the left. As such, its net momentum after stimulated emission is zero. On the other hand, if the excited atom was to interact with light traveling to the left, it would emit a photon that will travel to the left and receive a kick to the right. Thus, the net momentum of the atom after de-excitation would be $2\hbar\kappa_l$, and the atom will be seen moving to the right. A similar intuitive picture can be built for an atom moving to the left after the absorption–stimulated emission cycle. Higher-order diffraction can be achieved by repeating the whole process for each order increase in steps of $2\hbar\kappa_l$. Spontaneous emission is possible if the excited atom interacts with the vacuum mode of light instead. Of course, the atom will still receive a momentum kick in this case, but in a random direction that is not useful in the context of diffraction. We will, however, examine the absorption–spontaneous emission cycle employed in the cooling of atoms.

6.3.1 Evolution of Internal States of the Atom

The Hamiltonian for a two-level stationary atom interacting with standing wave light as described in Section 6.3 is under the dipole approximation (see Section 4.4):

$$H = \hbar\omega_g|g\rangle\langle g| + \hbar\omega_e|e\rangle|e\rangle - \mathbf{d}\cdot\mathbf{E}, \tag{6.3}$$

where $\mathbf{d} = \mathbf{d}_{ge}[|g\rangle\langle e| + |e\rangle\langle g|]$ is the dipole moment of the atom assumed to be real, $\mathbf{d}_{ge} = e\langle g|\mathbf{x}|e\rangle$, and $\omega_{g,e}$ are frequencies of the ground $|g\rangle$ and excited $|e\rangle$ of the atom, respectively. The classical standing wave laser field $\mathbf{E}$ experienced by the atom is

$$\mathbf{E} = \mathbf{E}_0(xt)\cos(\omega t + \phi_L(x)), \tag{6.4}$$

where $\mathbf{E}_0$ is the amplitude of the laser standing wave, ω is the frequency of the laser field, and $\phi_L(x)$ is the phase of the laser beam. The light field couples two internal states of the atom through dipole interactions, as shown in Fig. 6.3. The time evolution of the internal states of the atom,

$$|\psi(t)\rangle = a_g(t)|g\rangle + a_e(t)|e\rangle, \tag{6.5}$$

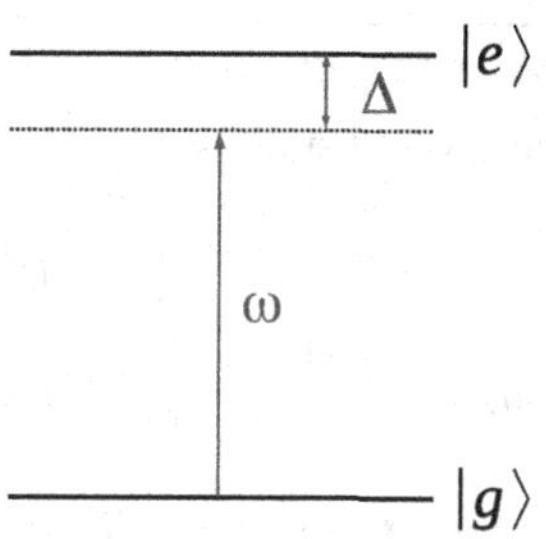

Fig. 6.3 Two-level atom with ground state $|g\rangle$ and excited state $|e\rangle$ is coupled by a laser field of frequency ω that is detuned from resonance by a frequency Δ, as defined in the text.

is governed by the Schrodinger equation,

$$i\hbar\frac{d}{dt}|\psi(t)\rangle = H|\psi(t)\rangle, \tag{6.6}$$

and the amplitudes $a_{g,e}$ are normalized to unity, $|a_g(t)|^2 + |a_e(t)|^2 = 1$. Substituting the state vector (6.5) into (6.6) gives coupled differential equations for the amplitudes

$$\begin{aligned} i\hbar\dot{a}_g(t) &= \hbar\omega_g a_g(t) + \hbar\Omega_{ge}\cos(\omega t + \phi_L)a_e(t), \\ i\hbar\dot{a}_e(t) &= \hbar\Omega_{ge}\cos(\omega t + \phi_L)a_g(t) + \hbar\omega_e a_e(t), \end{aligned} \tag{6.7}$$

where

$$\Omega_{ge} = -\frac{\mathbf{d}_{ge}\cdot\mathbf{E}_0(x,t)}{\hbar}. \tag{6.8}$$

The cosine function can be decomposed as $\cos(\omega t + \phi_L) = (e^{i(\omega t+\phi_L)} + e^{-i(\omega t+\phi_L)})/2$, which contains both fast and slow oscillatory terms. For instance, the field component $e^{-i\omega t}$ oscillates rapidly, so that its net effect on the ground state $|g\rangle$ is zero and vice versa. Transforming into the rotating frame according to

$$\begin{aligned} a_g(t) &= c_g(t)e^{-i\omega_g t - i\Delta t/2}, \\ a_e(t) &= c_e(t)e^{-i\omega_e t + i\Delta t/2}, \end{aligned} \tag{6.9}$$

where $\Delta = (\omega_e - \omega_g - \omega)$, and neglecting terms that oscillate rapidly, (6.7) becomes

$$\begin{aligned} i\dot{c}_g &= -\frac{\Delta}{2}c_g + \frac{\Omega_{ge}e^{i\phi_L}}{2}c_e, \\ i\dot{c}_e &= \frac{\Omega_{ge}e^{-i\phi_L}}{2}c_g + \frac{\Delta}{2}c_e. \end{aligned} \tag{6.10}$$

To solve (6.10), we assume $\Omega_{ge}(x,t)$ is constant when the light wave is interacting with the atom. This is true to a good approximation in most atom diffraction experiments. Defining the following parameters,

$$\tan\theta = \frac{\Omega_{ge}}{\Delta}, \quad \sin\theta = \frac{\Omega_{ge}}{\Omega_r}, \quad \cos\theta = \frac{\Delta}{\Omega_r}, \tag{6.11}$$

where $\Omega_r = \sqrt{\Delta^2 + \Omega_{ge}^2}$ and $0 < \theta < \pi$, the solution of amplitudes for an atom that is initially in the ground state is

$$\begin{aligned} a_g &= e^{-i(\omega_g+\Delta/2)t}\left(\cos^2\theta/2 e^{i\Omega_r t/2} + \sin^2\theta/2 e^{-i\Omega_r t/2}\right), \\ a_e &= \frac{\sin\theta}{2} e^{-i(\omega_e-\Delta/2)t}\left(e^{-i(\Omega_r t/2+\phi_L)} - e^{i(\Omega_r t/2-\phi_L)}\right). \end{aligned} \tag{6.12}$$

The energies $E_{g\pm}$ associated with the ground state in the presence of light are

$$E_{g\pm} = \hbar\left[\omega_g + \frac{\Delta}{2} \pm \frac{1}{2}\sqrt{\Delta^2 + |\Omega_{ge}|^2}\right]. \tag{6.13}$$

The detuning Δ can be controlled using the frequency ω of laser light. For large positive detuning $\Delta > 0$, θ is approximately zero and the state vector of the system becomes $|\psi\rangle \approx e^{-iE_{g-}t/\hbar}|g\rangle$, where $E_{g-} \approx \hbar\left(\omega_g - \frac{1}{4}\frac{|\Omega_{ge}|^2}{|\Delta|}\right)$. Similarly, for very large negative detuning $\Delta < 0$, θ is roughly π and the state vector of the system is given as $|\psi\rangle \approx e^{-iE_{g+}t/\hbar}|g\rangle$, where $E_{g+} \approx \hbar\left(\omega_g + \frac{1}{4}\frac{|\Omega_{ge}|^2}{|\Delta|}\right)$. Notice that in either detuning considered, the atoms are always found in the ground state $|g\rangle$, while the excited state $|e\rangle$ is negligibly occupied. The overall effect of large detuning is to raise or lower the ground state energy of the atoms by a constant amount, $\frac{1}{4}\frac{|\Omega_{ge}|^2}{|\Delta|}$. It also produces a periodic potential to the atoms since $\Omega_{ge} \sim \mathbf{d}_{ge} \cdot \mathbf{E}_0(t)\cos(\kappa_l x)$, which the ground state follows adiabatically. The discussion above is carried out without damping. Nevertheless, it captures the essential physics of the system even in the presence of decay, like spontaneous emission that leads to damping. For a treatment that includes damping and decay channels, see references and further reading at the end of this chapter.

Exercise 6.3.1 By solving (6.7), given that initially the atom is in the ground state $a_g = 1$, show that the solution is (6.12), and verify that the energies of the ground state $E_{g,\pm}$ are as given in (6.13).

Exercise 6.3.2 For zero detuning $\omega = \omega_e - \omega_g$, zero phases of the laser $\phi_L = 0$, and pulse duration $\Omega_r t = \pi/2$, show that the initial state $|\psi(0)\rangle = |g\rangle$ is transformed to the state $|\psi_{\pi/2}\rangle = \{|g\rangle - i|e\rangle\}/\sqrt{2}$ up to some global phase factors.

Exercise 6.3.3 If the pulse duration in the problem above is π instead, what is the effect of the laser pulse on the initial state $|\psi(0)\rangle = |g\rangle$? What would be the effect of having two π pulses act on the initial state $|\psi(0)\rangle = |g\rangle$?

6.3.2 Forces on Atom and Potential Due to Light Field

It is evident from the above discussion that there is a force acting on the atom due to the light that can be estimated from the Hamiltonian, and it must be in accordance with Newton's first law. The starting point to relate the quantum dynamics of the system with classical mechanics is to find the evolution of the quantum operators p, x in the Hamiltonian (6.3). The Heisenberg equation of motion for the dynamical operators x, p of (6.3) is $i\hbar\dot{x} = [x, H]$, $i\hbar\dot{p} = [p, H]$, where the dot denotes the time derivative and $[A, B]$ is the usual commutator. Averaging the Heisenberg equation of motion for the operators over the atomic wavefunction gives (Ehrenfest equations)

$$\langle \dot{x} \rangle = 0, \qquad (6.14)$$
$$\langle \dot{p} \rangle = \langle \mathbf{d} \rangle \cdot \frac{\partial}{\partial x} \mathbf{E}(x,t).$$

The first equation in (6.14) shows that the time evolution for the center of the atom wave $\langle \dot{x} \rangle$ is zero. This suggests three distinct possibilities for the atomic motion. The first case would be that the atom is stationary, and so is its wave packet. In this case, the center of the wave packet would not move for all time. This possibility is ruled out by the second equation of (6.14). Another scenario would be that the atom makes a random motion about some point, such that the net classical trajectory averages to zero. A third possibility, which is not intuitive classically, is for the center of the wave packet to be in a linearly superposed directed motion in opposite directions. As such, the atom is in an superposition of being at two places $-x$ and $+x$ at the same time, such that on averaging the velocity of the atom's center of the wave packet, one obtains zero. This ability to put the atoms in a directed motion in opposite directions will be exploited in the splitting of a wave packet of an atom.

When the spatial derivative is applied to the field as prescribed in the second equation of (6.14), it gives

$$\frac{\partial}{\partial x}\mathbf{E}(x,t) = \left(\frac{\partial \mathbf{E}_0(x,t)}{\partial x}\right)\cos(\omega t + \phi_L(x)) - \mathbf{E}_0(x,t)\sin(\omega t + \phi_L(x))\frac{\partial \phi_L(x)}{\partial x}. \qquad (6.15)$$

Thus, the force $\langle \dot{p} \rangle$ is composed of two terms: the dissipative force F_{diss} and the reactive force F_{react}. The dissipative force F_{diss} is proportional to the gradient of light phase $\frac{\partial \phi_L(x)}{\partial x}$ and is due to the spontaneous emission of light by the atom. Spontaneous emission is possible when there is a finite population of atoms in the excited state. An atom with a probability $|a_e(t)|^2 \approx \frac{|\Omega_{ge}|^2}{\Delta^2+|\Omega_{ge}|^2}$ of being in the excited state and with natural linewidth Γ will decay spontaneously to ground state at a rate $\Gamma|a_e(t)|^2 \approx \Gamma\frac{|\Omega_{ge}|^2}{\Delta^2+|\Omega_{ge}|^2}$. If the momentum of a photon lost in the spontaneous decay process is $\frac{\partial \phi_L}{\partial x} = \hbar\kappa_l$, then the dissipative force experienced by the atom is $F_{\text{diss}} \sim \Gamma\frac{|\Omega_{ge}|^2}{\Delta^2+|\Omega_{ge}|^2}\hbar\kappa_l$. This type of force is used in the cooling of atoms. However, for the purposes of directed motion of the atom, this process is not wanted but cannot be completely eliminated. It can, however, be controlled using light fields that are strongly detuned from atomic resonance transitions, $\Delta \gg 0$ and $|\Omega_{ge}|/\Delta \ll 1$.

The reactive force F_{react} is given by the term proportional to the gradient of field amplitude. To estimate this force, we will ignore the contributions from the gradient of the field phase as discussed in the preceding paragraph. Hence, we consider a standing wave field of the form $\mathbf{E}(x,t) = \mathbf{E}_0(x,t)\cos\omega t$, with spatial derivative

$$\frac{\partial}{\partial x}\mathbf{E}(x,t) = \left(\frac{\partial \mathbf{E}_0(x,t)}{\partial x}\right)\cos\omega t. \qquad (6.16)$$

The average dipole moment $\langle \mathbf{d} \rangle = \langle \psi|\mathbf{d}|\psi \rangle$ within the rotating frame defined in (6.9) gives

$$\langle \mathbf{d} \rangle = \mathbf{d}_{ge}\sin\theta\cos\theta\left[\cos(\Omega_r t) - 1\right]. \qquad (6.17)$$

This choice for calculating the dipole is a matter of convenience. Nevertheless, it gives the leading order term necessary to describe the effects of the reactive force. If the bare atom states $a_{g,e}$ are used instead of the rotating frame state $c_{g,e}$, the term not captured in this approximation is of the order $\left|\frac{\Omega_{ge}}{\Delta}\right|^3$, which is negligible in the regime of interest, $\Delta \gg 0$, and $|\Omega_{ge}|/\Delta \ll 1$. From (6.17), the dipole oscillates at frequency $\Omega_r/2$, so the atom responds to the field at frequencies lower than or equal to $\Omega_r/2$. For oscillations greater than $\Omega_r/2$, the atom will not have felt the effect of the field appreciably. Substituting (6.16) and (6.17) into (6.14) for $\langle \dot{p} \rangle$ and averaging over one period of the fastest oscillation $T = 2\pi/(\Omega_r + \omega)$ in the limit $(\Omega_r - \omega)T$, $\omega T \ll 1$ gives the reactive force

$$F_{\text{react}} = \frac{\hbar\Delta}{4} \frac{1}{\Delta^2 + \Omega_{ge}^2} \frac{\partial}{\partial x}(|\Omega_{ge}|^2), \tag{6.18}$$

where the definition of Ω_{ge}, (6.8), has been used and intensity of light is proportional to $|\Omega_{ge}|^2$. We see that the reactive force F_{react} depends on the detuning of laser light. For red-detuned light $\Delta > 0$, the atom is attracted to the region of high light intensity. Similarly, for blue-detuned light $\Delta < 0$, the atom is attracted to the region of low light intensity and will be repelled from the region of high light intensity. This points to the force $F_{\text{react}} = -\nabla V$ being conservative, thus derivable from a potential V given by

$$V = -\frac{\hbar\Delta}{4} \ln\left[1 + \left(\frac{\Omega_{ge}}{\Delta}\right)^2\right]. \tag{6.19}$$

For large detuning, $V \approx -\frac{\hbar\Delta}{4}\frac{|\Omega_{ge}|^2}{\Delta^2}$, which is exactly the shift in the energy of the atom calculated before in Section 6.3.1. For $\Delta > 0$, the atom will be attracted to the intensity maximum that appears to be a minimum of the potential V. Such potential minima can be used for trapping atoms.

Exercise 6.3.4 Let $|\psi\rangle = c_g|g\rangle + c_e|e\rangle$, where $c_{g,e}$ are the solutions of (6.10). Prove (6.17). Show that if the bare state solutions $a_{g,e}$ are used, there would be an additional term of order $|\Omega_{g,e}|^2/|\Delta|^2$.

Exercise 6.3.5 Prove (6.18) and obtain the potential V (6.19). Also, obtain the potential V in the large detuning limit, $\Delta > 0, |\Omega_{g,e}|$.

6.4 Bragg Diffraction by Standing Wave Light

The diffraction of atomic beams can be classified into two regimes: the Raman–Nath regime, where only the momentum is conserved; and the Bragg regime, where both the momentum and energy conservation are met. As such, an atomic beam in Raman–Nath diffraction can be diffracted into many orders, while in Bragg diffraction, only few diffraction orders are possible. In diffraction experiments, the momentum distribution of the atom shows a few narrow peaks. In this section, we examine Bragg diffraction using a square-wave pulse. Using a square-wave pulse removes the requirement of resonance and adiabaticity for the laser pulses. Instead, emphasis is placed on the proper timing of the square-wave pulses in order to achieve the desired goal.

As shown in the previous sections, an atom in the presence of an off-resonant standing wave of light experiences a periodic force and follows the light adiabatically. In the limit of large detuning and assuming the atom has no initial momentum, the Hamiltonian governing the dynamics of an atom is

$$i\hbar\frac{d}{dt}\psi(x,t) = \left[-\frac{\hbar}{2m}\frac{d^2}{dx^2} + \tilde{\Omega}(t)\cos(2\kappa_l x)\right]\psi(x,t), \tag{6.20}$$

where $\psi(x,t)$ is the ground state wavefunction of the atom and $\tilde{\Omega}(t)$ is the time-dependent amplitude of the applied laser pulses. Here, the optical potential $\tilde{\Omega}(t)\cos(2\kappa_l x)$ presents a grating of periodicity $\lambda_l/2$ to the atom's wavefunction, where κ_l and λ_l are the wavenumber and wavelength of the laser beam, respectively. The periodicity of the grating has a characteristic width of $2\hbar\kappa_l$ in momentum space and fulfills the Bragg condition,

$$p = 2n\hbar\kappa_l, \tag{6.21}$$

where n is the diffraction order and takes integer values only, and p is the momentum of the atom. Introducing the recoil frequency $\omega_{\text{rec}} = \hbar(2\kappa_l)^2/(2m)$, dimensionless coordinate $\nu = 2\kappa x$, and time $\tau = 2\omega_{\text{rec}}t$, (6.20) is written in dimensionless form,

$$i\frac{\partial}{\partial\tau}\psi(\nu,\tau) = \left[-\frac{1}{2}\frac{\partial^2}{\partial\nu^2} + \Omega(\tau)\cos\nu\right]\psi(\nu,\tau), \tag{6.22}$$

where $\Omega(\tau) = \tilde{\Omega}(t)/(2\omega_{\text{rec}})$. The above equation belongs to a special class of differential equations called the Mathieu equations, and its solutions cannot be written in terms of elementary functions in general. However, approximate solutions can be found if the properties of the equation are understood.

In Fourier space, if the width of an atom's wavefunction is much smaller than the length of the grating wavevector, then $\psi(\nu,\tau)$ consists of a series of narrow peaks. It is therefore convenient to expand $\psi(\nu,\tau)$ as

$$\psi(\nu,\tau) = \sum_{n=-\infty}^{\infty}\phi_n(\tau)e^{in\nu}, \tag{6.23}$$

where $\phi_n(t)$ are the time-varying amplitudes. Substituting (6.23) into the Schrodinger equation, (6.22) gives

$$i\dot{\phi}_n = \frac{n^2}{2}\phi_n + \left(\phi_{n-1} + \phi_{n+1}\right)\frac{\Omega(\tau)}{2}. \tag{6.24}$$

Equation (6.24) comprises an infinite set of coupled differential equations. In the Raman–Nath diffraction regime, the interaction time between the atoms and the laser pulses is short, resulting in the kinetic energy of the atoms being small in comparison with the optical potential energy presented by the laser pulses (second term of (6.24)). Hence, the kinetic energy term is neglected in solving (6.24), and the resulting solution is given in terms of Bessel functions.

In the Bragg limit, the kinetic energy term is no longer negligible. For some given diffraction order L, only diffraction orders $n = 0, \pm1, \pm2, \ldots, \pm(L-1)$ less than L are excited. Thus the series can be truncated if $L^2 \gg \Omega(\tau)$. For the lowest order diffraction $L = 2$, we have $n = 0, \pm1$, leading to the following three coupled differential equations:

$$i\begin{pmatrix} \dot{\phi}_{-1} \\ \dot{\phi}_0 \\ \dot{\phi}_1 \end{pmatrix} = \frac{1}{2}\begin{pmatrix} 1 & \Omega(\tau) & 0 \\ \Omega(\tau) & 0 & \Omega(\tau) \\ 0 & \Omega(\tau) & 1 \end{pmatrix}\begin{pmatrix} \phi_{-1} \\ \phi_0 \\ \phi_1 \end{pmatrix}. \tag{6.25}$$

The states $\phi_{\pm 1}$ correspond to atomic packets moving with momenta ± 1 in Fourier space, while ϕ_0 corresponds to an atomic wave packet with zero momentum in the Fourier space. The symmetric nature of the couplings in (6.25) means that we can eliminate one variable using the transformation $\eta_\pm = \frac{\phi_1 \pm \phi_{-1}}{\sqrt{2}}$, reducing it to two coupled differential equations and an uncoupled differential equation:

$$i\begin{pmatrix} \dot{\phi}_0 \\ \dot{\eta}_+ \end{pmatrix} = \frac{1}{2}\begin{pmatrix} 0 & \sqrt{2}\Omega(\tau) \\ \sqrt{2}\Omega(\tau) & 1 \end{pmatrix}\begin{pmatrix} \phi_0 \\ \eta_+ \end{pmatrix}, \tag{6.26}$$

$$i\dot{\eta}_- = \frac{1}{2}\eta_-. \tag{6.27}$$

The transformation η_-, which does not couple to ϕ_0, has a solution for the initial condition, $\eta_-(0)$:

$$\eta_-(\tau) = \eta_-(0)e^{-i\tau/2}. \tag{6.28}$$

This is the *dark state*, and it is often encountered in three-level systems. To solve (6.26), take the function $\Omega(\tau)$, a square-wave pulse of the form $\Omega(\tau) = \Theta(\tau) - \Theta(\tau + \tau_1)$, where $\Theta(\tau)$ is the Heaviside step function and the amplitude is Ω_0. The solution to (6.26) for the initial condition $\phi_0(0)$, $\eta_+(0)$ is

$$\begin{pmatrix} \phi_0(\tau) \\ \eta_+(\tau) \end{pmatrix} = e^{-i\frac{\tau}{4}}\begin{pmatrix} M_{11} & M_{12} \\ M_{12} & M_{11}^* \end{pmatrix}\begin{pmatrix} \phi_0(0) \\ \eta_+(0) \end{pmatrix}, \tag{6.29}$$

where

$$M_{11} = \cos\left(\frac{\sqrt{1+8\Omega_0^2}}{4}\tau\right) - \frac{i}{\sqrt{1+8\Omega_0^2}}\sin\left(\frac{\sqrt{1+8\Omega_0^2}}{4}\tau\right), \tag{6.30}$$

$$M_{12} = \frac{2\sqrt{2}\Omega_0 i}{\sqrt{1+8\Omega_0^2}}\sin\left(\frac{\sqrt{1+8\Omega^2}}{4}\tau\right). \tag{6.31}$$

Equation (6.29) gives the evolution of the atoms' wave packet in the presence of laser compound pulses for different initial conditions. A particular case of interest in many experiments using cold atoms is the coherent population transfer initially at zeroth order $n = 0$ to first order $n = \pm 1$, corresponding to atoms moving left and right of the atomic wave packet with momentum $2\hbar\kappa$. This can be realized using two compound laser pulses with pulse amplitude $\Omega_0 = 1/\sqrt{8}$, each with a duration of $\tau_d = \sqrt{2}\pi$. The first and the second pulses are separated by a period of free evolution lasting for a time $\tau_{\text{free}} = 2\pi$, during which the laser pulses are switched off and clouds are allowed to rephase. The evolution matrix for the splitting sequence is thus described as

$$U_{0\leftrightarrow +} = e^{-i\frac{\pi}{\sqrt{2}}}\begin{bmatrix} 0 & -1 \\ -1 & 0 \end{bmatrix}. \tag{6.32}$$

The expression (6.32) shows that the total population in ϕ_0 is transferred to the state η_+ at the end of the second pulse, as described. This transfer of population to η_+ happens in such a way that ϕ_+ and ϕ_- are equally populated. This is because according to (6.27) and (6.28), the evolution of the population difference has to be zero at all times if initially $\eta_-(0) = 0$.

The analytic expression for the solution of (6.25) can be obtained directly by assuming that the light amplitude $\Omega(\tau)$ is a square wave, as done in the preceding paragraph. The solutions obtained in this case are

$$\begin{pmatrix} \phi_{-1}(\tau) \\ \phi_0(\tau) \\ \phi_1(\tau) \end{pmatrix} = e^{-i\tau/4} \begin{pmatrix} A_{11} & A_{12} & A_{13} \\ A_{12} & A_{22} & A_{12} \\ A_{13} & A_{12} & A_{11} \end{pmatrix} \begin{pmatrix} \phi_{-1}(0) \\ \phi_0(0) \\ \phi_1(0) \end{pmatrix}, \tag{6.33}$$

where

$$A_{11} = \frac{1}{2}\left[\cos\left(\frac{\sqrt{1+8\Omega_0^2}\tau}{4}\right) + e^{-i\tau/4} - \frac{i}{\sqrt{1+8\Omega_0^2}}\sin\left(\frac{\sqrt{1+8\Omega_0^2}\tau}{4}\right)\right], \tag{6.34}$$

$$A_{12} = -2i\frac{\Omega_0}{\sqrt{1+8\Omega_0^2}}\sin\left(\frac{\sqrt{1+8\Omega_0^2}\tau}{4}\right), \tag{6.35}$$

$$A_{13} = \frac{1}{2}\left[\cos\left(\frac{\sqrt{1+8\Omega_0^2}\tau}{4}\right) - e^{-i\tau/4} - \frac{i}{\sqrt{1+8\Omega_0^2}}\sin\left(\frac{\sqrt{1+8\Omega_0^2}\tau}{4}\right)\right], \tag{6.36}$$

$$A_{22} = \cos\left(\frac{\sqrt{1+8\Omega_0^2}\tau}{4}\right) + \frac{i}{\sqrt{1+8\Omega_0^2}}\sin\left(\frac{\sqrt{1+8\Omega_0^2}\tau}{4}\right). \tag{6.37}$$

Evolving the initial state $[\phi_{-1}, \phi_0, \phi_1]^T = [0, 1, 0]^T$ under the same parameters considered in Section 6.3 gives the same result. In fact, it is easily seen that after Bragg pulses are applied in the sequence as described, two wave packets of equal amplitude in a superposition of mode-entangled states emerge.

Exercise 6.4.1 By solving (6.25), obtain the solution (6.33) for an atom that is initially stationary, $\phi_0 = 1$, $\phi_\pm = 0$.

Exercise 6.4.2 In diffraction experiment with atoms, it is desired to put atoms in the zero-momentum state $\phi_0 = 1$ into a superposition of state with atoms moving either to the left or to the right, $\phi_0 \to (\phi_{+1} + \phi_{-1})/\sqrt{2}$, and none in the stationary state $\phi_0 = 0$. Using the results of Exercise 6.4.1, find the splitting matrix that achieves this. Can the splitting be achieved by a single pulse? Compare the obtained result with your answer to Exercise 6.3.2.

Exercise 6.4.3 Laser pulses can be used as mirrors, whereby they reverse the motion of atoms in the moving state $\phi_\pm \to \phi_\mp$. Find the reflection matrix.

6.5 Bragg Diffraction by Raman Pulses

Another scheme for diffraction of matter waves uses the Raman pulses, as shown in Fig. 6.4. In this technique, two linearly polarized laser pulses with frequencies

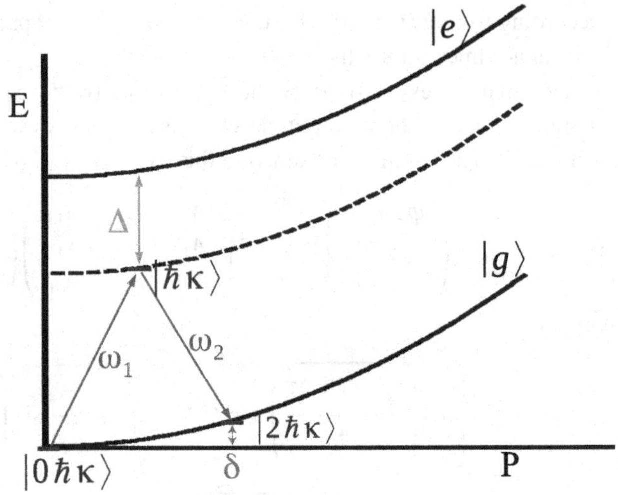

Fig. 6.4 The interaction of an atom with a detuned light of frequencies ω_1 and ω_2. Color version of this figure available at cambridge.org/quantumatomoptics.

$\omega_1 = \omega_0$ and $\omega_2 = \omega_0 + \delta$ detuned from the atomic resonance are incident on an atom. The frequency difference δ is chosen to be an integer multiple of the recoil frequency $2\hbar\kappa_l$, where κ_l is the wave number of the laser. An atom exposed to linearly polarized light makes a transition to the excited state allowed by the dipole transition rule. Since linearly polarized light consists of light with σ^+ and σ^- polarization, the two lasers provide channels for excitation and de-excitation of the atoms from virtual states. For instance, an atom absorbing a photon from light of any of frequency and de-exciting via stimulated emission into the same frequency gains no energy. Thus, its momentum change after the round trip is zero. However, an atom absorbing a photon from frequency ω_2 and de-exciting via stimulated emission into light with frequency ω_1 gains a net energy, $\hbar\delta$. The net energy $\hbar\delta$ gained by the atom changes the kinetic energy of the atom. Because the light at frequencies ω_1 and ω_2 couples to the same ground state, there is no change in the internal state of the atom at the end of the diffraction. Since an atom that is excited via ω_1 may return to the ground state via emission through the same frequency only because of energy conservation, the observed atomic motion is only in one direction when compared to the method of Section 6.4. This diffraction technique contrasts with a Raman transition, where both the internal and motional states of an atom are different from what they were at the beginning of the process. Because the transitions in momentum space resemble Raman transitions, this is also called a Raman transition in momentum space.

6.5.1 Evolution of the Ground State Momentum Family

Consider atoms interacting with a linearly polarized beam, as described above. When the lateral dimensions of the atoms are comparable or greater than the wavelength of light, each atom experiences a different relative phase. The external motional state of the atoms is no longer negligible. The atomic wavefunction is then described as a tensor product of the atom's internal energy state, and the external motional state represented by the momentum state. As such, the internal state of the atoms is associated with many different momenta. To see this, consider the Hamiltonian of a two-level atom,

$$H = \frac{\mathbf{p}^2}{2m} + \hbar\omega_e|e\rangle\langle e| + \hbar\omega_g|g\rangle\langle g| - \mathbf{d}\cdot\mathbf{E}(z,t), \tag{6.38}$$

where $\mathbf{p}$ operates on the momentum part of the atom's wavefunction, $\mathbf{d}$ is the electric dipole moment, and the external field $\mathbf{E}(z,t)$ on the atom is

$$\mathbf{E}(z,t) = \frac{1}{2}\left[\mathbf{E}_1 e^{i(k_1 z - i\omega_1 t)} + \mathbf{E}_2 e^{i(k_2 z + \omega_2 t)} + \text{c.c.}\right]. \tag{6.39}$$

Here, the incident laser beams have wave vectors $k_{1,2}$ and frequencies $\omega_{1,2}$, and c.c. stands for complex conjugate. The operator $e^{\pm ik_m z}$ ($m = 1, 2$) acts on the atom's momentum state $|p\rangle$ and shifts it as

$$e^{\pm ik_m z} = \int dp\, |p\rangle\langle p \mp \hbar k_m|. \tag{6.40}$$

Therefore, an atom absorbing or emitting a photon of wave vector k_m from the laser has its momentum changed by an amount $\hbar k_m$. Writing the basis state as a tensor product of internal energy and momentum $|g, p\rangle$ and $|e, p\rangle$, any state can be expanded in terms of these states as

$$|\psi(t)\rangle = \int dp\, [a_0(p,t)|g,p\rangle + a_1(p,t)|e,p\rangle]. \tag{6.41}$$

The state $|g, p\rangle$ is the ground state of an atom that has momentum p, while $|e, p\rangle$ is the excited state of an atom that has momentum p.

The evolution of the state $|\psi(t)\rangle$ is governed by the Schrodinger equation, $i\hbar\frac{d}{dt}|\psi(t)\rangle = H|\psi(t)\rangle$. By substituting (6.41) and (6.38) in the Schrodinger equation and equating coefficients in same basis $|g, p\rangle$, $|e, p\rangle$, we have the following coupled differential equations,

$$i\dot{c}_1(p) = \frac{1}{2}[\Omega_1^* e^{-i(\omega_1 - \omega_{eg} + \omega_{-k_1})t} c_0(p - \hbar k_1) + \Omega_2^* e^{-i(\omega_2 - \omega_{eg} + \omega_{k_2})t} c_0(p + \hbar k_2)], \tag{6.42}$$

$$i\dot{c}_0(p) = \frac{1}{2}[\Omega_2 e^{i(\omega_2 - \omega_{eg} - \omega_{-k_2})t} c_1(p - \hbar k_2) + \Omega_1 e^{i(\omega_1 - \omega_{eg} - \omega_{k_1})t} c_1(p + \hbar k_1)], \tag{6.43}$$

where $\omega_{eg} = \omega_e - \omega_g$ and the Rabi frequencies are $\Omega_i = -\langle g|\mathbf{d}\cdot\mathbf{E}_i|e\rangle/\hbar$, $i = 1, 2$. Note that for brevity in (6.42) and (6.43), $c_0(p,t)$ is written as $c_0(p)$. The variables $a_0(p,t)$, $a_1(p,t)$ are related to $c_0(p,t)$, $c_1(p,t)$ as

$$a_0(p,t) = c_0(p,t)\exp\left[-i\left(\frac{p^2}{2\hbar m} + \omega_g\right)t\right],$$

$$a_0(p \pm \hbar k, t) = c_0(p \pm \hbar k, t)\exp\left[-i\left(\frac{(p \pm \hbar k)^2}{2\hbar m} + \omega_g\right)t\right], \tag{6.44}$$

$$a_1(p,t) = c_1(p,t)\exp\left[-i\left(\frac{p^2}{2\hbar m}+\omega_e\right)t\right],$$
$$a_1(p\pm\hbar k,t) = c_1(p\pm\hbar k,t)\exp\left[-i\left(\frac{(p\pm\hbar k)^2}{2\hbar m}+\omega_e\right)t\right].$$

Terms that are fast rotating at frequencies $\omega_i + \omega_{eg}$ $(i = 1,2)$ were neglected in arriving at (6.42) and (6.43). The transformation (6.44) and the neglect of fast terms at frequency $\omega_i + \omega_{eg}$ is known as rotating wave approximation.

Equations (6.42) and (6.43) do not lend to easy analytical treatment because the coefficients depend on time. We begin our analysis by examining (6.42), which describes the coupling between the set of momenta $|g, p-\hbar k_1\rangle$, $|e, p\rangle$ and $|g, p+\hbar k_1\rangle$. The two ground states, $|g, p-\hbar k_1\rangle$ and $|g, p+\hbar k_2\rangle$, are coupled to the excited state $|e, p\rangle$. The coupling between the ground state $|g, p-\hbar k_1\rangle$ and the excited state $|e, p\rangle$ is mediated at a frequency ω_1. Similarly, the coupling between the excited state $|e, p\rangle$ state and the ground state $|g, p+\hbar k_2\rangle$ is mediated at the frequency ω_2. The detunings of laser light from the atomic resonance transition for the two transitions at wave vector of magnitude k_1, k_2 are given by

$$\Delta_{-k_1} = \omega_1 - \omega_{eg} + \frac{\hbar k_1^2}{2m} - \frac{pk_1}{m}, \tag{6.45}$$
$$\Delta_{k_2} = \omega_2 - \omega_{eg} + \frac{\hbar k_2^2}{2m} + \frac{pk_2}{m}.$$

In comparison with (6.10), the above detunings have additional terms. There is an additional photon-recoil frequency $\frac{\hbar k^2}{2m}$ term, which corresponds to the kinetic energy of an atom with momentum $\hbar k$, and a Doppler shift frequency $\frac{pk}{m}$.

It is then possible to eliminate the excited state adiabatically from the coupled differential equations (6.42) and (6.43), if the Rabi frequencies $\Omega_{1,2}$ are small in comparison with the detuning and the coefficients $c_0(p+\hbar k_2, t)$ and $c_0(p-\hbar k_1, t)$ in (6.42) are slow-varying functions of time within the interval specified by the detuning (6.45). Integrating (6.42) gives

$$c_1(p,t) = \frac{1}{2}\left[\frac{\Omega_1}{\Delta_{-k_1}}e^{-i\Delta_{-k_1}t}c_0(p-\hbar k_1,t) + \frac{\Omega_2^*}{\Delta_{k_2}}e^{-i\Delta_{k_2}t}c_0(p+\hbar k_2,t)\right]. \tag{6.46}$$

Substituting (6.46) in (6.43) gives a three-term recursive differential equation for the evolution of the ground state,

$$i\dot{c}_0(p,t) = \left[\frac{|\Omega_1|^2}{4\Delta_{-k_1}}e^{-i\frac{\hbar k_1^2}{m}t} + \frac{|\Omega_2|^2}{4\Delta_{k_2}}e^{-i\frac{\hbar k_2^2}{m}t}\right]c_0(p,t) \tag{6.47}$$
$$+\frac{\Omega_2\Omega_1^*}{4\Delta_{-k_1}}e^{-i\delta_1 t}c_0(p-\hbar K,t) + \frac{\Omega_1\Omega_2^*}{4\Delta_{k_2}}e^{i\delta_2 t}c_0(p+\hbar K,t),$$

where $K = k_1 + k_2$, and

$$\delta_1 = \omega_1 - \omega_2 + \frac{\hbar(k_1^2+k_2^2)}{2m} - \frac{pK}{m}, \tag{6.48}$$

$$\delta_2 = \omega_1 - \omega_2 - \frac{\hbar(k_1^2+k_2^2)}{2m} - \frac{pK}{m}. \tag{6.49}$$

Equation (6.47) shows coupling between three different ground state momentum families $|g, p\rangle$, $|g, p\pm\hbar K\rangle$ at frequencies $\delta_{1,2}$. Since the frequencies $\delta_{1,2}$ are different, only one of the $|g, p \pm \hbar K\rangle$ family is coupled to the $|g, p\rangle$ at resonance. To see this, if the lasers are detuned such that $\omega_1 - \omega_2 = \frac{p'K}{\hbar} + \frac{\hbar(k_1^2+k_2^2)}{2m}$, $\delta_1 = \frac{\hbar(k_1^2+k_2^2)}{m} + \frac{(p'-p)}{m}K$, and $\delta_2 = \frac{(p'-p)}{m}K$, then the frequency δ_1 oscillates fast compared to δ_2. At resonance, which is a special case where $p = p'$, the phases of the $c_0(p,t)$ and $c_0(p - \hbar K)$ terms in (6.47) are roughly the same order of magnitude, and their contribution to the evolution of $|g, p\rangle$ averages to zero. Then, only the state $|g, p + \hbar K\rangle$ contributes to the evolution of $|g, p\rangle$. Using similar analysis for evolution of the $|g, p + \hbar K\rangle$ state, one arrives at the following effective coupled differential equations:

$$i\dot{c}_0(p,t) = \frac{\Omega_1\Omega_2^*}{4\Delta_{k_2}} c_0(p + \hbar K, t), \tag{6.50}$$
$$i\dot{c}_0(p + \hbar K, t) = \frac{\Omega_2\Omega_1^*}{4\Delta_{-k_1}} c_0(p,t).$$

The solution of (6.50) for the initial condition $c_0(p, t = 0)$, $c_0(p + \hbar K, t = 0)$ is

$$\begin{pmatrix} c_0(p,t) \\ c_0(p + \hbar K, t) \end{pmatrix} = \begin{pmatrix} \cos\left(\frac{\Omega t}{2}\right) & -\frac{-iW}{\Omega}\sin\left(\frac{\Omega t}{2}\right) \\ -\frac{-iW}{\Omega}\sin\left(\frac{\Omega t}{2}\right) & \cos\left(\frac{\Omega t}{2}\right) \end{pmatrix} \begin{pmatrix} c_0(p,0) \\ c_0(p + \hbar K, 0) \end{pmatrix}, \tag{6.51}$$

where W and the effective Rabi frequency Ω are

$$\Omega^2 = \frac{1}{4}\frac{|\Omega_1|^2}{\Delta_{-k_1}}\frac{|\Omega_2|^2}{\Delta_{k_2}}, \tag{6.52}$$
$$W = \frac{\Omega_1\Omega_2^*}{2\Delta_{k_2}}. \tag{6.53}$$

Equation (6.51) gives the amplitudes of finding atoms in the states $|g, p\rangle$ and $|g, p + \hbar K\rangle$ moving with momenta p and $p + \hbar K$, respectively. Since the detuning depends sensitively on the momentum p, only a fraction of atoms that meet the resonance condition are transferred to $|g, p + \hbar K\rangle$. For $c_0(p, 0) = 1$ and $c_0(p + \hbar K, 0) = 0$, the probabilities of finding atoms in the state $|g, p\rangle$ and $|g, p + \hbar K\rangle$ are proportional to

$$|c_0(p)|^2 = \cos^2\left(\frac{\Omega t}{2}\right), \tag{6.54}$$
$$|c_0(p + \hbar K)|^2 = \left|\frac{W}{\Omega}\right|^2 \sin^2\left(\frac{\Omega t}{2}\right), \tag{6.55}$$

respectively. As inferred from (6.54), for $\Omega t = \frac{\pi}{2}$ half of the population initially in the ground state $|g, p\rangle$ moving with momentum p is transferred to the ground state momentum family $|g, p - \hbar K\rangle$, $|g, p + \hbar K\rangle$, and $|g, p + 2\hbar K\rangle$, moving with momenta $p - \hbar K$, $p + \hbar K$, and $p + 2\hbar K$, respectively. However, of the population transferred to the momentum state families, only the fraction $W^2/\Omega^2 = \frac{\Delta_{-k_1}}{\Delta_{k_2}}$ is found in the state $|p + \hbar K\rangle$. This is because $W^2/\Omega^2 = \frac{\Delta_{-k_1}}{\Delta_{k_2}}$ depends on the momentum of atoms, from (6.45). Unless this ratio is unity, the ground state momentum family $|g, p - \hbar K\rangle$, $|g, p + 2\hbar K\rangle$, other than the intended $|g, p + \hbar K\rangle$, are likely to be populated with finite probability. For a special case where $p = 0$ and $\omega_2 \approx \omega_1$, the ratio of laser detuning is roughly unity, and one-half of the atomic population is transferred to the target state

$|g, p + n\hbar K\rangle$, with atoms in that state having momentum $p + n\hbar K$, $(n = \pm 1, \pm 2, \ldots)$. This is routinely achieved with atomic Bose–Einstein condensate samples where the momentum distribution of the atoms is centered at $p = 0$. Similar analysis can be extended to the reflection of atoms using Bragg optical pulses.

Exercise 6.5.1 Substitute (6.40) and (6.41) in the Schrodinger equation and obtain the coupled differential equations (6.42) and (6.43).

Exercise 6.5.2 Apply the rotating wave approximation and adiabatic elimination to the results of Exercise 6.5.1 and obtain the differential equation for the evolution of the ground state probability amplitudes $c_0(p, t)$ and $c_0(p - \hbar K, t)$. See, for example, (6.47).

Exercise 6.5.3 Solve the coupled differential equations of Exercise 6.5.2 and use the solutions to find the splitting matrix that takes atoms initially in the state $|g, p\rangle$ to the state $(|g, p\rangle + |g, p - \hbar K\rangle)/\sqrt{2}$; $|g, p\rangle \rightarrow (|g, p\rangle + |g, p - \hbar K\rangle)/\sqrt{2}$. What can you say about the reflection matrix?

6.6 References and Further Reading

- Section 6.2: For a general introduction to the theory of diffraction, see [502, 176].
- Section 6.3: The first experiment demonstrating the diffraction of an atom, which used a mechanical grating, is detailed here [81]. For an introduction to interactions between atoms and radiation [104, 7, 428, 95, 330]. The text [330] is a very good resource on the topic of trapping and manipulating atoms using light.
- Section 6.4: Several theory proposals [321, 189, 494, 442] and experimental realizations are given in [322, 389, 346, 480, 161, 233, 68].
- Section 6.5: Bragg diffraction of atoms using Raman pulses [270, 461, 405, 255, 339, 367, 368, 240, 63, 228].
- A more detailed analysis of a two-level or multi-level atom interacting with light employing a semi-classical approach, including the effects of spontaneous emission, may be found in the books [7, 428].
- For a quantum treatment of atom interaction with light, see [95, 163, 152].
- Effect of the spontaneous emission on the Bragg diffraction is discussed in [258].

7 Atom Interferometry

7.1 Introduction

An interferometer is a measuring device that uses waves for its operation. It works on the principle of interference of waves; any object (not limited to physical objects) that can alter the path of the waves introduces a phase shift that leads to interference effects. The information about the object can then be estimated from the interference pattern. The most common type of interferometer uses light waves for its operation. Examples of interferometers are the Michelson interferometer, the Mach–Zehnder interferometer, and the Fabry–Perot interferometer. The Michelson interferometer played a significant role in the understanding of light at the turn of the nineteenth century.

Matter waves, too, can be used in the operation of an interferometer. Typical examples include neutron and atomic Bose–Einstein condensate (BEC) interferometers. As already discussed in Chapter 6, atoms can be manipulated using light to perform diffraction and other interference effects. The analogue of optical elements such as beam-splitters and mirrors are made of light. These elements are then used to split and diffract atomic BECs in an atom interferometer. This feat has been demonstrated in a number of BEC experiments that will be discussed in this chapter. We begin our discussion in Section 7.2 by first looking at optical interferometers. This is followed by the discussion of atomic BEC interferometry.

7.2 Optical Interferometry

We first describe two types of optical interferometers: the Michelson and Mach–Zehnder interferometers. Their principle of operation is closely related to the atom BEC interferometers described in Section 7.3. A Michelson interferometer is shown in Fig. 7.1(a). In a Michelson interferometer, coherent light enters a beam splitter and splits it into two parts that are allowed to evolve along different paths while being guided by mirrors. The split beams are then recombined by a beam splitter again to produce interference. The splitting and recombination of the light beam takes place at the same location, and the outgoing light is detected to observe the interference effect.

In a Mach–Zehnder interferometer as shown in Fig. 7.1(b), the splitting of the incident light beam and the recombination of the split light beams occur at different locations. If light of a particular phase is fed into the Mach–Zehnder interferometer,

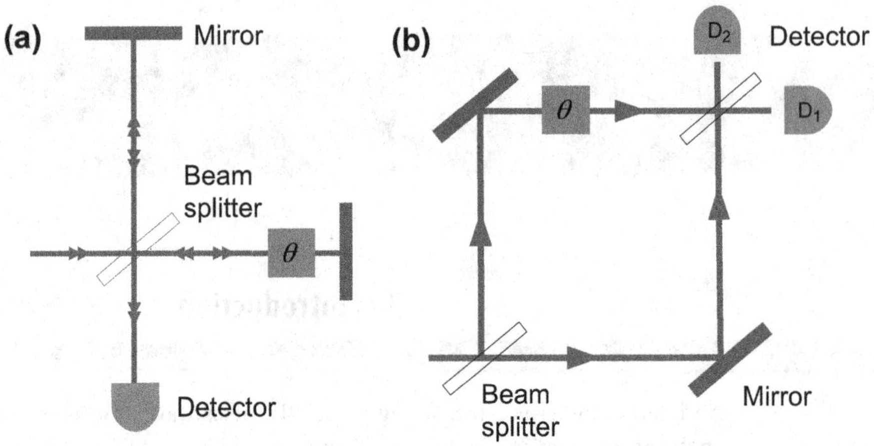

Fig. 7.1 A schematic representation of (a) a Michelson interferometer and (b) a Mach–Zehnder interferometer. The shaded box θ represents the relative phase between the two arms of the interferometer. Color version of this figure available at cambridge.org/quantumatomoptics.

and there is a relative phase θ between the two arms as shown, the light will be partly observed at detector D_1 with probability $P_1 = \frac{1+\cos\theta}{2}$, while the remaining light will be observed at detector D_2 with probability $P_2 = \frac{1-\cos\theta}{2}$. Notice that if the paths traversed by the split beams are identical, $\theta = 0$, only detector D_1 clicks and the light is not split between the two detectors.

In general, this result can be written as

$$P = \frac{1 + V\cos\theta}{2}, \tag{7.1}$$

where V is the visibility of the interference fringes and is usually defined with respect to the intensity maximum and minimum as follows:

$$V = \frac{I_{\max} - I_{\min}}{I_{\max} + I_{\min}}. \tag{7.2}$$

The visibility can be considered a measure of the coherence of the interfering beams. For $V = 1$, corresponding to perfect visibility as discussed above, one is able to measure accurately the interference signal. For $V < 1$, an interference signal can still be observed. However, if $V \ll 1$, it becomes difficult to observe anything (for $V = 0$, nothing is observed).

7.3 BEC Interferometry

In a BEC the matter waves of the bosons are coherent, forming a macroscopic quantum wave. For light, the formation of coherent light is the key to observing interference effects, as otherwise the final interference is a mixture of different phases, and no consistent interference pattern is obtained. We thus expect that it is possible to obtain similar interference effects as those discussed in Section 7.2 with BECs.

One of the most notable experiments showing matter wave interferometry is the demonstration in 1997 of interference between two independently prepared atomic condensates. It is striking because the atomic waves that were interfered did not share the same coherence initially. However, the interference was observed to vary from shot to shot (i.e., each run of the experiment). Since then, a number of atomic BEC interferometry experiments have been demonstrated in trapped-atom and guided-wave interferometers. In trapped-atom interferometers, a condensate in a potential well is divided into two condensates by deforming the well into a double-potential well. The split condensates are allowed to evolve before the double well is switched off, allowing the atomic BEC cloud to expand and interfere.

Interference of atomic BECs have also been demonstrated using guided-wave atom interferometers, where a condensate in trapping potential is split into two condensates using light. The arms of the potential well act as a guiding path for the split condensates. During the interferometric cycle, reflective laser pulses may be used to reverse the momentum of the split atomic condensates. At the end of the interferometric cycle, a light beam identical to the one used in the splitting of an atomic BEC cloud is used to recombine the split clouds.

The contrast observed in both the trapped-atom and guided-wave interferometers is degraded by atom-atom interaction within each split condensate. The quantum state of atomic BEC can be represented by a number state consisting of N atoms. When this state is split into two condensates with n_1 and n_2 atoms in each condensate, the presence of atom-atom interactions causes each number state to evolve at different rates and results in the accumulation of a relative time-dependent phase between the BECs. As a result, the interference fringes are lost at recombination. This effect is called phase diffusion. In addition, the spatially dependent phase, induced by the atom-atom interactions and the confining harmonic trap, degrades the observed contrast of guided-wave interferometers. In the rest of this chapter, we will analyze phase diffusion effects on guided-wave atom interferometry where the standing-wave light pulses of Sections 6.4 and 6.5 are used in the operation of the interferometer.

7.3.1 Splitting of Trapped Atomic Condensate

Consider an atomic BEC cloud with wavefunction $\psi_0(\boldsymbol{x})$ at rest in a confining harmonic potential. Applying the splitting laser pulses as discussed in Section 6.4, the initial cloud at rest $\psi_0(\boldsymbol{x})$ evolves to mode-entangled states $\psi_\pm(\boldsymbol{x})$ moving in opposite directions, where each state is a linear superposition of a number state with N atoms. The clouds are allowed to evolve for a duration T, after which they are subjected to recombination laser pulses that are identical to the splitting laser pulses. After applying the recombination pulses, atoms in general populate all the three modes $\psi_0(\boldsymbol{x})$, $\psi_\pm(\boldsymbol{x})$ as a result of of the relative phase accumulated by the split clouds during their evolution.

The many-body Hamiltonian describing the atomic BEC in the presence of an external potential V is

$$H(t) = \int d^2\boldsymbol{x}\, \Psi^\dagger \left[-\frac{\hbar^2}{2m}\nabla^2 + V + \frac{U_0}{2}\Psi^\dagger\Psi \right] \Psi, \tag{7.3}$$

where m is the atomic mass, $U_0 = 4\pi\hbar^2 a_s m^{-1}$ is the strength of the two-body interaction within the condensate, a_s is the s-wave scattering length, and $\Psi^\dagger$ is the creation field operator, which at a given time t creates an atom at position $\boldsymbol{x}$. Performing a transformation in a similar way to that done in (1.7), let $b_0^\dagger$, $b_{+1}^\dagger$, and $b_{-1}^\dagger$ be the operators that, upon acting on a vacuum state, create an atom belonging to a cloud at rest and moving to the right and left, respectively. They also satisfy the bosonic commutation relation $\left[b_i, b_j^\dagger\right] = \delta_{ij}$, $[b_k, b_k] = \left[b_k^\dagger, b_k^\dagger\right] = 0$, where $i, j, k = 0, \pm$. The field operator Ψ may be expanded in harmonics moving to the left and right as

$$\Psi(\boldsymbol{x}, t) = b_{+1}\psi_{+1}(\boldsymbol{x}, t) + b_{-1}\psi_{-1}(\boldsymbol{x}, t), \tag{7.4}$$

where $\psi_\pm(\boldsymbol{x}, t)$ are the wavefunctions of the BEC moving to right and left, respectively. They are solutions of the Gross–Pitaevskii equations, and are normalized to unity, $\int d\boldsymbol{x}\, |\psi_\pm(\boldsymbol{x}, t)|^2 = 1$.

Substituting (7.4) into (7.3) gives the Hamiltonian

$$H_{\text{eff}} = \frac{W}{2}(n_+ - n_-) + \frac{g}{2}\left[N^2 + (n_+ - n_-)^2 - 2N\right], \tag{7.5}$$

where $n_{\pm 1} = b_{\pm 1}^\dagger b_{\pm 1}$, $N = n_{+1} + n_{-1}$, $W = \varepsilon_+ - \varepsilon_-$ is the relative environment-introduced energy shift between the right- and left-propagating clouds $\varepsilon_\pm = \int d^3\boldsymbol{x}\, \psi_{\pm 1}^* \left[-\frac{\hbar^2}{2M}\nabla^2 + V\right]\psi_{\pm 1}$, and

$$g = \frac{U_0}{2}\int d^3\boldsymbol{x}\, |\psi_{+1}|^4 = \frac{U_0}{2}\int d^3\boldsymbol{x}\, |\psi_{-1}|^4 \tag{7.6}$$

is the coefficient characterizing the strength of atom-atom interaction within each cloud. In arriving at (7.5), a term that introduces constant energy shift has been omitted.

The initial state vector of the condensate, before the splitting laser pulses are applied, is well described by a number state with fixed atom number N,

$$|\Psi_{\text{ini}}\rangle = \frac{(b_0^\dagger)^N}{\sqrt{N!}}|0\rangle, \tag{7.7}$$

where $|0\rangle$ is the vacuum state. The splitting and recombination pulses couple the bosonic operators $b_0^\dagger$, $b_{\pm 1}^\dagger$ as described in (6.29) or (6.33):

$$\begin{aligned} b_{+1}^\dagger &\to -\frac{b_{+1}^\dagger}{2} + \frac{e^{i\pi/\sqrt{2}}}{\sqrt{2}} b_0^\dagger + \frac{b_{-1}^\dagger}{2}, \\ b_0^\dagger &\to \frac{b_{+1}^\dagger}{2} + \frac{b_{-1}^\dagger}{2}, \\ b_{-1}^\dagger &\to \frac{b_{+1}^\dagger}{2} + \frac{e^{i\pi/\sqrt{2}}}{\sqrt{2}} b_0^\dagger - \frac{b_{-1}^\dagger}{2}. \end{aligned} \tag{7.8}$$

The state vector (7.7), after the splitting pulses are applied, becomes

$$|\Psi_{\text{split}}\rangle = \frac{(b_{+1}^{\dagger} + b_{-1}^{\dagger})^N}{\sqrt{2^N N!}}|0\rangle,$$
$$= \frac{1}{2^{N/2}\sqrt{N!}} \sum_{n=0}^{N} \binom{N}{n} \left(b_{+1}^{\dagger}\right)^n \left(b_{-1}^{\dagger}\right)^{N-n} |0\rangle, \tag{7.9}$$

where $\binom{N}{n} = \frac{N!}{n!(N-n)!}$ is the binomial coefficient.

The state vector $|\Psi(t)\rangle$, at any other time after the splitting pulses have been applied, is obtained from the time evolution of $|\Psi_{\text{split}}\rangle$ under the Hamiltonian (7.5),

$$|\Psi(t)\rangle = e^{-i/\hbar \int H_{\text{eff}}\,dt}\,|\Psi_{\text{split}}\rangle, \tag{7.10}$$

and has the form

$$|\Psi(t)\rangle = \frac{1}{2^{N/2}\sqrt{N!}} \sum_{n=0}^{N} \binom{N}{n} e^{-i\frac{\theta}{2}(2n-N)-i\frac{\varphi}{2}[2n^2+2(n-N)^2]} \left(b_{+1}^{\dagger}\right)^n \left(b_{-1}^{\dagger}\right)^{N-n} |0\rangle, \tag{7.11}$$

where

$$\theta = \frac{1}{\hbar}\int_0^t d\tau\, W \tag{7.12}$$

is the accumulated phase difference between the left and right atomic clouds due to the environment, and

$$\varphi = \frac{1}{\hbar}\int_0^t d\tau\, g \tag{7.13}$$

is the accumulated nonlinear phase per atom due to interatomic interactions within each atomic cloud. In arriving at (7.11), the global phase term $\exp(iN\varphi)$ was neglected.

At the end of the interferometric cycle $t = T$, the recombination pulses act on $|\Psi(T)\rangle$ in accordance with (7.8), and the state afterward is

$$|\Psi_{\text{rec}}\rangle = \frac{1}{\sqrt{2^N N!}} \sum_{n=0}^{N} \binom{N}{n} e^{-i[\frac{\theta}{2}(2n-N)+\varphi(n^2+(n-N)^2)]} \left(-\frac{b_{+1}^{\dagger}}{2} + \frac{e^{i\pi/\sqrt{2}}b_0^{\dagger}}{\sqrt{2}} + \frac{b_{-1}^{\dagger}}{2}\right)^n$$
$$\times \left(\frac{b_{+1}^{\dagger}}{2} + \frac{e^{i\pi/\sqrt{2}}b_0^{\dagger}}{\sqrt{2}} - \frac{b_{-1}^{\dagger}}{2}\right)^{N-n} |0\rangle. \tag{7.14}$$

Exercise 7.3.1 Using the results of Exercise 6.4.2, verify (7.8).

Exercise 7.3.2 Verify (7.11).

7.3.2 Probability Density

After recombination, the particle numbers for the clouds at rest n_0, moving left n_- and moving right n_+ are measured. The state describing this result is given by

$$|n_+, n_-, n_0\rangle = \frac{\left(b_{+1}^{\dagger}\right)^{n_+}}{\sqrt{n_+!}} \frac{\left(b_{-1}^{\dagger}\right)^{n_-}}{\sqrt{n_-!}} \frac{\left(b_0^{\dagger}\right)^{n_0}}{\sqrt{n_0!}} |0\rangle. \tag{7.15}$$

The probability of measuring atoms in the state $|n_+, n_-, n_0\rangle$ after recombination is given by the modulus square of the probability amplitude $\langle n_+, n_-, n_0|\Psi_{\text{rec}}\rangle$. Using (7.14) and (7.15), the probability amplitude becomes

$$\langle n_+, n_-, n_0|\Psi_{\text{rec}}\rangle = \frac{1}{\sqrt{2^N N!}} \sum_{n=0}^{N} \binom{N}{n} e^{-i\left[\frac{\theta}{2}(2n-N)+\varphi(n^2+(n-N)^2)\right]}$$
$$\times \langle 0| \frac{\left(b_0^\dagger\right)^{n_0}}{\sqrt{n_0!}} \frac{\left(b_{-1}^\dagger\right)^{n_-}}{\sqrt{n_-!}} \frac{\left(b_{+1}^\dagger\right)^{n_+}}{\sqrt{n_+!}} \left(-\frac{b_{+1}^\dagger}{2} + \frac{e^{i\pi/\sqrt{2}} b_0^\dagger}{\sqrt{2}} + \frac{b_{-1}^\dagger}{2}\right)^n$$
$$\times \left(\frac{b_{+1}^\dagger}{2} + \frac{e^{i\pi/\sqrt{2}} b_0^\dagger}{\sqrt{2}} - \frac{b_{-1}^\dagger}{2}\right)^{N-n} |0\rangle, \tag{7.16}$$

where

$$\left(-\frac{b_{+1}^\dagger}{2} + \frac{e^{i\pi/\sqrt{2}} b_0^\dagger}{\sqrt{2}} + \frac{b_{-1}^\dagger}{2}\right)^n \left(\frac{b_{+1}^\dagger}{2} + \frac{e^{i\pi/\sqrt{2}} b_0^\dagger}{\sqrt{2}} - \frac{b_{-1}^\dagger}{2}\right)^{N-n}$$
$$= \sum_{j=0}^{n} \sum_{k=0}^{N-n} \binom{n}{j} \binom{N-n}{k} (-1)^{n-j} \left(\frac{b_0^\dagger e^{i\frac{\pi}{\sqrt{2}}}}{\sqrt{2}}\right)^{j+k} \left(\frac{b_{+1}^\dagger - b_{-1}^\dagger}{2}\right)^{N-k-j}.$$

7.3.2.1 Probability in the Absence of Atom-Atom Interactions

Here we consider the case where there are no atom-atom interactions among the atoms in the split clouds during evolution. In this case, $\varphi = 0$, the probability amplitude (7.16) takes the form

$$\langle n_+, n_-, n_0|\Psi_{\text{rec}}\rangle = \sqrt{\frac{N!}{2^N n_+!\, n_-!\, n_0!}} (-1)^{n_-} \left(-i \sin\frac{\theta}{2}\right)^{N-n_0} \left(\sqrt{2}\cos\frac{\theta}{2}\right)^{n_0}, \tag{7.17}$$

and the probability density $P(n_+, n_-, n_0) = |\langle n_+, n_-, n_0|\Psi_{\text{rec}}\rangle|^2$ is given by

$$P(n_+, n_-, n_0) = \frac{1}{2^N} \frac{N!}{n_+!\, n_-!\, n_0!} \left(\sin^2\frac{\theta}{2}\right)^{N-n_0} \left(2\cos^2\frac{\theta}{2}\right)^{n_0}. \tag{7.18}$$

Equation (7.18) is a binomial distribution and can be written as a product of two probability density functions,

$$P(n_+, n_-, n_0) = P_\pm(n_+, n_-) P(n_0), \tag{7.19}$$

where

$$P_\pm(n_+, n_-) = \frac{(N-n_0)!}{2^{N-n_0} n_+!\, n_-!} \tag{7.20}$$

and

$$P(n_0) = \frac{N!}{n_0!\,(N-n_0)!} \left(\sin^2\frac{\theta}{2}\right)^{N-n_0} \left(\cos^2\frac{\theta}{2}\right)^{n_0}. \tag{7.21}$$

The function $P_\pm$ describes the probability of observing n_+ and n_- atoms in the right- and left-moving clouds, respectively, for a fixed number of atoms in the cloud at rest. This function is independent of the phase angle θ and is normalized to unity. The function P is the probability of observing n_0 atoms in the cloud at rest. It is normalized to unity and depends on the phase angle θ introduced by the environment.

For a very large population of atoms $N \gg 1$, the factorials may be approximated using Stirling's formula, $n! = \sqrt{2\pi n}n^n e^{-n}$, and the probability densities that correspond to P and $P_\pm$ become

$$P(n_0) = \frac{2}{\sqrt{2\pi N}\sin\theta}\exp\left[-\frac{2}{N}\frac{(n_0 - N\cos^2(\theta/2))^2}{\sin^2\theta}\right] \tag{7.22}$$

and

$$P_\pm(n_+, n_-) = \sqrt{\frac{2}{\pi(n_+ + n_-)}}\exp\left[-\frac{2}{n_+ + n_-}\left(\frac{n_+ - n_-}{2}\right)^2\right], \tag{7.23}$$

where $n_+ + n_- \gg 1$.

Both the probability densities $P_\pm$ and P are Gaussian. For a fixed value of n_0 atoms in the stationary cloud, the peak of the probability function $P_\pm$ is located at $(N-n_0)/2$, with average values of n_+ and n_- given by

$$\langle n_+\rangle = \langle n_-\rangle = \frac{1}{2}(N - n_0) \tag{7.24}$$

and standard deviations

$$\sigma_{n_+} = \sigma_{n_-} = \frac{1}{2}\sqrt{N - n_0}. \tag{7.25}$$

The number of atoms in the right and left clouds are anticorrelated according to

$$\text{cov}(n_+, n_-) = \langle n_+ n_-\rangle - \langle n_+\rangle\langle n_-\rangle = -\frac{1}{4}(N - n_0). \tag{7.26}$$

The maximum of the probability density function $P(n_0)$ is located at $n_0 = N\cos^2(\theta/2)$. Since n_0 takes values in the interval $[0, N]$, θ takes values in the interval $0 < \theta < \pi$. The end points $\theta = 0$ and $\theta = \pi$ are excluded because the probability density (7.22) is not defined at the end points. However, the values of $P(n_0)$ at the end points may be obtained using (7.21). This gives $P(n_0) = 1$ for $\theta = 0, \pi$. The probability density $P(n_0 = N) = 1$ for $\theta = 0$ means that all the atoms are in the cloud at rest after recombination, and $P(n_0 = 0) = 1$ for $\theta = \pi$ implies that no atom is observed in the cloud at rest after recombination; all the atoms are found in the clouds moving to the left and right after recombination.

Figure 7.2(a) shows a plot of the probability density (7.22) at three different values of θ. The width of each peak on the graph scales roughly as $\sqrt{(N\sin^2\theta)/4}$ so that the relative width of the distribution scales roughly as $\sqrt{\sin^2\theta/(4N)}$. Because of the dependence of the width of the distribution function on θ, the width of the probability density is largest at $\theta = \pi/2$ and vanishes at $\theta = 0, \pi$. The changing values of θ move the peak of the probability density $P(n_0)$ from $n_0 = N$, corresponding to the situation where more atoms are in the stationary cloud towards $n_0 = 0$ that corresponds to situations where a fewer number of atoms are in the stationary cloud.

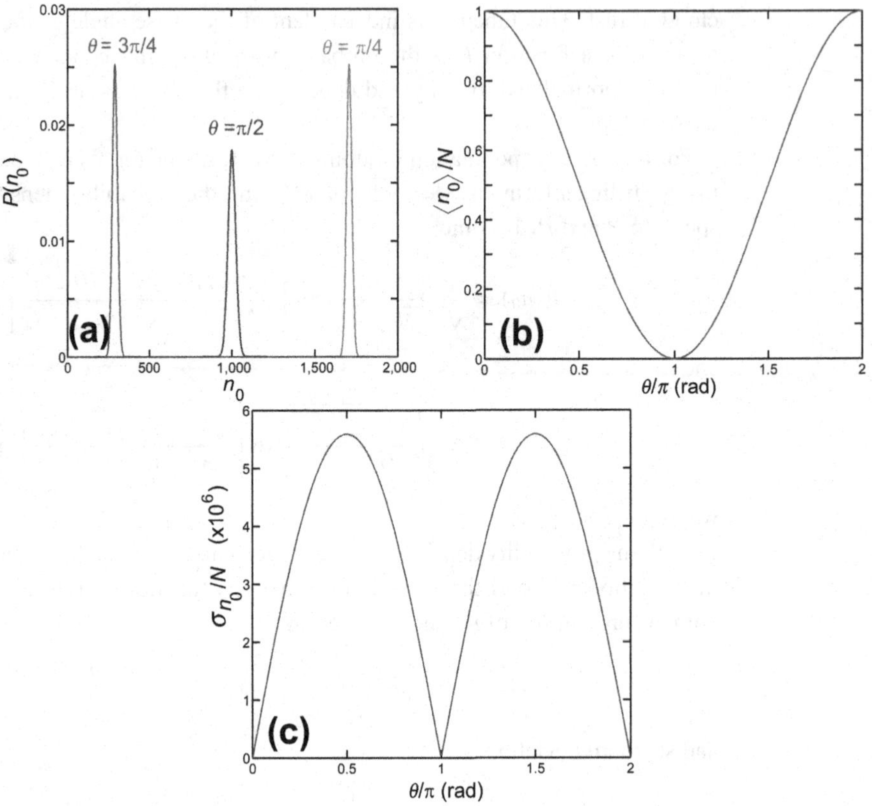

Fig. 7.2 (a) The probability density $P(n_0)$ as a function of n_0 at three different values of θ. (b) The relative mean value $\langle n_0 \rangle /N$ as a function of θ. (c) The relative standard deviation σ_{n_0}/N as a function of θ. Color version of this figure available at cambridge.org/quantumatomoptics.

The mean value and variance of the probability density $P(n_0)$ are

$$\langle n_0 \rangle = N \cos^2 \frac{\theta}{2} \tag{7.27}$$

and

$$\sigma_{n_0}^2 = N \cos^2 \frac{\theta}{2} \sin^2 \frac{\theta}{2}. \tag{7.28}$$

Figure 7.2(b) and (c) shows the plots of the relative mean value and relative standard deviation, respectively. In Fig. 7.2(b), the contrast is unity and the visibility as defined in (7.2) is maximum, taking the value of unity. Thus, for noninteracting condensates, full fringes would be observed in every run of the experiment. The error associated in counting the number of atoms in the stationary cloud shows sinusoidal oscillations with a periodicity of π, as shown in Fig. 7.2(c). At $\theta = 0, m\pi$ (where m is any integer value), the standard deviation is zero and corresponds to situations where all the atoms are known with absolute certainty to be either in the cloud at rest or in the moving clouds. At this point, the width of the probability density vanishes as previously described above. Even values of m correspond to the case when all the atoms are in the cloud at rest, while odd values of m correspond to the case when all the atoms are in the moving clouds. The standard deviation is maximum at odd multiples of $\pi/2$, as shown in Fig. 7.2(c). This can also be seen from the width of the distribution at $\theta = \pi/2$ in Fig. 7.2(a), which occurs when equal population of atoms are found in the moving clouds and the cloud at rest.

7.3.2.2 Probability in the Presence of Atom-Atom Interactions

Calculating the probability density when $\varphi \neq 0$ seems a daunting task. However, due to orthogonality of the number states, all terms in (7.16) vanish except for the term $j + k = n_0$, giving

$$\langle n_{+1}, n_{-1}, n_0 | \Psi_{\text{rec}} \rangle = \sqrt{\frac{N!\, n_0!}{2^{(3N-n_0)} n_{+1}!\, n_{-1}!}} (N - n_0)!\, e^{i n_0 \frac{\pi}{\sqrt{2}}} (-1)^{n_{-1}} \times \sum_{n=0}^{N} e^{-i\theta(n-N/2) - i\varphi[n^2 + (n-N)^2]} S(n_0, n), \tag{7.29}$$

where

$$S(n_0, n) = \sum_{j=\max(0, n_0+n-N)}^{\min(n, n_0)} \frac{(-1)^{n-j}}{j!\,(n-j)!\,(n_0-j)!\,(N-n-n_0+j)!}. \tag{7.30}$$

Comparing (7.29) for $\varphi = 0$ and (7.17) shows that $S(n_0, n)$ may be obtained from a Fourier transform of the trigonometric functions in (7.17):

$$S(n_0, n) = \frac{1}{2\pi} \frac{2^N}{(N-n_0)!\, n_0!} \int_0^{2\pi} d\theta\, e^{i(n-N/2)\theta} \left(\cos\frac{\theta}{2}\right)^{n_0} \left(-i \sin\frac{\theta}{2}\right)^{N-n_0}. \tag{7.31}$$

Substituting (7.31) into (7.29), the probability

$$P(n_+, n_-, n_0) = |\langle n_{+1}, n_{-1}, n_0 | \Psi_{\text{rec}} \rangle|^2 \tag{7.32}$$

of observing n_- atoms in the clouds moving to the left, n_+ atoms in the cloud moving to the right, and n_0 atoms in the stationary cloud after recombination may be written as a product of two functions $P(n_+, n_-, n_0) = P_\pm(n_+, n_-) P_0(n_0, \theta, \varphi)$, where $P_\pm$ defined in (7.20) gives the probability of finding n_+ and n_- atoms in the clouds that are moving to the right and left, respectively, after recombination. The probability density function $P_0(n_0, \theta, \varphi)$ describes the probability of finding n_0 atoms in the stationary clouds after recombination,

$$P_0(n_0, \theta, \varphi) = \frac{N!}{n_0!\,(N - n_0!)} |f(n_0, \theta, \varphi)|^2. \tag{7.33}$$

The function $f(n_0, \theta, \varphi)$ is

$$f(n_0, \theta, \varphi) = \frac{e^{-iN^2\varphi/2}}{\sqrt{1 - 2iN\varphi}} \left[(N - n_0) \ln \sqrt{1 - \frac{n_0}{N}} + n_0 \ln \sqrt{\frac{n_0}{N}} \right] \times \left(e^{-\eta_-^2} + (-1)^{N-n_0} e^{-\eta_+^2} \right), \tag{7.34}$$

where

$$\eta_\pm = \frac{N \left(\arccos \sqrt{\frac{n_0}{N}} \pm \frac{\theta}{2} \right)^2}{1 - 2iN\varphi}. \tag{7.35}$$

Unlike the probability density function $P_\pm$, P_0 has a nontrivial dependence on its arguments and contains parameters φ, θ that affect the fringes observed in the interferometer. Section 7.3.3 is devoted to understanding the dependence of P_0 on its arguments. The dependence of $P_\pm$ on its arguments was described earlier in Section 7.3.2.1.

Exercise 7.3.3 Show that powers of operators $b_0^\dagger\, b_{+1}^\dagger$ and $b_{-1}^\dagger$ in (7.16) can be expanded as written.

Exercise 7.3.4 For $\varphi = 0$, verify that the probability (7.16) takes the form (7.17). Show that (7.17) can be decomposed into two probability density functions (7.20) and (7.21).

Exercise 7.3.5 Assuming $N \gg 1$ and using Stirling's approximation, prove (7.21) and (7.23). What is the assumption on n_0 or $n_+ + n_-$?

Exercise 7.3.6 Verify (7.24), (7.25), and (7.26).

Exercise 7.3.7 Verify (7.27) and (7.28).

Exercise 7.3.8 Show that the Fourier transform of (7.30) is (7.31).

7.3.3 Features of the Probability Density

In the limit of very large numbers of atoms, $N, n_0 \gg 1$, the factorial function can be approximated by Stirling's approximation. The probability density function $P_0(n_0, \theta, \varphi)$ is proportional to the modulus square of two terms,

$$P_0(n_0, \theta, \varphi) = \frac{1}{\sqrt{(1 + 4N^2\varphi^2)}} \sqrt{\frac{N}{2\pi n_0(N - n_0)}} |e^{-\eta_-^2} + (-1)^{N-n_0} e^{-\eta_+^2}|^2, \quad (7.36)$$

where $\eta_\pm$ is given in (7.35). The relative phase difference between the two terms of $P_0(n_0, \theta, \varphi)$ changes rapidly with n_0 due to the multiplier $(-1)^{N-n_0}$. Thus, the interference terms do not contribute to the averages and are neglected in calculating the mean $\langle n_0 \rangle = \int_0^N dn_0\, n_0 P_0(n_0, \theta, \varphi)$ and standard deviation of the probability density function. Evaluating the integral gives

$$\langle n_0 \rangle = \frac{N}{2} \left[1 + \exp\left(-\frac{1 + 4N^2\varphi^2}{2N} \right) \cos\theta \right]. \quad (7.37)$$

Similarly, the variance is

$$\sigma_{n_0}^2 = \frac{N^2}{2} \left[\frac{1}{4} + \frac{\exp\left(-2\frac{1+4N^2\varphi^2}{N}\right) \cos 2\theta}{4} - \frac{\exp\left(-\frac{1+4N^2\varphi^2}{N}\right) \cos^2\theta}{2} \right]. \quad (7.38)$$

These results are understood by studying the dependence of the function $P_0(n_0, \theta, \varphi)$ on the number of atoms n_0 for different values of the strength of two-body atom interactions φ. At relatively small values of φ, such that $\varphi \ll 1/\sqrt{N}$, the term $\exp(-\eta_-)$ in (7.36) dominates the other. The probability is then in the form of a Gaussian,

$$P_0(n_0, \theta, \varphi) \approx \frac{1}{\sqrt{1 + 4N^2\varphi^2}} \sqrt{\frac{N}{2\pi n_0(N - n_0)}} \exp\left[-\frac{2N\left(\theta/2 - \arccos\sqrt{n_0/N}\right)^2}{1 + 4N^2\varphi^2} \right], \quad (7.39)$$

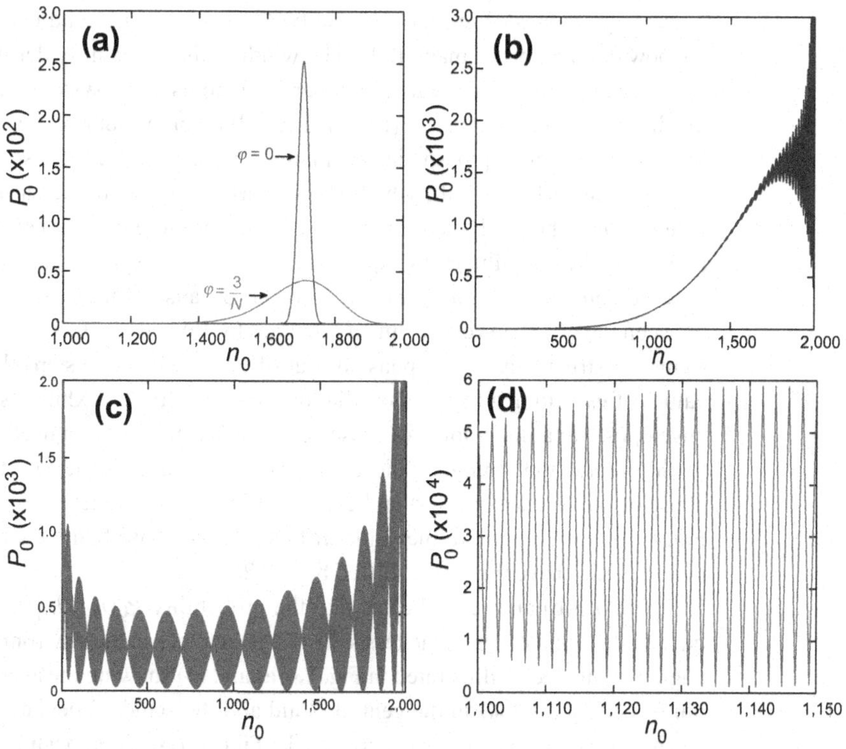

Fig. 7.3 (a) The probability density function $P_0(n_0, \theta, \varphi)$ as a function of n_0 for $\varphi = 0$ and $\varphi = 3/N$. (b) The probability density function $P_0(n_0, \theta, \varphi)$ as a function of n_0 for $\varphi = 0.2/\sqrt{N}$. (c) The probability density function $P_0(n_0, \theta, \varphi)$ as a function of n_0 for $\varphi = 1/\sqrt{N}$. (d) An enlargement of the plot in (c), showing the fast-scale spatial oscillations of the probability density function $P_0(n_0, \theta, \varphi)$. For all plots, $\theta = \pi/4$ and $N = 2{,}000$. Color version of this figure available at cambridge.org/quantumatomoptics.

with a maximum located at $n_0 = N\cos^2\theta/2$. This situation is shown in Fig. 7.3(a). The two curves in the figure are plots of the probability density $P_0(n_0, \theta, \varphi)$ given by (7.39) versus n_0 for two different values of two-body atomic interaction strength φ. Both curves correspond to the same value of angle θ. The noticeable feature of Fig. 7.3(a) is the increase in the width of the probability distribution with φ. This behavior is explained by (7.38), which in the limit $\varphi \ll 1/\sqrt{N}$ reduces to

$$\sigma_{n_0} = \frac{\sqrt{N}}{2}\sqrt{1 + 4N^2\varphi^2}\sin\theta. \tag{7.40}$$

For very small values of $\varphi \ll 1/N$, the influence of interatomic interactions on the operations of the atom interferometer is negligible. The relative standard deviation of the number of atoms in the central cloud is inversely proportional to the square root of the total number of atoms in the system: $\sigma_{n_0} \propto 1/\sqrt{N}$. For $1/N \ll \varphi \ll 1/\sqrt{N}$, the width of the distribution grows linearly with the increases in φ. The mean value of n_0 for $\varphi \ll 1/\sqrt{N}$ reasonably corresponds to the position of the peak. Equation (7.37) for $\langle n_0 \rangle$ in this limit gives

$$\langle n_0 \rangle \approx \frac{N}{2}(1 + \cos\theta). \tag{7.41}$$

As is seen, n_0 depends on θ but not on φ.

For large values of $\varphi \approx 1/\sqrt{N}$, the two terms $\exp(-\eta_-)$ and $\exp(-\eta_+)$ in (7.36) are now comparable in magnitude. The width of the probability density $P_0(n_0, \theta, \varphi)$ becomes of the order of the total number N of atoms in the systems. The transition to this limit is shown in Figs. 7.3(b)–(c). The solid regions not resolved in Figs. 7.3(b)–(c) correspond to rapid spatial oscillations with period 2. These oscillations are clearly seen in Fig. 7.3(d), which shows part of Fig. 7.3(c) for a narrow range of values of n_0. The oscillations are caused by the interference between the two terms in (7.36). As the magnitude of φ approaches $1/\sqrt{N}$, the $\exp(-\eta_-)$ and $\exp(-\eta_+)$ terms become comparable in magnitude. However, because of the nearly π-phase change between the two terms every time n_0 changes by 1 due to the factor $(-1)^{N-n_0}$, the two terms constructively add in phase or out of phase when one steps through different values of n_0. Along with rapid oscillations, Figs. 7.3(b)–(c) exhibit oscillations of the envelopes – at a much longer timescale – that are more pronounced for large values of the interaction strength. These oscillations are due to the fact that the relative phase of the terms $\exp(-\eta_-)$ and $\exp(-\eta_+)$ in (7.36) changes with n_0. The nodes in Fig. 7.3(c) correspond to the value of this relative phase being equal to 0 or π and the antinodes have the phase shifted by $\pm\pi/2$.

Figure 7.3(b) and (c) indicates that the probability $P_0(n_0, \theta, \varphi)$, and as a consequence, $\langle n_0 \rangle$ and σ_{n_0} become less sensitive to changes in the environment-introduced angle θ. This fact is illustrated in Fig. 7.4(a) and (b) showing the average value of the number of atoms $\langle n_0 \rangle$ in the central cloud and the standard deviation σ_{n_0} versus θ, as given by (7.37) and (7.38), respectively. Figure 7.4(a) demonstrates that increased interatomic interactions eventually lead to the loss of contrast of interference fringes. Additionally, larger interatomic interactions cause large shot-to-shot fluctuations in the number of atoms in each of the three ports, as seen in Fig. 7.4(b). The loss of interference fringe contrast can be quantified by writing (7.37) as

$$\langle n_0 \rangle = \frac{N}{2}\left(1 + V\cos\theta\right), \tag{7.42}$$

where the contrast V is

$$V = \exp\left(-\frac{1 + 4N^2\varphi^2}{N}\right). \tag{7.43}$$

Figure 7.4(c) shows the fringe contrast (7.43) as a function of φ and demonstrates that the values of φ approaching $1/\sqrt{N}$ results in the washing out of interference fringes.

Exercise 7.3.9 Using (7.31) in (7.29), show that the probability of atoms in the stationary cloud may be written as (7.37), and state all your assumptions. Hint: First use the steepest descent method to get (7.34) and apply Stirling's approximation to simplify your result.

Exercise 7.3.10 Calculate the variance and expectation value of n_0. Will there be any changes to expectation value and variance of atoms if the moving clouds change?

Exercise 7.3.11 Show that in the limit $\varphi\sqrt{N} \ll 1$, the variance and expectation value of n_0 reduce to (7.41) and (7.42), respectively.

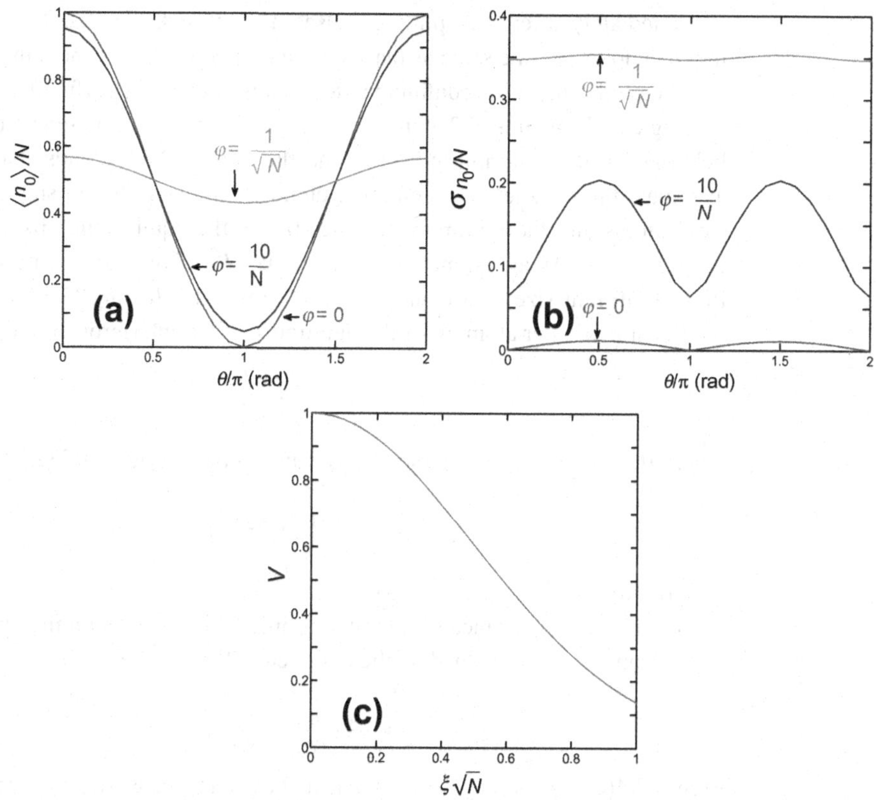

Fig. 7.4 (a) The normalized mean value of the number of atoms in the central cloud $\langle n_0 \rangle / N$ as a function of θ. (b) The normalized standard deviation σ_{n_0}/N as functions of θ. (c) Interference fringe contrast $\mathcal{V}$ as a function of the interatomic in interactions $\varphi\sqrt{N}$. All plots use $N = 2{,}000$. Color version of this figure available at cambridge.org/quantumatomoptics.

7.3.4 Controlling Nonlinear Phase per Atom

As shown in Section 7.3.3, limited interference fringes were observed when the nonlinear phase per atom φ due to two-body interactions is about the order of $1/\sqrt{N}$. In order to quantify the effect due to two-body interactions within the condensate, the phase φ is calculated in terms of experimental parameters. We consider experiments that are conducted in parabolic traps with confining potential of the form

$$V = \frac{M}{2}\left(\omega_x^2 x^2 + \omega_y^2 y^2 + \omega_z^2 z^2\right). \tag{7.44}$$

The density profiles of the moving clouds are well described by the Thomas–Fermi approximation,

$$|\psi_\pm|^2 = \frac{\mu_n}{U_0 n}\left(1 - \frac{x^2}{R_x^2} - \frac{y^2}{R_y^2} - \frac{z^2}{R_z^2}\right), \quad |\psi_\pm|^2 \geq 0, \tag{7.45}$$

where R_i is the radial sizes of the cloud for the ith dimension, μ_n is the chemical potential of the BEC cloud with n atoms, and $U_0 = 4\pi\hbar^2 a_{sc}/M$ is the strength of the two-body interactions within the cloud.

Immediately after the splitting pulses are applied, the density profiles of the moving clouds are the same as that of the initial BEC cloud containing N atoms and are in equilibrium in the confining potential given by (7.44). After the splitting, each moving cloud contains $N/2$ atoms on average. The repulsive nonlinearity is no longer balanced by the confining potential and the radii of both clouds start to oscillate. The maximum size of the oscillating clouds is the equilibrium size corresponding to N atoms and the minimum size lies below the equilibrium size corresponding to $N/2$ atoms. As an estimate, the radial size R_i of the condensate is taken to be the equilibrium size of a cloud with $N/2$ atoms. Evaluating (7.13), we obtain the nonlinear phase per atom φ for the duration T of the interferometric cycle,

$$\varphi = \frac{2}{7}\frac{\mu_n T}{n\hbar}, \tag{7.46}$$

where $\mu_n = 2^{-3/5}\mu$, μ being the equilibrium chemical potential [32, 100],

$$\mu = \frac{\hbar\bar{\omega}}{2}\left(15N\frac{a_{sc}}{\bar{a}}\right)^{2/5}, \tag{7.47}$$

$\bar{\omega} = (\omega_x\omega_y\omega_z)^{1/3}$, $\bar{a} = \sqrt{\hbar/(M\bar{\omega})}$.

The relative importance of interatomic interaction effects on the operation of the interferometer is determined by the parameter $P = \varphi\sqrt{N} \ll 1$,

$$P = 0.64\left(\frac{a_{sc}}{\bar{a}}\right)^{2/5}\bar{\omega}TN^{-1/10}. \tag{7.48}$$

Figure 7.4(c) shows that the contrast of the interference fringes decreases with P. The condition of good contrast is in the regime $P < 1/2$; for $P = 0.5$, the contrast is $V = 0.6$.

Equation (7.48) shows that $P \propto T\bar{\omega}^{6/5}N^{-1/10}$. The dependence of P on the total number N of atoms in the BEC cloud is very weak. Hence, the parameter P is primarily dependent on the duration T of the interferometric cycle and averaged frequency $\bar{\omega}$ of the trap. Equation (7.48) is handy in quantifying the amount of nonlinear phase per atom that would be present in an experiment. For example, consider an experiment where a condensate consisting of about 10^5 ^{87}Rb atoms are used to perform interferometry, say in a Michelson geometry with transverse and longitudinal frequencies of the trap being 177 Hz and 5 Hz, respectively. If the interferometry time is 10 ns and the s-wave scattering length is 5.2×10^{-9} m, one obtains $P \approx 1.6\times10^{-2}$. This value of P is very small in comparison to unity. As such, the interatomic interactions will not limit the visibility of the interference fringes obtained in the experiment.

It is of importance to quantify the amount of error one makes in estimating parameters of the measurement, such as the θ parameter in this chapter. One would expect that the large number of atoms in BEC would easily allow one to estimate θ errors with scaling better than the standard quantum limit, $1/\sqrt{N}$. However, this is not the case, as one can see from the probability density distribution before and after the interferometry. This is not surprising, since the initial input state is a linear superposition of the atomic coherent state, which in the limit of large N is Gaussian. Indeed, the self-interaction within each atomic condensate during propagation belongs to the family of correlations that produce squeezing – the one-axis twisting squeezed states of Section 5.7.2. Incidentally, the measured observables $n_\pm$ or n_0 of interest in these experiments commute with the generator of the correlations. To observe the squeezing, one would have to find ways to calculate averages of operators that do not commute with the generators of the correlations in the condensate. Most importantly, the self-interaction that is responsible for the nonlinear evolution would only lead to phase diffusion. We will be studying ways to utilize squeezing contained in quantum states for improved estimation in Chapter 8.

Exercise 7.3.12 One of the principal conditions required to arrive at (7.37) is that $\varphi\sqrt{N} \ll 1$. For $\varphi\sqrt{N} = 1$, verify (7.48).

Exercise 7.3.13 A trap holding ^{87}Rb atom condensate containing 3×10^3 atoms was used in an interferometry experiment. The trap has radial frequency fixed at 60 Hz, while the axial frequency of the trap varied. For a propagation time T of 60 ms, the axial frequency was 17 Hz, whereas for a propagation time of 97 ms, the axial frequency was 10.29 Hz. Comment on the effect of nonlinear phase per atom in each experiment.

7.4 References and Further Reading

- Section 7.2: For a general introduction to interferometry, especially in optics, see [207, 317].
- Section 7.3: Experiments describing BEC interferometry in trapped and guided-wave interferometers detailing the effect of two-body interactions [16, 427, 418, 247, 480, 227, 161, 228, 67]. Theoretical works describing BEC atom interferometry and the effects of two-body interactions [242, 236, 354, 442, 253, 234, 235]. A theory paper giving the scattering coefficients of an atomic BEC [251]. For more review articles on the subject, see [37, 173, 99]. For texts describing the steepest descent method and asymptotic evaluation of integrals, see [33, 20].
- Other interferometers involving fermions, for instance, and other methods for splitting the atom BEC [396, 193, 412, 243, 401, 298, 149].
- Applications of interferometers to atomtronic devices and gyroscopes [423, 10, 84, 129, 395, 112, 216, 190, 191, 119].
- Other applications to precision measurements [486, 42, 347].

8 Atom Interferometry Beyond the Standard Quantum Limit

8.1 Introduction

As pointed out in Chapter 7, the error in interferometry depends sensitively on the input state that is used in the measurement. The types of input states used in the interferometer of Chapter 7 are linear superpositions of atoms in different modes. Using such states composed of N independent particles in interferometry results in phase estimation that cannot be better than the standard quantum limit scaling as $1/\sqrt{N}$. However, nonlinear interactions among the particles can modify the linear superposed states such that the particles become correlated. Such correlations give rise nonclassical states such as twin-Fock states, NOON, and squeezed states. The NOON state is a linear superposition of two quantum states with each state consisting of two modes. The first quantum state has N particles in the first mode and 0 particles in the second mode, while the second quantum state has 0 particles in the first mode and N particles in the second mode. The arrangement of the particle numbers gives the name NOON. These nonclassical states have special features in their distributions, where the width of the distribution in one of the degrees of freedom is less than the square root of N. An interferometer using these nonclassical states at the input realizes phase estimation error that is better than the standard quantum limit scaling as $1/\sqrt{N}$.

In optics, passing coherent light through nonlinear materials allows for the generation of nonclassical states such as the squeezed and twin-Fock states. Similarly, by using collisions, dipole interactions, or optical field modes, nonclassical states are generated in condensed atoms like Bose–Einstein condensates (BECs) and atomic ensembles. The collisions in atomic systems, such as in two-component BECs, dipole interactions in Rydberg atoms, or spin-1 (or spinor) condensates, produce nonlinear interactions that have a similar form as those found in their optics counterpart. These types of states have been extensively studied and have been suggested to provide improvements in rotation sensing and the precision of atomic clocks. We have already seen examples of squeezed states in BECs in Section 5.7. In this chapter, we further study the generation of squeezed states in two-component atomic condensates and its characterization in an interferometric context.

8.2 Two-Component Atomic Bose–Einstein Condensates

Multicomponent atomic condensates, such as two-component BECs or spinor condensates, consist of atoms with different degrees of freedom. The most common way of realizing this is using internal states of the atom, such as the hyperfine ground states as seen in Section 4.2. When the components are put in a linear superposition, the nonlinear two-body atom interactions cause fluctuations in the relative number and phase between the components. The net effect is that a nonclassical state emerges. In this section, we derive the nonclassical state generated by collisions in a two-component atomic condensate.

8.2.1 Two-Mode Model Hamiltonian

Consider a condensate consisting of different atomic species labeled by a, b, that are in the same trap. In the case of ^{87}Rb, the two species may be realized by the hyperfine ground states $|F = 1, m_f = -1\rangle$ and $|F = 2, m_f = 1\rangle$, for instance. This can be achieved by population transfer after initially preparing the condensate in one of the atomic states. As discussed in Section 4.3, the atoms interact, experiencing interspecies and intraspecies collisions. Assuming that the interspecies collisional interactions are energetically insufficient to cause spin flips from one species to the other, the number of atoms in any species is then constant, and there is no weak link or Josephson oscillations between the species. The many-body Hamiltonian governing their dynamics during the nonlinear interactions is

$$
\begin{aligned}
H = \sum_{k=1,2} & \left[\int d\mathbf{r} \Psi_k^\dagger(\mathbf{r},t) H_0 \Psi_k(\mathbf{r},t) + \frac{U_k}{2} \int d\mathbf{r} \Psi_k^\dagger(\mathbf{r},t)\Psi_k^\dagger(\mathbf{r},t)\Psi_k(\mathbf{r},t)\Psi_k(\mathbf{r},t) \right] \\
& + U_{12} \int d\mathbf{r} \Psi_1^\dagger(\mathbf{r},t)\Psi_2^\dagger(\mathbf{r},t)\Psi_1(\mathbf{r},t)\Psi_2(\mathbf{r},t), \qquad (8.1)
\end{aligned}
$$

where H_0 is a single-particle Hamiltonian, $U_k = 4\pi\hbar^2 a_{\text{sc}}^k/m$, m is the atomic mass, a_{sc}^k is the kth species s-wave scattering length, $U_{12} = 4\pi\hbar^2 a_{\text{sc}}^{ab}/m$, and a_{sc}^{ab} is the interspecies scattering length. The single-particle Hamiltonian includes the confining potential that traps the atomic condensate, and it also includes the effect of the environment that results in the different dynamics for condensates in different hyperfine states.

The condensate wavefunctions $\psi_k(\mathbf{r},t)$ for each component k in the BEC are found by solving two coupled Gross–Pitaevskii equations,

$$\mu_1\psi_1(\mathbf{r},t) = \left(H_0 + U_1 n_1|\psi_1(\mathbf{r},t)|^2 + U_{12}n_2|\psi_2(\mathbf{r},t)|^2\right)\psi_1(\mathbf{r},t), \qquad (8.2)$$

$$\mu_2\psi_2(\mathbf{r},t) = \left(H_0 + U_2 n_2|\psi_2(\mathbf{r},t)|^2 + U_{12}n_1|\psi_1(\mathbf{r},t)|^2\right)\psi_2(\mathbf{r},t), \qquad (8.3)$$

where μ_k is the chemical potential per particle, and n_k is the total number of particles in the kth component. The condensate wavefunction $\psi_k(\mathbf{r},t)$ satisfies the normalization condition $\int d\mathbf{r}\, |\psi_k(\mathbf{r},t)|^2 = 1$.

Let $a_1^\dagger$ and $a_2^\dagger$ be operators that act on the vacuum state to create an atom in components 1 and 2, respectively. They satisfy the commutation relation $[a_k, a_j] = 0$

and $[a_k, a_j^\dagger] = \delta_{kj}$. These operators are defined in (1.8). The field operators in terms of the $a_1^\dagger$ and $a_2^\dagger$ are

$$\Psi_1(\mathbf{r},t) = \psi_1(\mathbf{r},t)a_1, \qquad \Psi_2(\mathbf{r},t) = \psi_2(\mathbf{r},t)a_2. \tag{8.4}$$

Using (8.2), (8.3), and (8.4) in (8.1), we have

$$\begin{aligned} H &= \mu_+ N + \mu_-(n_1 - n_2) - g_+\frac{N^2}{4} - g_- N\left(\frac{n_1 - n_2}{2}\right) \\ &\quad + \frac{1}{2}\left[g_+ N + g_-\left(n_1 - n_2\right) - 2\left(g_+ - g_{12}\right)\left(\frac{n_1 - n_2}{2}\right)^2 + \frac{g_{12}N^2}{2}\right], \end{aligned} \tag{8.5}$$

where $N = n_1 + n_2$ is the total number of atoms in the two condensates, $n_1 = a_1^\dagger a_1$, $n_2 = a_2^\dagger a_2$, $\mu_\pm = (\mu_1 \pm \mu_2)/2$, $g_\pm = (g_{11} \pm g_{22})/2$, and

$$g_{kk} = U_k \int d\mathbf{r}\, |\psi_k|^4, \qquad k = 1, 2 \tag{8.6}$$

$$g_{12} = U_{12} \int d\mathbf{r}\, |\psi_1|^2 |\psi_2|^2. \tag{8.7}$$

The terms of the form $(a_1^\dagger a_1 - a_2^\dagger a_2)$ in (8.5) imprint a linear phase on the atoms. The g_-, g_+, and g_{12} are the two-body interactions terms. The g_- imprints a linear phase that is enhanced by a factor $N - 1$ on the atoms. Similarly, the chemical potential μ_- imprints a linear phase on the atoms. The term $(g_+ - g_{12})$ imprints a nonlinear phase on the atoms. It is this phase that will determine if the distribution is squeezed and gives the degree of the squeezing. For instance, if g_{12} vanishes, then the two-component condensates are separated and do not overlap. Hence, there are no interparticle collisions between them. This is akin to the system studied in Section 7.3 where self-interactions (i.e., the intraparticle collisions) dominate, leading to phase diffusion. For $g_{12} \neq 0$ and $g_{12} < g_+$, the phase diffusion dominates the dynamics of the two-component condensates, but some fingerprints of the squeezing effect due to the nonnegligible value of g_{12} may be seen on the distribution of the atomic condensate. In the case where the phase diffusion term $g_{12} = g_+$, the intraparticle collisions in each condensate cancel the interparticle collisions between the two components. As a result, there is no nonlinear phase imprinted on the atom. For $g_{12} > g_+$, the interparticle collision between the two-component condensates dominates and drives the dynamics of the condensates. In this case, the atoms' distribution would show strong squeezing effects. We point out that none of these situations is static. The two-component states starting out with $g_{12} > g_+$ will evolve in such a manner that at some point in time $g_{12} \leq g_+$, where the phase diffusion effect dominates the dynamics or the interparticle and intraparticle collisions in the two-component condensates completely cancel each other out. Thus, generating a squeezed state would require optimizing the interactions such that the squeezing effect is not destroyed or washed out by phase diffusion.

From (8.5), the Hamiltonian H, N, n_1, and n_2 all commute. Hence, these four operators have a common set of eigenstates $|N, n_1, n_2\rangle$. In what follows, we assume to work within the total particle sector of N and ignore all terms in the Hamiltonian only involving the total particle number N, as they merely add a constant energy shift to the Hamiltonian. Hence, we write (8.5) as

$$H_{\text{eff}} = \mu_-(n_1 - n_2) - g_-(\mathcal{N} - 1)\left(\frac{n_1 - n_2}{2}\right) - (g_+ - g_{12})\left(\frac{n_1 - n_2}{2}\right)^2. \tag{8.8}$$

The operators $a_k^\dagger$ and a_k may be related to the angular momentum operators in the same way as (5.14). For our current notation, this is

$$S_x = \frac{1}{2}\left(a_1^\dagger a_2 + a_2^\dagger a_1\right), \tag{8.9}$$

$$S_y = \frac{i}{2}\left(a_2^\dagger a_1 - a_1^\dagger a_2\right), \tag{8.10}$$

$$S_z = \frac{1}{2}\left(a_1^\dagger a_1 - a_2^\dagger a_2\right). \tag{8.11}$$

We note that to make the connection to standard conventions of angular momentum, we include here the factor of 1/2 in the definitions of the spin. The raising angular momentum operator is defined as $S_+ = S_x + iS_y = a_1^\dagger a_2$ and the lowering angular momentum operator is $S_- = S_x - iS_y = a_2^\dagger a_1$. The Hamiltonian (8.8) in terms of the angular momentum operators is

$$H_{\text{eff}} = 2\mu_- S_z - g_-(\mathcal{N} - 1)S_z - (g_+ - g_{12})S_z^2. \tag{8.12}$$

It can be shown that these operators obey the angular momentum commutation algebra with the Casimir invariant operator S^2 as

$$S^2 = S_x^2 + S_y^2 + S_z^2 = \frac{N}{2}\left(\frac{N}{2} + 1\right), \tag{8.13}$$

The Hamiltonian (8.12), S^2, N, and S_z all commute. Another set of orthogonal basis vectors for this Hamiltonian is $|N, s, m\rangle$. It is then evident that $S^2|N, s, m\rangle = N/2(N/2 + 1)|N, s, m\rangle = s(s + 1)|N, s, m\rangle$, from which we immediately conclude that $s = N/2$. For brevity, $|N, s, -s\rangle$ is henceforth written as $|s, -s\rangle$.

The angular momentum states can be represented in terms of the Fock states of the Hamiltonian (8.8) by expanding either of the angular momentum states $|s, m = \pm s\rangle$ in terms of Fock states, $|N, n, N - n\rangle$ defined earlier. Take, for example, the angular momentum state $|s, -s\rangle$, which is expanded in the Fock states as

$$|s, -s\rangle = \sum_{n=0} C_n |N, n, N - n\rangle, \tag{8.14}$$

where $C_n = \langle N, n, N - n|s, -s\rangle$, $n_1 = n$, and $n_2 = N - n$. To determine the coefficients C_n, the lowering angular momentum operator $S_- = a_2^\dagger a_1$ is applied to (8.14) and set to zero, giving

$$\begin{aligned} S_-|s, -s\rangle &= 0 \\ &= \sum_n C_n \sqrt{(N - n + 1)n}\,|N, n - 1, N - n + 1\rangle. \end{aligned} \tag{8.15}$$

It is evident that, except for $n = 0$, C_n must vanish for every other term. Thus, $|s, m = -s\rangle = |N, 0, N\rangle$, where $C_0 = 1$ has been chosen to satisfy the normalization condition. The state corresponds to a situation where all the atoms are in the second component of the condensate $n_2 = N$. To get the rest of the angular momentum states in terms of the Fock state, the raising angular momentum operator $S_+ = a_1^\dagger a_2$ is applied repeatedly to $|s, -s\rangle$:

$$|s, m\rangle = \sqrt{\frac{1}{(2s)!}\frac{(s-m)!}{(s+m)!}}\, S_+^{l+m}\, |s, -s\rangle, \tag{8.16}$$

where $|s, -s\rangle = |N, 0, N\rangle$. This immediately gives

$$|s, m\rangle = |N, n, N-n\rangle. \tag{8.17}$$

Thus, there is a one-to-one correspondence between the angular momentum states and the Fock states.

Exercise 8.2.1 Using (8.4) in (8.1), verify (8.5).

Exercise 8.2.2 By applying the lowering operator S_- to (8.14), show that $C_{n\neq 1} = 0$, and verify (8.17).

Exercise 8.2.3 A three-level atom has three (energy) levels, such as the three hyperfine energy levels of the $F = 1$ of the alkali metals like ^{87}Rb: $m_F = -1, 0, 1$. The operators $a_{+1}^\dagger$, $a_0^\dagger$, and $a_{-1}^\dagger$ act on a vacuum state to create an atom in the corresponding level, respectively. The operators describing the interactions in the system can be expressed in terms of angular momentum operators: $S_z = a_{-1}^\dagger a_{-1} - a_{+1}^\dagger a_{+1}$, $S_+ = \sqrt{2}(a_{-1}^\dagger a_0 + a_0^\dagger a_{+1})$, and $S_- = \sqrt{2}(a_{+1}^\dagger a_0 + a_0^\dagger a_{-1})$. Show that the operators S_z, S_x, and S_+ are angular momentum operators. Hint: Verify that the operators satisfy angular momentum commutation relations.

Exercise 8.2.4 If the angular momentum state $|s, -s\rangle$ is related to the three-level Fock states as

$$|s, -s\rangle = \sum_k^n C_k |2k, n+s-k, n-k\rangle,$$

determine C_k. How is the Fock state representation of the angular momentum state for a two-level atom different from that of a three-level atom?

8.2.2 Evolution of the Initial State

The atoms are initially prepared with N atoms in one of the internal states such as the $|F = 1, m_f = -1\rangle$ of ^{87}Rb. This is represented by the state

$$|\Psi_{\text{ini}}\rangle = \frac{\left(a_1^\dagger\right)^N}{\sqrt{N!}}|0\rangle, \tag{8.18}$$

where $|0\rangle$ is the vacuum state. To create a linear superposition of two-component atomic condensates, a $\pi/2$ pulse is used (see Section 5.5) to couple the bosonic operators $a_1^\dagger$ and $a_2^\dagger$ that act on the vacuum state to create an atom in each state, respectively, according to the following rules:

$$a_1^\dagger \rightarrow \frac{1}{\sqrt{2}}\left(a_1^\dagger - i a_2^\dagger\right), \tag{8.19}$$

$$a_2^\dagger \rightarrow \frac{1}{\sqrt{2}}\left(-i a_1^\dagger + a_2^\dagger\right). \tag{8.20}$$

A single-atom state is transformed as

$$a_1^\dagger|0\rangle \to \frac{1}{\sqrt{2}}\left(a_1^\dagger - ia_2^\dagger\right)|0\rangle, \tag{8.21}$$

so that the product state of the N-particle system after the $\pi/2$ pulse has been applied becomes

$$\begin{aligned}|\Psi_{\text{split}}\rangle &= \frac{1}{\sqrt{2^N N!}}\left(a_1^\dagger - ia_2^\dagger\right)^N|0\rangle,\\ &= \frac{1}{\sqrt{2^N}}\sum_{n=0}^{N}\sqrt{\frac{N!}{n!\,(N-n)!}}(-i)^{N-n}|N,n,N-n\rangle,\end{aligned} \tag{8.22}$$

where

$$|N,n,N-n\rangle = \frac{(a_1^\dagger)^n}{\sqrt{n!}}\frac{(a_2^\dagger)^{N-n}}{\sqrt{(N-n)!}}|0\rangle \tag{8.23}$$

is the state vector having $n_1 = n$ atoms in one of the hyperfine states $|F=1,m_f=-1\rangle$ and $n_2 = N-n$ atoms in the other hyperfine state $|F=2,m_f=1\rangle$.

The time evolution of the state vector is governed by the Hamiltonian (8.8)

$$|\Psi(t)\rangle = \exp\left[-\frac{i}{\hbar}\int_0^t H_{\text{eff}}dt'\right]|\Psi_{\text{split}}\rangle. \tag{8.24}$$

The states $|N,n_1=n,n_2=N-n\rangle$ given by (8.23) are eigenstates of the Hamiltonian, with eigenvalues

$$E(n_1,n_2) = \left(2\mu_- - g_-(N-1)\right)\frac{(n_1-n_2)}{2} - (g_+ - g_{12})\left(\frac{n_1-n_2}{2}\right)^2. \tag{8.25}$$

The state vector of the system at time $t=T$ is

$$|\Psi(T)\rangle = \frac{1}{\sqrt{2^N N!}}\sum_{n=0}^{N}\frac{N!}{n!\,(N-n)!}e^{-i\phi_T(n-N/2)+i\varphi(n-N/2)^2}\left(a_1^\dagger\right)^n\left(-ia_2^\dagger\right)^{N-n}|0\rangle, \tag{8.26}$$

where

$$\phi_T = \frac{1}{\hbar}\int_0^T dt(2\mu_- - g_-(N-1)) \tag{8.27}$$

is the linear phase difference accumulated due to the relative difference in the nonlinear self-interactions in each component and environment effects that are coupled to the condensate through the chemical potential. Meanwhile,

$$\varphi = \frac{1}{\hbar}\int_0^T (g_+ - g_{12})\,dt \tag{8.28}$$

is the nonlinear phase per atom due to self-interactions in each BEC component and the mutual interactions between the two components.

Exercise 8.2.5 Verify that the state $|N,n,N-n\rangle$ is an eigenstate of (8.12) with eigenvalue (8.25). Verify that the state of the system at any time is as given in (8.26).

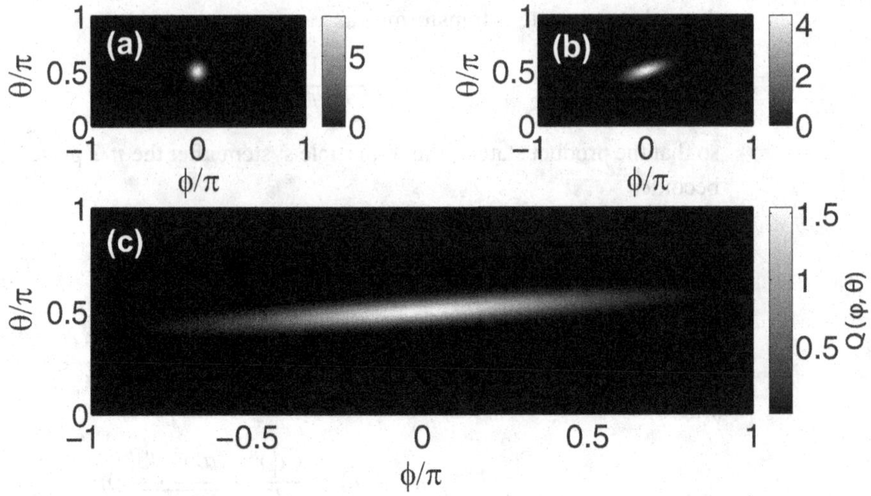

Fig. 8.1 The Q-distribution as a function of ϕ and θ for varying interaction strengths (a) $\varphi = 0$; (b) $\varphi = 3/N$; and (c) $\varphi = 1/\sqrt{N}$. The parameters used are $N = 1{,}00$, $\phi_T = -3\pi/2$. Color version of this figure available at cambridge.org/quantumatomoptics.

8.3 Husimi Q-function

To visualize the state (8.26), we plot the associated Q-function as defined in (5.128). The numerical calculations of the Q-function are shown in Fig. 8.1. It is seen that various strengths of the nonlinear interaction parameter φ give different shapes of the Q-function. To make sense out of this, we begin analysis in the simplest case scenario, which is $\varphi = 0$. The Q-function in this case is given as

$$Q = \frac{N+1}{4\pi} \cos^{2N}\left(\frac{\theta}{2} - \frac{\theta_0}{2}\right) e^{-Nab(\phi - \frac{3}{2}\pi - \phi_T)^2}, \tag{8.29}$$

where

$$a = \frac{\cos\theta/2 \cos\theta_0/2}{\cos(\theta/2 - \theta_0/2)}, \tag{8.30}$$

$$b = \frac{\sin\theta/2 \sin\theta_0/2}{\cos(\theta/2 - \theta_0/2)}, \tag{8.31}$$

and $\theta_0 = \pi/2$ specifically for the state (8.26). It is clear from (8.29) that the dominant term in the Q-function comes from points $\theta = \theta_0$ and $\phi = \frac{3}{2}\pi + \phi_T$. Close to the maximum of the Q-function, the width of the distribution in either the ϕ or θ directions is roughly the same and is of the order $1/\sqrt{N}$. Hence, the Q-function appears symmetric about its maximum, as expected, since there are no nonlinear interactions.

As the nonlinear interaction parameter is increased, the width of the Q-function changes. This can be seen from the analytical form of the Q-function in the limit $N \gg 1$:

$$Q = \frac{N+1}{4\pi}\sqrt{\frac{1}{1+4N^2a^2b^2\varphi^2}}\cos^{2N}\left(\frac{\theta}{2}-\frac{\theta_0}{2}\right) \times \exp\left[-\frac{Nab}{1+4N^2a^2b^2\varphi^2}\left(\phi-\frac{3}{2}\pi-\phi_T-N\varphi(b-a)\right)^2\right]. \tag{8.32}$$

Notice that the location and width of the maximum along θ do not change. However, the location of width and the maximum along the ϕ axis changes with φ. At small values of φ ($\varphi \ll 1/N$), the location of the distribution's peak along ϕ has dependence on both ϕ_T and φ. Also, its width along ϕ grows linearly, with the largest contribution coming from points around $\theta = \theta_0$. Since the width along the θ dimension remains fixed while the width along the ϕ axis grows by some amount proportional to φ, as shown in Fig. 8.1(b) and (c), the distribution becomes rotated about the location of its maximum such that it is tilted at an angle with the ϕ axis. Large values of φ ($1/N < \varphi \ll 1/\sqrt{N}$) further increases the width along ϕ and decreases the angle of inclination of the distribution with the ϕ axis, resulting in a tilted ellipse on the (θ, ϕ) plane. The net effect is that a measurement along the semi-minor axis of the ellipse results in a reduced error that is less than the shot noise. States with this property are referred to as squeezed states and are exploited in metrology.

Exercise 8.3.1 Verify that the Q-function for a coherent state is as given in (8.29). Also, verify (8.32).

Exercise 8.3.2 Verify that for $\varphi \ll 1/\sqrt{N}$, the width of the Q-function along ϕ grows at a rate of $4Nab\varphi^2$.

8.4 Ramsey Interferometry and Bloch Vector

This section applies the results of Section 6.3 to describe the working principles of Ramsey interferometry. This technique is often used to measure the transition frequency of atoms, as done in magnetic resonance imaging. The process involves interacting an atom with electromagnetic radiation for a time τ_r, which is very short in comparison with its decay time (or lifetime), $1/\gamma$. To illustrate the working principle, we will consider the interaction of two square pulses of electromagnetic wave with an atom.

8.4.1 Pure State Evolution

Consider a two-level atom with eigenstates $|\psi_g\rangle, |\psi_e\rangle$ of the Hamiltonian H_0, with energies E_g and E_e, respectively. The two-level atom is interacting with two square pulses of an electromagnetic wave. Each square pulse of the radiation has a constant amplitude of duration τ_r and is separated by time interval T. During the interaction of the atom and the radiation, the total Hamiltonian of the atom and radiation is $H = H_0 + H_I$. If the interaction strength of the atom and radiation described by H_I is sufficiently small, the eigenstates of H can be approximated as a linear superposition of the eigenstate of the bare Hamiltonian H_0. The state can then be written as $|\psi\rangle = a_g(t)|\psi\rangle + a_e(t)|\psi\rangle$, where $a_g(t)$ is the probability amplitude to populate the lower of the two eigenstates and $a_e(t)$ is the probability to populate the upper level of the two eigenstates. The evolution of the amplitudes $a_g(t)$ and $a_e(t)$ is described by the solution (6.12). Moving into the frame that is rotating with eigenfrequencies of the bare atom states and assuming that the detuning $\Delta = \omega_e - \omega_g - \omega$ is large compared to the Rabi frequency Ω_{ge}, $\Delta \gg \Omega_{ge}$, as well as ignoring the phases of the radiation $\phi_L = 0$, the solution for $a_e(t)$ in this limit becomes

$$c_e(t) = -i\frac{\Omega_{ge}}{\Delta}e^{it\Delta/2}\sin\left(\frac{t\Delta}{2}\right). \tag{8.33}$$

Applying (8.33) to the square pulses for $t = 0$ to $t = \tau_r$ and from $t = T$ to $t = T + \tau_r$ gives

$$c_e(t) = -2i\frac{\Omega_{ge}}{\Delta}\sin\left(\frac{\tau_r\Delta}{2}\right)\cos\left(\frac{T\Delta}{2}\right)e^{i\tau_r\Delta/2}e^{iT\Delta/2}, \tag{8.34}$$

The probability that the atom is excited after applying the pulses is

$$|c_e(t)|^2 = \Omega_{ge}^2\tau_r^2\left[\frac{\sin(\tau_r\Delta/2)}{\tau_r\Delta/2}\right]^2\cos^2\left(\frac{T\Delta}{2}\right). \tag{8.35}$$

Figure 8.2 shows the Ramsey fringes that are given in (8.35). The interference exhibited in the figure may be understood as follows. Each pulse taken independently (as in (8.33)) gives a sinc^2 function with a maximum at $\tau_r\Delta = 0$ and minimum

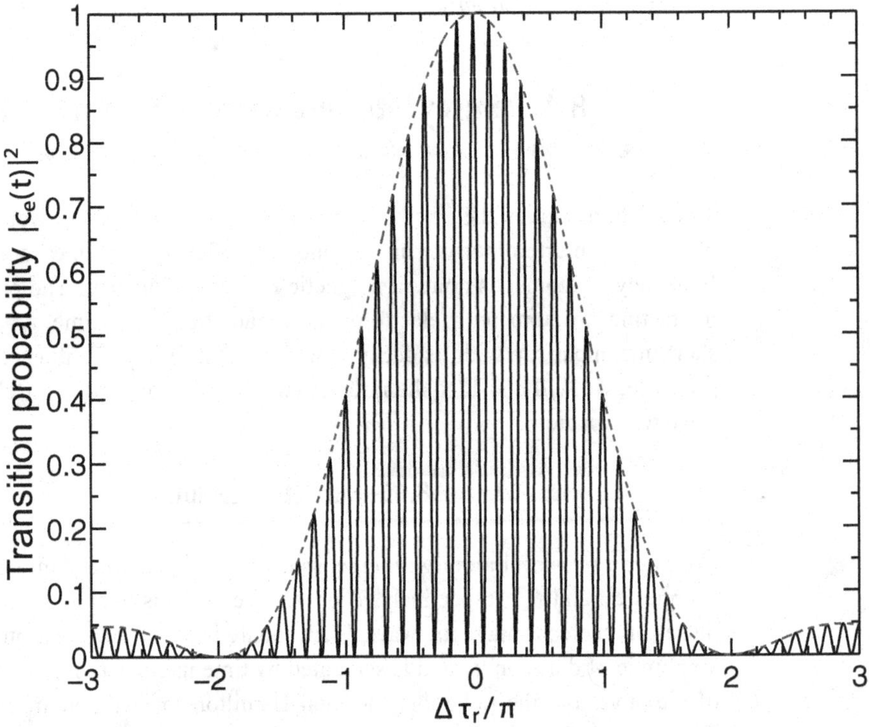

Fig. 8.2 Ramsey interference. The dashed line is the envelope from a single square pulse, $\Omega_{ge}^2\tau_r^2\left[\frac{\sin(\tau_r\Delta/2)}{\tau_r\Delta/2}\right]^2$, while the solid line is the interference pattern from two square pulses separated by a time interval T, (8.35). The parameters used are $\tau_r = \pi/4$, $T = 4\pi$, $\Omega_{ge} = 1/\tau_r$.

at $\tau_r\Delta = 2\pi$. This gives the width of the central maximum to be $\sigma_{\Delta,c} = 2\pi/\tau_r$. However, for the two square pulses, there is a superposition of two effects: interference and diffraction as result of applying the first pulse in the interval $t = 0$ to $t = \tau_r$, and applying the pulse from $t = T$ to $t = T + \tau_r$. The maximum of a fringe is dictated by the cosine term and occurs when the $T\Delta$ is an integral multiple of 2π. The central peak or maximum occurs for $\Delta = 0$, and goes to zero at $T\Delta = \pi$, giving the width of the central peak as $\sigma_{\Delta,p} = \pi/T$. The sinc^2 function is an envelope for the interference pattern and controls the amplitude of the transition probability with the width of each band given by $\sigma_{\Delta,c} = 2\pi/\tau_r$. Comparing with the single square pulse, one can see from (8.33) that the sinc term is being sliced by the cos^2 term such that a better resolution is obtained. This is a classic two-slit diffraction, where the slits of size τ_r are separated in time by an interval T. Hence, the benefit of the Ramsey spectroscopy is that it resolves the width of the frequency band $2\pi/\tau_r$ of a single square pulse into finer intervals of π/T due to interference.

8.4.2 Mixed-State Evolution

Thus far, the atom interaction with radiation has been treated using the wavefunction approach. In order to take into account the interaction with the environment, we must extend this to a density matrix approach. Working in the rotating frame, the wavefunction describing the evolution of an atom interacting with radiation, as described at the beginning of this section, is given by $|\psi\rangle = c_g(t)|\psi\rangle + c_e(t)|\psi\rangle$. The evolution of the probability amplitudes $c_{g,e}(t)$ is as described in (6.10). To calculate the evolution of the density matrix ρ, we need to know the elements of the matrix $\rho = |\psi\rangle\langle\psi|$,

$$\rho = \begin{pmatrix} c_g(t) \\ c_e(t) \end{pmatrix} \begin{pmatrix} c_g^*(t) & c_e^*(t) \end{pmatrix} = \begin{pmatrix} c_g(t)c_g^*(t) & c_g(t)c_e^*(t) \\ c_e(t)c_g^*(t) & c_e(t)c_e^*(t) \end{pmatrix} = \begin{pmatrix} \rho_{11} & \rho_{12} \\ \rho_{21} & \rho_{22} \end{pmatrix}. \tag{8.36}$$

The diagonal elements ρ_{11}, ρ_{22} of the density matrix are called populations and are real. The off-diagonal elements, ρ_{12} and ρ_{21}, are coherences and are complex, having a time-dependent phase factor that describes the frequency response of the atom to radiation field.

The evolution of the density matrix are calculated using (6.10). Noting, for instance, $\dot{\rho}_{11} = \dot{c}_g(t)c_g^*(t) + c_g(t)\dot{c}_g^*(t)$, $\dot{\rho}_{12} = \dot{c}_g(t)c_e^*(t) + c_g(t)\dot{c}_e^*(t)$, and assuming that Ω_{ge} is real as well as ignoring the phase of the radiation field, $\phi_L = 0$, we find from these that the evolution of the matrix elements are

$$\frac{d\rho_{11}}{dt} = -i\frac{\Omega_{ge}}{2}(\rho_{21} - \rho_{12}) = -\frac{d\rho_{22}}{dt}, \tag{8.37}$$

$$\frac{d\rho_{12}}{dt} = i\Delta\rho_{12} + i\frac{\Omega_{ge}}{2}(\rho_{11} - \rho_{22}), \tag{8.38}$$

$$\frac{d\rho_{21}}{dt} = -i\Delta\rho_{21} - i\frac{\Omega_{ge}}{2}(\rho_{11} - \rho_{22}). \tag{8.39}$$

From (8.37), it is easy to deduce that $\frac{d\rho_{11}}{dt} + \frac{d\rho_{22}}{dt} = 0$ is a constant of motion and expresses the conservation of the probability, $c_g(t)c_g^*(t) + c_e(t)c_e^*(t) = 1$. The conservation of the probability allows three variables u, v, and w to be defined as follows. Decomposing the coherences into their real and imaginary parts $\rho_{12} = u + iv$, $\rho_{21} = u - iv$,

$$u = \frac{\rho_{12} + \rho_{21}}{2}, \tag{8.40}$$

$$v = i\frac{\rho_{21} - \rho_{12}}{2}. \tag{8.41}$$

Finally, due to the conservation of the probability, the population gives only one variable, the relative population difference, which we define as

$$w = \rho_{11} - \rho_{22}. \tag{8.42}$$

From (8.37) to (8.39), the evolutions of the variables u, v, and w are

$$\frac{du}{dt} = -\Delta v, \tag{8.43}$$

$$\frac{dv}{dt} = \Delta u + \Omega_{ge} w, \tag{8.44}$$

$$\frac{dw}{dt} = -\Omega_{ge} v. \tag{8.45}$$

Defining a Bloch vector $\mathbf{r} = u\hat{\mathbf{i}} + v\hat{\mathbf{j}} + w\hat{\mathbf{k}}$, and $\mathbf{Q} = -\Omega_{ge}\hat{\mathbf{i}} + \Delta\hat{\mathbf{k}}$, (8.43)–(8.45) can be written in vector notation as

$$\frac{d\mathbf{r}}{dt} = \mathbf{Q} \times \mathbf{r}. \tag{8.46}$$

The solution of (8.46) may be obtained by using (6.9) and (6.12) in the definitions of u, v and w. Equation (8.46) can be solved more generally under various conditions, including damping and losses. Furthermore, the Bloch vector allows for easy visualization on a sphere and gives an intuitive interpretation of the dynamics. For instance, $\dot{\mathbf{r}} \cdot \mathbf{r} = 0$ immediately tells us that $\dot{\mathbf{r}} \cdot \mathbf{r} + \mathbf{r} \cdot \dot{\mathbf{r}} = 0$, so that $\mathbf{r} \cdot \mathbf{r} = u^2 + v^2 + w^2 = 1$ is a constant of motion. This is not surprising, since $\mathbf{Q} \times \mathbf{r}$ gives a vector that is perpendicular to the plane of $\mathbf{Q}$ and $\mathbf{r}$, so the dot product of this vector $\mathbf{Q} \times \mathbf{r}$ with either $\mathbf{Q}$ or $\mathbf{r}$ must vanish. It is then evident that $\mathbf{Q} \cdot \frac{d\mathbf{r}}{dt} = 0$ or $\mathbf{Q} \cdot \mathbf{r}$ is also a constant of motion. This implies that motion of $\mathbf{r}$ must be such that the projection of $\mathbf{r}$ along $\mathbf{Q}$ must remain invariant. Supposing the vector $\mathbf{Q}$ is held fixed in space, then $\mathbf{Q} \cdot \mathbf{r} = Qr\cos\theta$, where θ is some angle between $\mathbf{Q}$ and $\mathbf{r}$ measured from $\mathbf{Q}$. It then means that $\mathbf{r}$ is a vector on the surface of the cone traced out by θ measured from $\mathbf{Q}$ (with $\mathbf{Q}$ at the center of the cone); that is, $\mathbf{r}$ precesses around $\mathbf{Q}$. In interferometry, as we shall see later, the goal is to be able to estimate this rotation angle θ with the best precision possible. Our discussion so far does not account for loss mechanisms in the dynamics of the density matrix, such as spontaneous emission. Such loss mechanisms and decoherence can be incorporated, for example, using master equation methods as seen in Sections 4.7 and 4.8.

8.5 Error Propagation Formula and Squeezing Parameter

In Section 8.3, visual inspection of the Q-function for the state (8.32) showed that the noise in the state can be affected and controlled by the strength of the interaction parameter. The amount of noise, which originates from quantum mechanical noise in this case, affects the accuracy of estimating particular parameters in an experiment. For example, it may be of interest to know the parameter ϕ that influences the relative population difference S_z. But the parameter ϕ cannot be directly measured in an experiment. By measuring the relative population difference S_z, one hopes to get the best possible estimate of the unknown parameter ϕ. The way that S_z changes with ϕ may not necessarily be a linear function and may be different for various parameter choices of ϕ. That is, for a change from ϕ to $\phi + \Delta\phi$, the relative population difference changes from $S_z(\phi)$ to $S_z(\phi + \Delta\phi)$. Assuming that the change in the parameter ϕ is very small, we can expand $S_z(\phi + \Delta\phi) \approx S_z(\phi) + \frac{dS_z(\phi)}{d\phi}\Delta\phi$ to get the error propagation formula, whose inversion gives the error in estimation of ϕ as

$$\Delta\phi = \frac{\Delta S_z(\phi)}{\left|\frac{dS_z(\phi)}{d\phi}\right|}, \tag{8.47}$$

where $S_z(\phi + \Delta\phi) - S_z(\phi) = \Delta S_z(\phi)$ and the modulus sign has been put on the derivative in recognition that it could sometimes be negative.

To give a concrete example, consider using the state prepared in (8.26) in sensing rotation or measuring an unknown field. The expectation values of the operators S_x, S_y, and S_z are as follows:

$$\begin{aligned} \langle\Psi(T)|S_x|\Psi(T)\rangle &= \frac{N}{2}\sin\phi_T\cos^{N-1}\varphi, \\ \langle\Psi(T)|S_y|\Psi(T)\rangle &= -\frac{N}{2}\cos\phi_T\cos^{N-1}\varphi, \\ \langle\Psi(T)|S_z|\Psi(T)\rangle &= 0. \end{aligned} \tag{8.48}$$

This implies that at time T, the state $|\Psi(T)\rangle$ lies on the x–y plane. Without loss of generality and setting $\phi_T = \pi$, the state $|\Psi(T)\rangle$ points along the y axis at $t = T$. This state can be used to measure the rotation of a radiation field in any of the other two orthogonal directions, say, the x axis. The field tips the mean spin and causes it to rotate about the x axis in the y–z plane. The Hamiltonian describing the applied field within the rotating wave approximation is $H_I = \frac{\hbar\Delta}{2}S_z + \hbar\Omega_{ge}S_x/2$. The dynamics of the spin are governed by the Heisenberg equations, which are (8.43)–(8.45), if u, v, and w of the Bloch vectors are replaced with S_x, S_y, and S_z, respectively: $\Delta = 0$ and $\Omega_{ge} = -\Omega_{ge}/2$.

The time evolution of the spins is given by

$$S_x(t) = S_x, \tag{8.49}$$

$$S_y(t) = S_y\cos\frac{\Omega_{ge}t}{2} + S_z\sin\frac{\Omega_{ge}t}{2}, \tag{8.50}$$

$$S_z(t) = S_y\sin\frac{\Omega_{ge}t}{2} - S_z\cos\frac{\Omega_{ge}t}{2}, \tag{8.51}$$

where $S_i(0) = S_i$, $i = x, y, z$. To estimate the angle $\phi = \Omega_{ge}t/2$ at which the spin is rotating about the x axis, one calculates the spin in z direction (8.51),

$$\langle S_z(t)\rangle = \langle S_y\rangle\sin\phi - \langle S_z\rangle\cos\phi = \langle S_y\rangle\sin\phi, \tag{8.52}$$

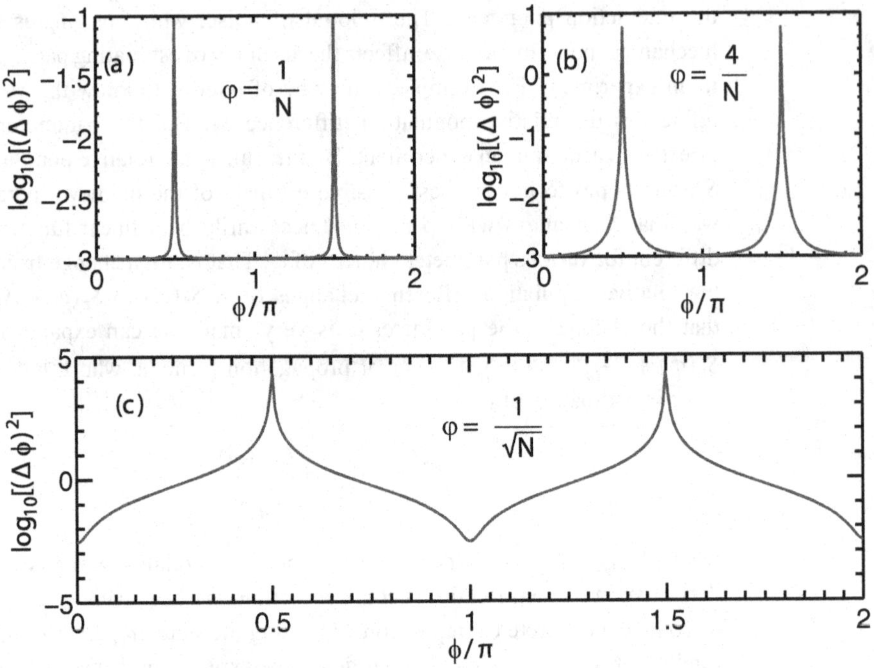

Fig. 8.3 The error in estimating ϕ using the error-propagation formula (8.53). For the state $|\Psi(T)\rangle$, the following parameters were used: $N = 1,000$ and $\phi_T = 0$. The values of φ are as shown in the figure.

where $\langle S_i \rangle = \langle \Psi(T)|S_i|\Psi(T)\rangle$, $i = x, y, z$. In the following discussion, we will approximate the change in S_z (i.e., ΔS_z) with the standard deviation σ_{S_z} and similarly $\Delta\phi$ will be approximated with σ_ϕ. Then, the error in estimating ϕ in accordance with (8.47) becomes

$$\sigma_\phi^2 = \frac{(\langle (S_y)^2\rangle - \langle S_y\rangle^2)\sin^2\phi + \langle S_z^2\rangle \cos^2\phi}{\langle S_y\rangle^2 \cos^2\phi}, \tag{8.53}$$

where all the expectations are taken with respect to $|\Psi(T)\rangle$. The cross terms $\langle S_y S_z\rangle$ in the variance vanish, hence they are not present. Note that the denominator of (8.53) is the expectation of $\langle S_y(t)\rangle$, while the numerator is the variance of $\langle S_z^2(T)\rangle$. The error σ_ϕ^2 is plotted in Fig. 8.3. As seen from the plots, as φ remains small, $0 \le \varphi \le 1/N$, the error is relatively low except at $\phi = n\pi/2$, n being odd. The points $\phi = n\pi/2$ give a value of ϕ where the spin is aligned along the z axis, thus $\langle S_y(t)\rangle$ is zero. Increasing the value of φ increases the error in estimating ϕ. However, from Fig. 8.3(c) there exist error minima at values of $\phi = 0, m\pi$ (m even), corresponding to the spins aligned along the y axis. At these points, the variance of $S_z(t)$ is small, while the mean value of $S_y(t)$ has its maximum value. Thus, the error σ_ϕ is minimized for $\phi = 0, m\pi$, giving

$$\sigma_{\phi,\min}^2 = \frac{\langle S_z^2\rangle}{\langle S_y\rangle^2}. \tag{8.54}$$

The variance of S_z is shot-noise limited, $\langle S_z^2\rangle = N/4$, while $\langle S_y\rangle^2$ is given in (8.48) with $\phi_T = \pi$. In summary, (8.54) states that if the direction of the mean spin $\langle \mathbf{S}\rangle$ of a state is along some axis, then the variance $\sigma_{S_\perp}$ that minimizes the error propagation formula is on a plane that is perpendicular to the mean spin $\langle \mathbf{S} \rangle$, $\sigma_\phi = \frac{\sigma_{S_\perp}}{|\langle \mathbf{S} \rangle|}$. Further, when this error is benchmarked against the error obtained from a spin coherent state $\sigma_{\phi,\mathrm{s}} = 1/\sqrt{N}$, one gets a parameter $\xi_{\mathrm{R}} = \frac{\sigma_\phi}{\sigma_{\phi,\mathrm{s}}}$ that estimates the improvement of using the given state over the coherent state, which is given in this case by

$$\xi_{\mathrm{R}} = \frac{\sqrt{N}\sigma_{S_\perp}}{|\langle \mathbf{S} \rangle|}. \tag{8.55}$$

The failure of (8.54) to obtain the desired improvement from using spin squeezing stems from the fact that even though the error σ_{S_z} along the z axis is perpendicular to the mean spin S_y direction, which is along the y axis, the direction z is not that of the minimum variance. In order to utilize the squeezing effect contained in the state $|\psi(T)\rangle$, one can observe from the Q-function that the dominant spin direction of the atoms is in the y axis. Any rotation that is not along the mean spin tends to shift the mean spin to another position. It then suggests that the best sensing will be obtained if the atoms are rotated about their dominant spin direction. For the phase of the laser light $\phi_L = \pi/2$ and detuning $\Delta = 0$ in (6.10), one obtains within the rotating wave approximation the Hamiltonian $H = \hbar\Omega_{ge} S_y$. This Hamiltonian rotates spin about the dominant spin direction $\langle \mathbf{S} \rangle$. The evolution of the spin operators is

$$S_x(t) = S_x \cos\Omega_{ge}t + S_z \sin\Omega_{ge}t, \tag{8.56}$$

$$S_y(t) = S_y, \tag{8.57}$$

$$S_z(t) = -S_x \sin\Omega_{ge}t + S_z \cos\Omega_{ge}t. \tag{8.58}$$

The mean spin $|\langle \mathbf{S} \rangle|$ direction has not changed, hence the variance on the plane perpendicular to the direction of the mean spin that points along y can be found using either (8.56) or (8.58). Using (8.56), the variance along x,

$$\langle S_x^2(t) \rangle = \langle S_x^2 \rangle \cos^2\phi + \langle S_z S_x + S_x S_z \rangle \sin\phi\cos\phi + \langle S_z^2 \rangle \sin^2\phi, \tag{8.59}$$

where $\phi = \Omega_{ge}t$. This result is shown in Fig. 8.4, normalized with respect to the shot noise $N/4$. From Fig. 8.4(b), we see that there exists a ϕ value for which the variance is minimum and is well below the shot noise for a given φ with $\varphi > 0$. Minimizing $\langle S_x^2(t) \rangle$ with respect to ϕ gives the optimum ϕ value that optimizes the variance,

$$\tan(2\phi) = \frac{\langle S_x S_z + S_z S_x \rangle}{\langle S_x^2 \rangle - \langle S_z^2 \rangle}. \tag{8.60}$$

Using the property of the trigonometric functions, $\tan x = \sin x / \cos x$, then

$$\sin(2\phi_{\mathrm{opt}}) = \frac{\langle S_x S_z \rangle + \langle S_z S_x \rangle}{\sqrt{(\langle S_x S_z \rangle + \langle S_z S_x \rangle)^2 + (\langle S_x^2 \rangle - \langle S_y^2 \rangle)^2}}, \tag{8.61}$$

$$\cos(2\phi_{\mathrm{opt}}) = \frac{\langle S_x^2 \rangle - \langle S_z^2 \rangle}{\sqrt{(\langle S_x S_z \rangle + \langle S_z S_x \rangle)^2 + (\langle S_x^2 \rangle - \langle S_y^2 \rangle)^2}}. \tag{8.62}$$

Thus the variance is minimized if $\sin(2\phi) = -\sin(2\phi_{\mathrm{opt}})$ and $\cos(2\phi) = -\cos(2\phi_{\mathrm{opt}})$, which is obtained for $\phi = \pi/2 + \phi_{\mathrm{opt}}$ in (8.59), giving $\langle S_x^2(t) \rangle = \sigma_{S_\perp}$ as

$$\sigma^2_{S_\perp,\mathrm{min}} = \frac{\langle S_x^2 \rangle + \langle S_z^2 \rangle}{2} - \frac{\sqrt{(\langle S_x^2 \rangle - \langle S_z^2 \rangle)^2 + (\langle S_x S_z \rangle + \langle S_z S_x \rangle)^2}}{2}. \tag{8.63}$$

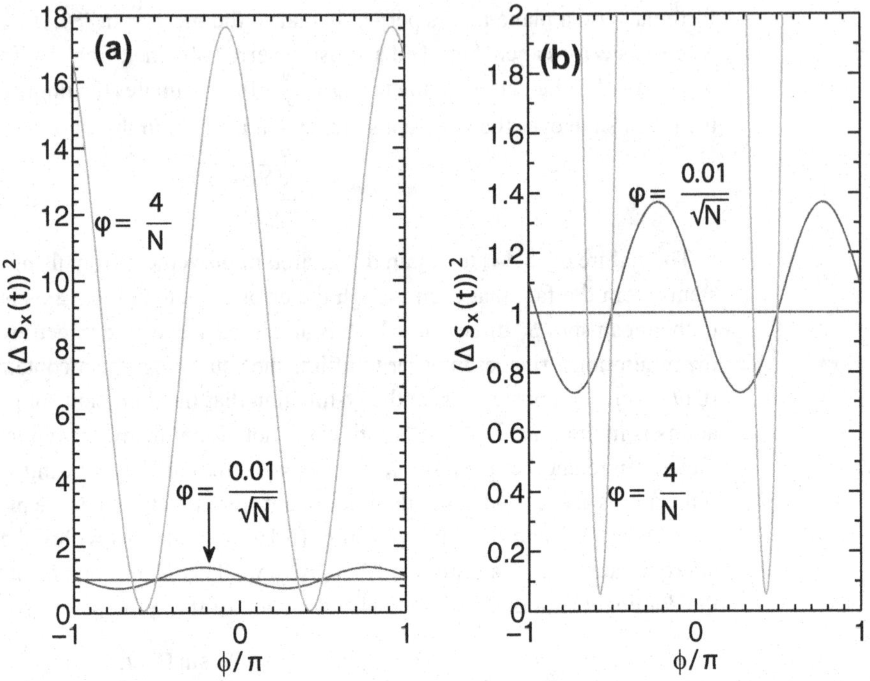

Fig. 8.4 The normalized variance of $S_x(t)$, (8.59). (b) Shows a zoomed-in region of (a) to show the minimum values of variance. The parameters are $N = 1{,}000$ and $\phi_T = 0$. The values of φ are as shown in the figure, except for the straight line, which is the shot noise with $\varphi = 0$. Color version of this figure available at cambridge.org/quantumatomoptics.

Similarly, for $\phi = \pi + \phi_{\text{opt}}$ the variance is maximized $\sigma_{S_\perp,\max}$. This is often referred to the antisqueezing, the opposite of the squeezing effect. Other scenarios are also possible, such as having $\phi = \pi/2 - \phi_{\text{opt}}$ or $\phi = \pi - \phi_{\text{opt}}$. In this case, $\phi = \pi/2 - \phi_{\text{opt}}$ will minimize the error $S_z(t)$, while $\phi = \pi - \phi_{\text{opt}}$ will maximize the error in $S_z(t)$.

Exercise 8.5.1 Calculate the variance of $S_z(t)$ (8.51) and expectation value of $S_y(t)$ (8.50), and verify (8.53).

Exercise 8.5.2 Calculate the variance of $S_z(t)$ (8.58) and find the angle that minimizes $\langle S_z^2 \rangle$. For what value of ϕ is ξ_R antisqueezed?

8.6 Fisher Information

In many practical situations, the parameters of physical interest are not directly accessible due to experimental circumstances or from the fact that they can only be inferred from a measurement of a conjugate parameter. This scenario abounds in quantum mechanics, where some parameters like phase do not have defined operators. Thus, inference about the unknown parameter relies on indirect measurement of a conjugate observable, which adds additional uncertainty for the estimated parameter, even for optimal measurements. The estimation of the parameter usually occurs in a two-step process, namely the parameter to be estimated is first encoded in the state of the system (to be probed), and then the system that reveals information about the parameter is measured.

Fisher information is the amount of information that can be extracted about an unknown parameter encoded in the state of any system from a measurement of an observable of the system. The amount of information extracted depends of course on the procedure that is used in obtaining the information, and hence the procedure needs to be typically optimized for better and improved estimation. The best estimators are those that saturate the Cramer–Rao inequality,

$$\sigma_\theta^2 \geq \frac{1}{MF(\theta)}, \tag{8.64}$$

where M is the number of measurements. Equation (8.64) establishes a lower bound on the variance σ_θ^2 of any estimator of the parameter θ, which is given by the Fisher information,

$$F(\theta) = \int dx p(x|\theta) \left(\frac{\partial \ln p(x|\theta)}{\partial \theta} \right)^2, \tag{8.65}$$

where $p(x|\theta)$ is the conditional probability of obtaining the value of x when the parameter has a value θ.

In quantum mechanics, the probability $p(x|\theta)$ of obtaining the value x given that the quantum state of the system is parametrized by θ is $p(x|\theta) = \mathrm{Tr}\left[\Pi_x \rho_\theta\right]$, where Π_x is a positive operator satisfying the normalization condition $\int dx\, \Pi_x = \mathbb{1}$, and ρ_θ is the density matrix parametrized by the quantity θ that is to be estimated. Introducing the symmetric logarithmic derivative L_θ as a Hermitian operator satisfying the equation

$$\frac{L_\theta \rho_\theta + \rho_\theta L_\theta}{2} = \frac{\partial \rho_\theta}{\partial \theta}, \tag{8.66}$$

then

$$\frac{\partial p(x|\theta)}{\partial \theta} = \mathrm{Re}\left(\mathrm{Tr}\left[\rho_\theta \Pi_x L_\theta\right]\right). \tag{8.67}$$

The Fisher information (8.65) is then written as

$$F(\theta) = \int dx \frac{\mathrm{Re}\left(\mathrm{Tr}\left[\rho_\theta \Pi_x L_\theta\right]\right)}{\mathrm{Tr}\left[\Pi_x \rho_\theta\right]}. \tag{8.68}$$

For a given quantum measurement, (8.68) gives a classical bound on the precision that can achieved through data processing. Optimizing the measurements Π_x gives the ultimate bound on the Fisher information,

$$F(\theta) \leq \mathrm{Tr}\left[\rho_\theta L_\theta^2\right]. \tag{8.69}$$

The quantum Fisher information $F_Q(\theta) = \mathrm{Tr}\left[\rho_\theta L_\theta^2\right]$ places a limit on the amount of information regarding θ that can be gained in any measurement. Using (8.69) in (8.64) gives the quantum Cramer–Rao bound,

$$\sigma_\theta^2 \geq \frac{1}{MF_Q(\theta)}, \tag{8.70}$$

for the variance of any estimator. This implies that any measurement can produce a minimum acceptable error that is determined by the quantum Fisher information, and this bound does not depend on the measurement.

For a pure state, the symmetric logarithmic derivative is $L_\theta = 2\frac{\partial \rho_\theta}{\partial \theta}$. Using the fact that $\rho_\theta = \rho_\theta^2$, then

$$\frac{\partial \rho_\theta}{\partial \theta} = \frac{\partial \rho_\theta}{\partial \theta}\rho_\theta + \rho_\theta \frac{\partial \rho_\theta}{\partial \theta}. \tag{8.71}$$

From (8.71) and the symmetric logarithmic derivative, the quantum Fisher information for a pure state is

$$F_Q = 4\left(\left\langle \frac{\partial \psi_\theta}{\partial \theta} \middle| \frac{\partial \psi_\theta}{\partial \theta} \right\rangle + \left(\left\langle \psi_\theta \middle| \frac{\partial \psi_\theta}{\partial \theta} \right\rangle\right)^2\right). \tag{8.72}$$

For a pure state $|\psi_\theta\rangle$ parametrized by θ, which is related to the initial state $|\psi_0\rangle$ by the unitary transform $|\psi_\theta\rangle = e^{-i\theta S}|\psi_0\rangle$, the quantum Fisher information is then given by

$$F_Q = 4\sigma_S^2 = 4(\langle \psi_0|S^2|\psi_0\rangle - \langle \psi_0|S|\psi_0\rangle^2), \tag{8.73}$$

where S is any of the spin operators S_x, S_y, or S_z. Equation (8.73) immediately tells us that the quantum Fisher information is independent of the parameter θ and is given by the variance in the operator S that generates the parameter θ.

We now apply the above concepts to find the component of the total angular momentum operator S_k, $k = x, y, z$ that gives the maximum information on the estimation of the rotation angle ϕ from a state $|\psi_\phi\rangle$. Here, (8.26) is the initial state

$$|\psi_\phi\rangle = e^{-i\phi S_k}|\psi(T)\rangle, \tag{8.74}$$

where $\phi_T = 0$. According to (8.73), finding the optimum operator amounts to calculating the variances which using (8.48) are

$$\sigma_{S_x}^2 = \frac{N[(N+1) - (N-1)e^{-2(N-2)\varphi^2}]}{8}, \tag{8.75}$$

$$\sigma_{S_y}^2 = \frac{N[(N+1) + (N-1)e^{-2(N-2)\varphi^2}]}{8} - \frac{N^2 e^{-(N-1)\varphi^2}}{4}, \tag{8.76}$$

$$\sigma_{S_z}^2 = \frac{N}{4}. \tag{8.77}$$

The quantum Fisher information (8.73) is plotted in Fig. 8.5 using (8.75)–(8.77). The Fisher information F_Q^z for a rotation about the z axis remains constant throughout the variation of atom-atom interactions φ. This is not surprising, as the state $|\psi_\phi\rangle$ is on the x–y plane. Thus, the fluctuations along the z axis remain fixed, on average, and the rotations about the z axis would not produce any information better than N. However, rotations about the x axis and y axis produce an increase in the amount of information that can be obtained. At $\varphi = 0$, we have $F_Q^x = N$ and $F_Q^y = 0$. This is because the $|\psi_\phi\rangle$ is an eigenstate of S_y with eigenvalue $N/2$. Thus, all the atoms in this state would simply produce an absolute phase for the rotation angle ϕ. Since the atoms are known with certainty to be in the eigenstate of S_y with eigenvalue $N/2$, there is no fluctuation and the variance is thus zero. As such, this state will not provide any useful information in sensing rotation, as the

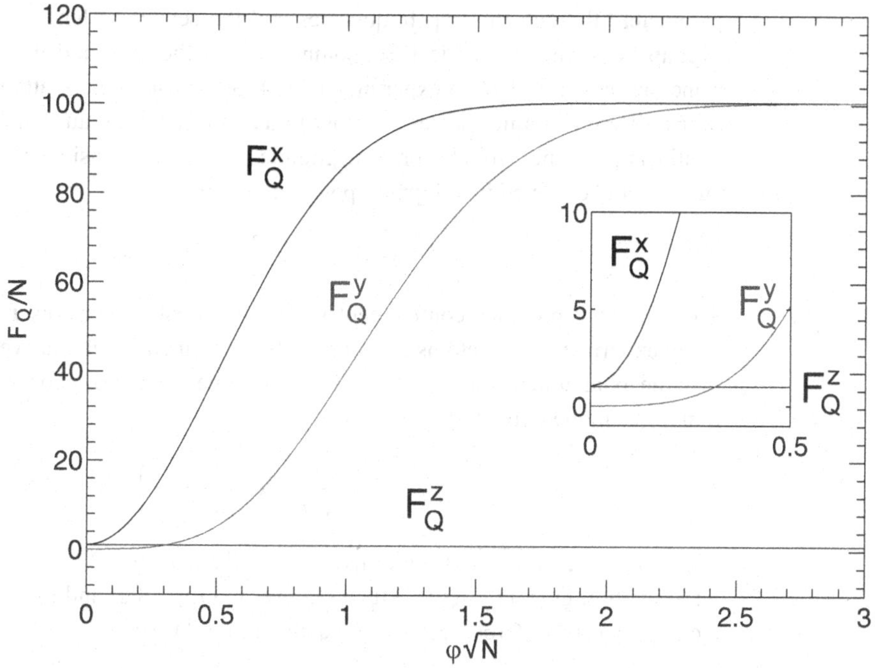

Fig. 8.5 The quantum Fisher information as a function of the nonlinear phase per atom φ. The inset shows a zoomed-in region near origin. Parameters are $N = 200$. Color version of this figure available at cambridge .org/quantumatomoptics.

$\sigma_{\theta=\phi}$ is infinite in (8.64). However, a rotation about the x axis at $\varphi = 0$ produces a variance of $N/4$, giving $F_Q^x = N$. As φ increases, the fluctuations about the x and y axes increase; hence, more information can be estimated about ϕ. Also, at every φ value, $F_Q^x > F_Q^y$. Consequently, the variance in the estimation of ϕ (see (8.64)), using a rotation about the x axis, is smaller compared to the variance in estimation of ϕ using a rotation about the y axis for the same φ value. For large values of φ, the quantum Fisher information is approximately the same for a rotation about the x and y axes, such that $F_Q^y = F_Q^x \approx N(N+1)/2$. We thus see that S_x is the operator that gives maximal information in estimating the rotation angle ϕ.

Exercise 8.6.1 Using the initial state $|\Psi(T)\rangle$ with $\phi_T = 0$, verify (8.75), (8.76), and (8.77).

Exercise 8.6.2 The Fisher information is as defined in (8.65). Given the state (8.26), calculate the Fisher information. Hint: First calculate the probability density of having the kth atom in a state, say, $|k, N-k\rangle$. Next, take the derivative of the probability with the parameter ϕ_T and complete the calculation.

8.7 Controlling the Nonlinear Phase per Atom

From Sections 8.5 and 8.6, we have seen that the nonlinear phase per atom φ controls the amount of squeezing that is useful for measurement, and dictates the lower bound on the Fisher information. The parameter φ is related to experimental parameters through the trap frequencies and the scattering lengths, the degree of overlap between the atomic BEC components in the trap, and therefore could in principle be controlled. In experiments [324, 329], where the scattering lengths a_k of the different atomic species are close to the interspecies scattering length a_{12}, the nonlinear phase per atom becomes negligible. To see this, consider a two-component atomic BEC confined in a trapping potential of the form

$$V(x, y, z) = \frac{m_i}{2}(\omega_x^2 x^2 + \omega_\perp^2 r_\perp^2). \tag{8.78}$$

The $m\omega_\perp^2 r_\perp^2/2$ provides confinement in the transverse dimension, $\mathbf{r}_\perp = (y, z)$. In many experiments, the atoms are more tightly confined in the transverse dimension than the axial dimension, $\omega_x \ll \omega_\perp$. Let us assume that the BEC occupies the lowest transverse mode in the trap

$$\psi_\perp(r_\perp) = \frac{1}{\sqrt{\pi} a_\perp} \exp\left(-\frac{r_\perp^2}{2a_\perp^2}\right), \tag{8.79}$$

where $a_\perp = \sqrt{\hbar/(m\omega_\perp)}$ is the transverse oscillator length. Writing the condensate wavefunction $\psi_i(\mathbf{r}, t) = \psi_i(x, t)\psi_\perp(r_\perp)$, we find that (8.2) and (8.3) are reduced to a one-dimensional Gross–Pitaevskii equation in $\psi_i(x, t)$:

$$\mu_1\psi_1(x,t) = \left[-\frac{\hbar^2}{2m}\frac{\partial^2}{\partial x^2} + \frac{1}{2}m\omega_x^2 x^2 + \frac{U_1 n_1}{2\pi a_\perp^2}|\psi_1(x,t)|^2 + \frac{U_{12} n_2}{2\pi a_\perp^2}|\psi_2(x,t)|^2\right]\psi_1(x,t), \tag{8.80}$$

$$\mu_2\psi_2(x,t) = \left[-\frac{\hbar^2}{2m}\frac{\partial^2}{\partial x^2} + \frac{1}{2}m\omega_x^2 x^2 + \frac{U_2 n_2}{2\pi a_\perp^2}|\psi_2(x,t)|^2 + \frac{U_{12} n_1}{2\pi a_\perp^2}|\psi_1(x,t)|^2\right]\psi_1(x,t). \tag{8.81}$$

Within the Thomas–Fermi approximation (n_1, $n_2 \gg 1$), the kinetic energy terms (8.80) and (8.81) can be neglected, and the density profile of the condensates becomes

$$|\psi_1(x)|^2 = 2\pi a_\perp^2 \frac{(\mu_1 - V(x))U_2 - (\mu_2 - V(x))U_{12}}{n_1(U_1U_2 - U_{12}^2)}, \tag{8.82}$$

$$|\psi_2(x)|^2 = 2\pi a_\perp^2 \frac{(\mu_2 - V(x))U_1 - (\mu_1 - V(x))U_{12}}{n_2(U_1U_2 - U_{12}^2)}, \tag{8.83}$$

where $|\psi_i(x)|^2 \geq 0$. The boundary of the cloud where the density vanishes, $|\psi_i(x)|^2 = 0$, determines the spatial extent, $R_{k,x}$, $k = 1, 2$, of the cloud. For $\psi_1(x)$,

$$V(x = R_{1,x}) = \frac{1}{2}m\omega_x R_{1,x}^2 = \frac{U_2\mu_1 - U_{12}\mu_2}{U_2 - U_{12}}. \tag{8.84}$$

The length $R_{1,x}$ of the cloud at the point where $|\psi_1(x)|^2$ vanishes then becomes

$$R_{1,x}^2 = \frac{2}{m\omega_x^2}\frac{U_2\mu_1 - U_{12}\mu_2}{U_2 - U_{12}}. \tag{8.85}$$

A similar calculation for $|\psi_2(x)|^2$ gives

$$R_{2,x}^2 = \frac{2}{m\omega_x^2}\frac{U_1\mu_2 - U_{12}\mu_1}{U_1 - U_{12}}. \tag{8.86}$$

Note that for noninteracting components $U_{12} = 0$, the length for a uniform one-dimensional cloud is recovered. Hence, the normalized density function becomes

$$|\psi_1(x)|^2 = \frac{3}{4R_{1,x}}\left(1 - \frac{x^2}{R_{1,x}^2}\right), \tag{8.87}$$

$$|\psi_2(x)|^2 = \frac{3}{4R_{2,x}}\left(1 - \frac{x^2}{R_{2,x}^2}\right). \tag{8.88}$$

Also, the normalization conditions,

$$\frac{3}{4} = 2\pi a_\perp^2 R_{1,x}\frac{U_2\mu_1 - U_{12}\mu_2}{U_1U_2 - U_{12}^2}, \tag{8.89}$$

$$\frac{3}{4} = 2\pi a_\perp^2 R_{2,x}\frac{U_1\mu_2 - U_{12}\mu_1}{U_1U_2 - U_{12}^2}, \tag{8.90}$$

allow for the calculation of the chemical potentials.

Consider a situation where the density of condensate in state $|2\rangle$ is displaced with respect to the condensate in state $|0\rangle$. Denoting the position of the center of the condensate in state $|2\rangle$ with respect to the condensate in $|0\rangle$ by x_0, the density of the condensate in state $|2\rangle$ is $|\psi_2(x)|^2 = \frac{3}{4R_{2,x}}\left(1 - \frac{(x-x_0)^2}{R_{2,x}^2}\right)$. The self-interaction energy terms g_{kk} (8.6) are then

$$g_{kk} = 3U_k/(10\pi a_\perp^2 R_{k,x}). \tag{8.91}$$

Meanwhile, the interspecies interaction energy is

$$g_{12} = \frac{3U_{12}(R_{1,x} + R_{2,x} - x_0)^3}{320\pi a_\perp^2 R_{1,x}^3 R_{2,x}^3}\left(x_0^2 + 3(R_{2,x} + R_{1,x})x_0 - 4(R_{1,x}^2 + R_{2,x}^2) + 12R_{1,x}R_{2,x}\right). \tag{8.92}$$

Equation (8.92) shows that g_{12} depends on the separation x_0 between the centers of the condensate densities and the ratio of condensate lengths, $R_{2,x}/R_{1,x}$. The point $x_0 = R_{1,x} + R_{2,x}$ is where the condensate in state $|1\rangle$ and $|2\rangle$ separate, and the interspecies interaction vanishes, $g_{12} = 0$. Thus, $g_{12} = 0$ for $x_0 \geq R_{1,x} + R_{2,x}$, and the dominant contribution to the nonlinear phase per atoms φ then comes from the self-interaction terms g_{kk}. On the other hand, when $x_0 = 0$, g_{12} becomes

$$g_{12} = \frac{3U_{12}}{80\pi a_\perp^2 R_{1,x}}\left(1 + \frac{R_{1,x}}{R_{2,x}}\right)^3\left(3\frac{R_{2,x}}{R_{1,x}} - \left(\frac{R_{2,x}}{R_{1,x}}\right)^2 - 1\right). \tag{8.93}$$

If the ratio of the lengths of condensates $R_{2,x}/R_{1,x}$ are roughly the same, $R_{2,x}/R_{1,x} \approx 1$, then g_{11}, g_{22}, and g_{12} all would have same form. Then both self-interaction terms and the intraspecies interaction term contribute to φ. In atomic species such as ^{87}Rb BEC, where the scattering lengths (a_1, a_2, and a_{12}) are all rather similar, then $\varphi \propto (a_1 + a_2 - 2a_{12})$ approximately vanishes.

One method to generate a nonlinear phase is to use spontaneous nonequilibrium dynamics. This is done by condensing the atoms into one of the hyperfine levels, say $|1\rangle$. A combination of radio frequency and microwave pulses are then used to transfer some of the condensate in hyperfine level $|1\rangle$ to another state $|2\rangle$, thus producing condensates that are in a linear superposition of state, $(|1\rangle + |2\rangle)/\sqrt{2}$, and interacting with each other. Immediately after the transfer of atomic population to $|2\rangle$, the density of the two atomic species containing different numbers of atoms n_1 and n_2 are not in equilibrium. This is because the nonlinear repulsive interaction energy due to the difference in the scattering length of the two different hyperfine levels is no longer balanced by the confining potential. As such, the equilibrium density distributions oscillate, driving each spin component to different regions of the trap. The decrease in overlap between the atomic densities at separation increases the nonlinear phase per atom, which leads to squeezing [329, 291].

Another approach is to use state-dependent potentials and Feshbach resonance [348, 394] to achieve deterministic control of the interactions. Different hyperfine levels have different magnetic moments and thus can be localized in different regions of a trap. This is exploited to produce state-dependent potentials, by physically separating the trap minima for the two-component species that coalesced during the production of the two-component atomic condensate. As a result, deterministic control of the overlap–and hence nonlinear phase per atom–is achieved. Also, Feshbach resonance [185] has been used to tune the scattering rates, thereby controlling the nonlinear phase per atom.

Exercise 8.7.1 Using (8.79) in (8.2) and (8.3), verify that the one-dimensional coupled Gross–Pitaevskii equation for a two-component atom is as given in (8.80) and (8.81). Also verify using the Thomas–Fermi approximation that the normalized solutions are given by (8.87) and (8.88).

8.8 References and Further Reading

- Section 8.1: For precision interferometry beyond the standard quantum limit, see [223, 54, 119, 52, 118, 262, 470, 78, 406, 139]. These references discuss spin-1 atom condensates and their components: [219, 194, 195, 342, 218, 352, 436]. The first trapping of all the spin components of an atomic BEC in a trap is found in [31].
- Section 8.2: These references discuss the dynamics of two-component atom Bose–Einstein condensates: [324, 194, 140]. These references discuss the tunneling of atoms at a Josephson junction: [335, 432, 385, 440, 85]. The splitting of atom BEC in a trap is discussed in [241]. The effect of two-body interactions on the operation of an atom interferometer are discussed in [242, 236]. The application of the spin-1 atoms to interferometry are discussed in [503]. Review articles on two-component atomic and spinor condensates by [257, 254] are excellent resources.
- Section 8.3: These references are good resources for applications of the Husimi Q-function [317, 421, 333]. Refs. [19, 183] are review papers on the atomic coherent state. For an application of the Q-function on the atom BEC interferometry, see [234, 235].
- Section 8.4: For an introduction to diffraction from a double slit, see [197]. Ramsey interferometry and the depiction of particle evolution on a Bloch sphere may be found in [387, 388, 7, 428, 29].
- Section 8.5: Discussions on the error propagation formula and the comparison with the standard quantum limit are provided in [491, 492, 262, 86].

- Section 8.6: The following materials give a detailed introduction to Fisher information: [211, 57, 361, 370].
- Section 8.7: For controlling the nonlinear phase per atom using state-dependent potentials and Feshbach resonances, see [394, 183, 348]. The following references controlled the nonlinear phase per atom by manipulating the spatial overlap between different spin components: [329, 291]. Ref. [324] gives an experimental method to calculate the scattering lengths by controlling the nonlinear phase per atom. Ref. [344] is an experiment describing how to realize two-atom BECs in different internal states with overlapping densities. Ref. [32] is a theory paper detailing how to estimate parameters of an atomic BEC in a trap.

9 Quantum Simulation

9.1 Introduction

It has long been appreciated that simulating quantum many-body problems is a difficult computational task. Take, for example, an interacting system of N spin 1/2 particles. The Hilbert space dimension grows exponentially as 2^N, and therefore to find the ground state would generally require diagonalization of a Hamiltonian of matrix dimension $2^N \times 2^N$. Even with the most powerful supercomputers available today, only about $N = 40$ spins can handled to perform exact calculations. Typically, in condensed matter physics one is interested in the behavior of large-scale systems. The number of atoms in a typical crystalline material could be more in the region of Avogadro's number, $N \approx 6 \times 10^{23}$. This makes the study of realistic materials far beyond the reach of direct simulation. In condensed matter physics, this has necessitated the development of sophisticated numerical techniques such as quantum Monte Carlo, density matrix renormalization group (DMRG), density functional theory, dynamical mean field theory, and series expansion methods to calculate the properties of the quantum many-body problem of interest. However, each of these methods has various shortcomings that prohibit certain quantum many-body problems to be applied with reliable accuracy, particularly those involving fermions in more than one dimension.

The above problem is not one that is limited to condensed matter physics. In high-energy physics, the fundamental theory of the strong force, quantum chromodynamics (QCD), is formulated as an SU(3) gauge-invariant quantum field theory, where the underlying fermions are quarks. Particles such as protons and neutrons are emergent quasiparticle excitations of the underlying quantum field theory. This again involves solving a quantum-many-body problem involving fermions and gauge fields. Many decades of effort have been dedicated to verify that QCD is the correct theory of the strong force, particularly in the form of lattice gauge theories, which are discrete lattice versions of QCD used to simulate the systems. This again runs into the same issues that are raised above, where only a relatively small number of sites can be simulated due to the exponential complexity. Likewise, in quantum chemistry one aims to understand the nature of molecules that are fundamentally described by a Schrodinger equation for the many-electron system. Due to the computational complexity, various approximations are typically made, such as density functional methods. Similarly, in atomic physics, numerous approximations are made to understand the nature of heavy atoms, as they involve large numbers of interacting electrons and nucleons.

Richard Feynman originally proposed the quantum computer as a method of solving such quantum many-body problems. The basic idea for this is rather simple. He noted that computers that we utilize today are based on classical logic and do not utilize any aspect of quantum physics. So, the idea is that if the computers themselves used quantum physics, then they might be better at solving quantum problems. While it was shown later that indeed it is possible to gain an exponential speedup over classical computers using these methods, attaining this in practice is difficult due to all the same problems as realizing a quantum computer: decoherence, scalability, and errors occurring during qubit manipulations. The trend recently has been to instead fabricate special-purpose systems that are manipulated so that they have the Hamiltonian of interest. This approach is called *analogue quantum simulation* and is distinguished from the former algorithmic approach, which is termed *digital quantum simulation*.

In this chapter, we will briefly describe the various techniques involved in performing quantum simulation. After introducing the problem in more detail, we will describe the difference between analogue and digital simulation. We will focus more on analogue simulation, because digital simulation is more suited for performing quantum simulations when a quantum computer is available, which is out of the scope of this book. There are already huge numbers of demonstrations for performing analogue quantum simulation on various quantum systems. Rather than describe each of these, we will describe the main techniques that are used to perform such quantum simulations, which can be used together to construct various Hamiltonians as desired. To illustrate these techniques, we also describe one of the classic experiments that demonstrates a quantum simulation of the Bose–Hubbard model with cold atoms. For a comprehensive list of systems that have been simulated using these ideas, we refer the reader to the excellent review articles listed at the end of this chapter.

9.2 Problem Statement: What is Quantum Simulation?

The first ideas for quantum simulation were originally motivated by Richard Feynman, who proposed the quantum computer for simulating quantum many-body problems. Feynman made this conjecture in 1982, but it remained unproven until Lloyd gave an explicit argument that this is true in 1996. Lloyd showed that it is possible to construct the Hamiltonian of a chosen physical system out of a limited set of controls–for example, in a quantum computer. While conceptually this was a breakthrough, this did not mean that one can easily perform quantum simulations. Lloyd's method relies upon performing a series of gates to simulate the desired Hamiltonian. This *digital quantum simulation* approach requires the technological capability of something rather close – or even equivalent – to a full quantum computer.

As the field of quantum technology developed since these early days, it became evident that quantum computers were a type of device that is one of the most challenging in terms of engineering. In a quantum computer, one needs to be able to perform a completely arbitrary unitary evolution and readout of a very large number of qubits. Other quantum technological applications such as quantum cryptography are closer to being used in practice, since only few-qubit manipulations need to be performed. For this reason, alternative directions were investigated where the technological demands are less. The approach was to construct purpose-built systems that have the same Hamiltonian as the system of interest. This has been a popular approach, since it is technologically more achievable, rather than first implementing a fully programmable quantum computer. In this *analogue quantum simulation* approach, one typically combines several techniques to build a specified Hamiltonian. Then one experimentally studies the artificially created system, which often can be measured using advanced techniques that may not be available to other systems. This gives a unique and practical way of studying complex quantum many-body systems.

To understand the task of quantum simulation, it is illustrative to look at a simple example. An archetypical quantum many-body problem from condensed matter physics is the transverse Ising model (see Fig. 9.1), which is specified by the Hamiltonian

$$H = -\frac{J}{2} \sum_{j,k \in \text{n.n.}} \sigma_j^z \sigma_k^z - \Gamma \sum_j \sigma_j^x, \tag{9.1}$$

where J, Γ are some constants, and the sum runs over nearest neighbors (n.n.) of a lattice. The lattice could have any geometry or dimensionality; we can consider it to be a square lattice, for instance. The Hamiltonian encodes any number of lattice sites: for example, for $N = 2$ sites it can be equivalently represented as a matrix:

$$H_{N=2} = -\begin{pmatrix} J/2 & \Gamma & \Gamma & 0 \\ \Gamma & -J/2 & 0 & \Gamma \\ \Gamma & 0 & -J/2 & \Gamma \\ 0 & \Gamma & \Gamma & J/2 \end{pmatrix}. \tag{9.2}$$

Here the basis vectors are taken as $\{|00\rangle, |01\rangle, |10\rangle, |11\rangle\}$, where $|0\rangle, |1\rangle$ are the eigenstates of the σ^z Pauli matrix. A typical condensed matter physics problem would then involve diagonalizing the Hamiltonian to find the eigenstates. In particular, one is often interested in the ground state and low-lying excitations. These states possess particular properties, which determine the physics of the model. For example, in the one-dimensional transverse Ising model, there is a quantum phase transition when $J/2 = \Gamma$, where the ground state is a ferromagnet for $J/2 > \Gamma$ and a paramagnet for $J/2 < \Gamma$. One can identify such phases using *order parameters* that give characteristic behavior in the region of a quantum phase transition. For instance, the magnetization has the relation

$$\langle \sigma_i^z \rangle = \begin{cases} (1 - (\frac{2\Gamma}{J})^2)^{1/8} & \frac{J}{2\Gamma} > 1 \\ 0 & \frac{J}{2\Gamma} \leq 1, \end{cases} \tag{9.3}$$

which is plotted in Fig. 9.1(b). This shows a sharp transition at the quantum phase transition point.

While diagonalizing (9.2) is straightforward, such small systems unfortunately do not usually exhibit the phase transition physics that one would like to observe. The sharp transition occurs when the system size is large $N \to \infty$, usually called the *thermodynamic limit*. In Fig. 9.1(c), the energy gap between the first excited state

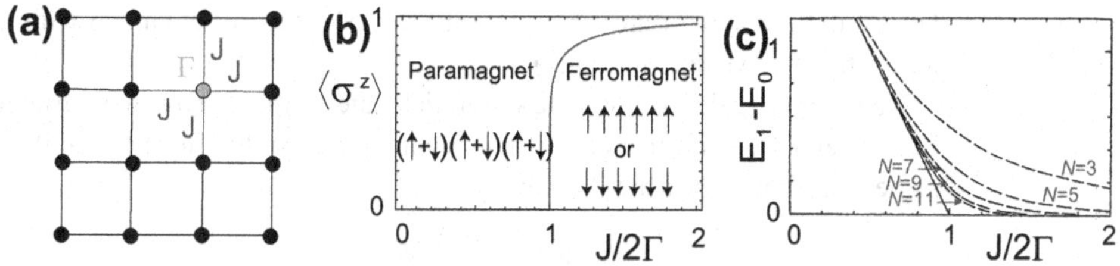

Fig. 9.1 The transverse Ising model. (a) Schematic description of the transverse Ising model in two dimensions. The couplings J indicate the interactions between nearest neighbor sites, and a magnetic field in the transverse direction of strength Γ is applied across all sites. (b) Phases of the one-dimensional transverse Ising model. The magnetization result of (9.3) taken from [371] is shown. (c) The energy gap reproduced from [198]. Dashed lines show finite size results; the solid line is for the thermodynamic limit.

and the ground state is shown. The sharp transition only occurs for large systems, and for finite systems the precise transition is less visible. In this way, one is usually more interested in larger systems. However, this is a computationally expensive task, hence one typically performs small-scale simulations and extrapolates to larger-scale systems. Furthermore, there are often many generalizations of the above scenario that one would like to handle. Looking at the ground state alone corresponds to a physical system at the absolute zero of temperature. Realistically, one would like to look also at the effects of nonzero temperature. However, this is another nontrivial computational task, since one must also find the spectrum of states. Another interesting but nontrivial task is to understand nonequilibrium physics of such systems, which do not follow a Boltzmann distribution of states. This could involve time-dynamical problems, which are quite difficult from a computational point of view, hence they have been studied to a lesser extent.

9.3 Digital Quantum Simulation

We first describe the approach of digital quantum simulation, which is historically the first method that was found to have a quantum speedup over classical methods. Suppose that we wish to study a Hamiltonian that takes the general form

$$H = \sum_j H_j, \tag{9.4}$$

where the H_j is a locally acting Hamiltonian. This is an extremely common structure for physical Hamiltonians of interest, since the terms in a Hamiltonian are usually derived from reducing particular forces that tend to be locally acting. For example, in the transverse Ising model of (9.1), the $\sigma_j^z \sigma_k^z$ originates from a typically short-ranged exchange interaction, and the σ_j^x arises from an applied magnetic field that only acts one spin at a time. The key point that makes (9.4) difficult to solve is that not all the terms in the Hamiltonian commute:

$$[H_j, H_{j'}] \neq 0. \tag{9.5}$$

For example, for the transverse Ising model, σ_j^x does not commute with all $\sigma_j^z \sigma_k^z$ terms that involve the same site j.

Now, suppose that we prepare the initial state of the system in a particular state $|\psi(0)\rangle$, and we want to find the time evolution according to the Hamiltonian. At some later time t, the state is given by

$$\begin{aligned} |\psi(t)\rangle &= e^{-iHt/\hbar}|\psi(0)\rangle \\ &= \exp\left(-\frac{it}{\hbar}\sum_j H_j\right)|\psi(0)\rangle. \end{aligned} \tag{9.6}$$

Now, further suppose that we have in our possession a quantum computer that can perform any combination of one and two qubit gates. For example, such gates can be written respectively as

$$\begin{aligned} U_l(\boldsymbol{n},\theta) &= \exp(-i\theta \boldsymbol{n}\cdot\boldsymbol{\sigma}_l/2) \\ U_{ll'}(\phi) &= \exp(-i\phi\sigma_l^z\sigma_{l'}^z) \end{aligned} \tag{9.7}$$

where $\boldsymbol{n}$ is a unit vector specifying the axis of the single qubit gate and θ, ϕ are freely choosable angles. A well-known result in quantum information states that an arbitrary unitary evolution can be made by combining the one- and two-qubit operations. Thus, the unitary evolution $e^{-iHt/\hbar}$ should also be written in terms of the elementary gates. The question is, then, how to perform this decomposition?

The decomposition can be achieved by a Suzuki–Trotter expansion. The first-order Suzuki–Trotter expansion is written

$$e^{(A+B)} = \lim_{n\to\infty}\left(e^{\frac{A}{n}}e^{\frac{B}{n}}\right)^n, \tag{9.8}$$

where n is the Trotter number. In practice, the Trotter number is kept at a finite number, so that one expands the unitary evolution in (9.6) according to

$$\begin{aligned} e^{-iHt/\hbar} &= \exp\left(-\frac{it}{\hbar}\sum_j H_j\right) \\ &= \left[\exp\left(-\frac{i\Delta t}{\hbar}\sum_j H_j\right)\right]^{t/\Delta t} \\ &\approx \left[\prod_j \exp\left(-\frac{i\Delta t}{\hbar}H_j\right)\right]^{t/\Delta t}, \end{aligned} \tag{9.9}$$

This appears as a sequence of terms of the form $\exp\left(-\frac{i\Delta t}{\hbar}H_j\right)$, which is suitable to perform a gate decomposition. For example, for the case of the transverse Ising model (9.1), the sequence of gates would involve a sequence of operations including

$$U_l(\boldsymbol{x}, \lambda\Delta t/\hbar) = \exp\left(-\frac{i\lambda\Delta t}{\hbar}\sigma_l^x\right) \tag{9.10}$$

$$U_{ll'}(J\Delta t/\hbar) = \exp\left(-\frac{iJ\Delta t}{\hbar}\sigma_l^z\sigma_{l'}^z\right). \tag{9.11}$$

The error of the operation by performing the Trotter decomposition (9.9) for one of the time steps Δt (inside the square brackets) scales as $O((\Delta t)^2)$. This means that the error of the whole operation (9.9) scales as $O(t\Delta t)$.

The key point here is that the number of terms in the decomposition (9.9) is only proportional to the number of terms in the original Hamiltonian (9.4). Taking the transverse Ising model as an example again, the total number of terms in a periodic one-dimensional chain with N sites is $2N$. The number of gates to evolve the system forward in time by δt is N units of (9.10) and N units of (9.11). In fact, many of these gates can be done in parallel, since terms that do not involve the same site commute. Repeating this procedure $t/\Delta t$ times, one achieves a time evolution of the state with a time t. In comparison, time-evolving a quantum state on a classical computer is a computationally intensive task. On a classical computer, one way to perform this would be to find the eigenstates of the Hamiltonian $|E_n\rangle$ and perform an expansion (9.6) according to

$$|\psi(t)\rangle = \sum_n e^{-iE_n t/\hbar}\langle E_n|\psi(0)\rangle|E_n\rangle. \tag{9.12}$$

The part that is computationally expensive is the evaluation of the eigenstates, which requires diagonalization of a matrix of dimension $2^N \times 2^N$. This is far less efficient, and hence using a quantum computer has an exponential speedup over classical methods.

The above shows that time evolution can be performed more efficiently. But what if one is interested in other aspects of a quantum many-body model? We saw in Section 9.2 that often one is interested in finding the eigenstates and energies of the Hamiltonian, such as the ground and excited states. For these tasks, one can also show that these can be performed with a quantum speedup. For example, an eigenstate can be obtained by performing the quantum phase estimation algorithm on the operator

$$e^{-iHt/\hbar} = \sum_n e^{-iE_n t/\hbar}|E_n\rangle\langle E_n|. \tag{9.13}$$

The time-evolution operator applies a phase of $\phi = E_n t/\hbar$ on the eigenstates, which is the phase to be estimated. To perform the time-evolution operator, we again use the Trotter expansion (9.9). Since the Trotter expansion can be performed efficiently, it also shows that the quantum phase estimation algorithm can be performed efficiently. The eigenstates are found by a by-product of the same algorithm, where after a measurement of the energy is made, the initial state is projected into an eigenstate of the Hamiltonian. Numerous other methods to examine various quantities have been investigated in the past in the digital quantum simulation approach. We do not discuss them here; we refer the reader to several excellent review articles as given in Section 9.6.

Exercise 9.3.1 Expand $\left(e^{\frac{A}{n}}e^{\frac{B}{n}}\right)^n$ to second order and verify that it agrees with the expansion of $e^{(A+B)}$. What is the form of the error that occurs due to the Suzuki–Trotter expansion? What happens if $[A, B] = 0$?

9.4 Toolbox for Analogue Quantum Simulators

The digital quantum simulation approach described in Section 9.3 assumes the availability of a highly controllable quantum system consisting of a large number of qubits. To perform the time evolution, one- and two-qubit gates were applied in sequence, in possible combination with quantum algorithms such as the phase estimation algorithm. This requires the capabilities of an experimental system that is close or equivalent to a full quantum computer. While great progress has been made in recent years, at the time of writing only relatively small numbers of qubits can be fully controlled, hence the digital quantum simulation approach can only be applied to small quantum systems.

One of the features of many quantum–many problems is that there are usually symmetries that can be exploited. For example, in the transverse Ising model (9.1), the couplings J, Γ are translationally invariant. This means that the gates such as (9.10) and (9.11) are the same for all the qubits. This means that they can be applied in parallel across all qubits, as long as the gates commute. In fact, depending upon the particular experimental system, it may be possible to even perform several Hamiltonians at the same time, even if they do not commute. For example, one might be able to apply a one- and two-qubit operation,

$$H_j = -\frac{J}{4} \sum_{k \in \text{n.n.}} \sigma_j^z \sigma_k^z - \Gamma \sigma_j^x, \tag{9.14}$$

on the jth qubit. Then, performing this on all qubits, $H = \sum_j H_j$ reproduces (9.1), directly implementing the desired Hamiltonian. The only reason that we separated the two types of gates in (9.10) and (9.11) is because of the assumption that these gates must be performed separately. If experimentally it is in fact possible to perform them at the same time, then the procedure involving the Trotter decomposition of the last section is clearly unnecessary and it is possible to directly perform the time evolution.

What kind of Hamiltonians can be applied, and whether they can be done simultaneously, is largely an experimental question that depends upon the particular system. The cold atomic system is one of the most versatile systems with a variety of experimental techniques to construct various Hamiltonians. In addition to creating the Hamiltonian of interest, it is necessary to measure the system in an appropriate way to find the physics of interest. For example, some observables such as the order parameter need to be extracted from the state of interest. In this section, we outline the various techniques available for analogue quantum simulation with cold atoms.

9.4.1 Optical Lattices

As we have seen in Chapter 3, the spatial wavefunction of Bose–Einstein condensates (BECs) can be described by an order parameter in continuous space. Meanwhile, many of the models of interest that are discussed in condensed matter physics and other areas of physics are discrete lattice systems. Therefore, if we are to use BECs to simulate the physics of a lattice model, some way of discretizing space is required. This is performed experimentally using optical lattices and has proved to be an immensely powerful tool to manipulate cold atom systems.

The simplest configuration of an optical lattice consists of two counterpropagating laser beams that produce a standing wave. The counterpropagating lasers may be conveniently produced, for example, by using mirrors to reflect the laser light

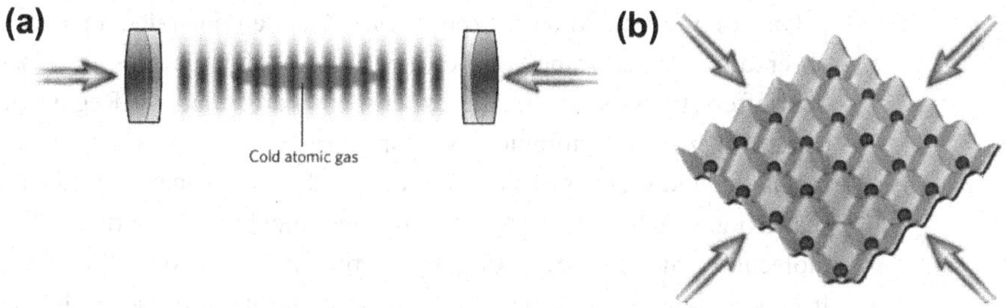

Fig. 9.2 An optical lattice for creating a periodic potential with cold atoms. (a) Two counterpropagating laser beams produce a standing wave. (b) The effective potential produced by the optical lattice. Diagrams reproduced from [44]. Color version of this figure available at cambridge.org/quantumatomoptics.

entering from one side (Fig. 9.2(a)). This produces a light intensity that varies along the axis of the laser beams as

$$I(x) = I_0 \cos^2(\kappa x) \tag{9.15}$$

where $\kappa = 2\pi/\lambda$ and λ is the wavelength of the light.

As we have already encountered in Section 4.5, an off-resonant light field will create an ac Stark shift. From (4.38), we observe that the Rabi frequency Ω is proportional to the amplitude of the light, hence the energy shift (4.51) is proportional to the intensity of the light. Since the laser beam in the standing wave configuration is of spatially varying intensity, there will be spatially varying energy due to the ac Stark shift according to

$$V(x) \propto V_0 \cos(2\kappa x), \tag{9.16}$$

where $V_0 = -\frac{|\Omega|^2}{\Delta}$ and has removed an irrelevant constant, which contributes to a global energy offset. This periodic potential is what is experienced by the atoms and is the effect of applying the optical lattice. Generally, the parameter V_0 is relatively easily controlled, for example by changing the laser intensity, so one can produce a potential depth as desired. Although the above potential is only in one dimension here, by applying more than one laser it is possible to produce two- and three-dimensional lattices by combining other lasers in standing wave configurations (Fig. 9.2(b)). The simplest types of lattices are square lattices, but other lattice geometries are also possible using combinations of laser beams in other orientations.

9.4.2 Feshbach Resonances

The optical lattice manipulates the single-particle Hamiltonian (1.31), specifically the potential energy $V(x)$. On the other hand, as discussed in Section 1.5, there is another important component to the atomic Hamiltonian, which is the interactions between the atoms. As we have already discussed in Section 4.6, by applying a magnetic field it is possible to change the strength of the atom-atom interaction, and even change the sign from repulsive to attractive. Feshbach resonances can therefore be used to manipulate the interaction part of the Hamiltonian (4.10) to a desired value.

One of the best-known experiments that demonstrated this technique was the observation of the transition between a BEC of molecules to a Bardeen–Cooper–Schrieffer (BCS) state. In the experiment performed by Regal, Greiner, and Jin, a trapped gas of fermionic ^{40}K atoms was cooled to quantum degeneracy and a magnetic field was applied to produce a Feshbach resonance. In the regime where the scattering length corresponds to repulsive interactions, there exists a weakly bound molecular state, where the binding energy is controlled by the Feshbach resonance. In this regime, the state is a BEC of bosonic diatomic molecules, and there are no fermion degrees of freedom. When the Feshbach resonance is such that interactions are attractive, no molecules form, but there is still a condensation of Cooper pairs that can be described by a BCS theory. Thus, by controlling the Feshbach resonances, it is possible to observe both phenomena in the same atomic system.

9.4.3 Artificial Gauge Fields

In condensed matter physics, a common situation that one encounters is a charged particle in a magnetic field. The archetypical experiment is the quantum Hall effect, where electrons in a two-dimensional plane have a magnetic field applied to them. In terms of the Schrodinger equation, this can be taken into account by a vector potential $\boldsymbol{A}(\boldsymbol{x})$, where the single-particle Hamiltonian (1.2) becomes

$$H_0(\boldsymbol{x}) = -\frac{(\boldsymbol{p} - \boldsymbol{A}(\boldsymbol{x}))^2}{2m} + V(\boldsymbol{x}). \tag{9.17}$$

where $\boldsymbol{p} = -i\hbar\boldsymbol{\nabla}$ is the momentum operator. One issue with using cold atoms is that they are charge neutral – this means that applying a magnetic field will not produce such a vector potential. How can one produce a magnetic field, and more generally, a gauge field? Here we will introduce two methods that can be used to make an effective vector potential $\boldsymbol{A}(\boldsymbol{x})$.

The first method takes advantage of the fact that there is a very close relationship between rotation and a magnetic field. To illustrate this, consider the single-particle Hamiltonian in a harmonic oscillator rotating around the z axis,

$$H_{\text{rot}}(\boldsymbol{x}) = \frac{p^2}{2m} + \frac{1}{2}m\omega^2(x^2 + y^2 + z^2) - \Omega L^z, \tag{9.18}$$

where Ω is the rotation frequency, and $L^z = \hat{z} \cdot (\boldsymbol{r} \times \boldsymbol{p})$ is the angular momentum. The above can be rewritten as

$$H_{\text{rot}}(\boldsymbol{x}) = \frac{(\boldsymbol{p} - \boldsymbol{A}(\boldsymbol{x}))^2}{2m} + \frac{1}{2}m(\omega^2 - \Omega^2)(x^2 + y^2) + \frac{1}{2}m\omega^2 z^2, \tag{9.19}$$

where $\boldsymbol{A}(\boldsymbol{x}) = m\Omega\hat{z} \times \boldsymbol{r}$. Thus, the rotation has the effect of producing an effective vector potential, as well as an antitrapping in the perpendicular plane to the axis of rotation.

The second method takes advantage of Berry phases that are produced in an adiabatic evolution. The simplest configuration that shows the effect is when there is a spatially varying spin Hamiltonian. Consider a similar situation to what was discussed in Section 4.4, where a laser is applied to atoms such that it produces the effective Hamiltonian (4.44). The difference that we will consider here is that the

Rabi frequency Ω and detuning Δ will have a more general spatial dependence, so that the overall Hamiltonian is

$$H_{\text{Berry}}(\boldsymbol{x}) = \left(\frac{p^2}{2m} + V(\boldsymbol{x})\right)\begin{pmatrix} 1 & 0 \\ 0 & 1 \end{pmatrix} + \hbar \begin{pmatrix} 0 & \Omega(\boldsymbol{x})/2 \\ \Omega^*(\boldsymbol{x})/2 & \Delta(\boldsymbol{x}) \end{pmatrix}. \tag{9.20}$$

Here, the first term is the spin-independent part of the Hamiltonian due to the kinetic and potential energy, and the second term is due to the laser creating transitions between the spin states $|1\rangle$ and $|2\rangle$. The spin-dependent part of the Hamiltonian can be diagonalized and is given by

$$|E_s^{\pm}(\boldsymbol{x})\rangle = \Delta(\boldsymbol{x}) \mp \sqrt{\Delta^2(\boldsymbol{x}) + |\Omega(\boldsymbol{x})|^2}|1\rangle - \Omega^*(\boldsymbol{x})|2\rangle, \tag{9.21}$$

where the spin-dependent energy is

$$E_s^{\pm}(\boldsymbol{x}) = \frac{\Delta(\boldsymbol{x}) \pm \sqrt{\Delta^2(\boldsymbol{x}) + |\Omega(\boldsymbol{x})|^2}}{2}. \tag{9.22}$$

The above only considers the spin part of the Hamiltonian. Let us now obtain an effective Hamiltonian including the spatial degrees of freedom. The most general form of the wavefunction including the spatial degrees of freedom takes the form

$$|\Psi(\boldsymbol{x},t)\rangle = \psi^+(\boldsymbol{x},t)|E_s^+(\boldsymbol{x})\rangle + \psi^-(\boldsymbol{x},t)|E_s^-(\boldsymbol{x})\rangle. \tag{9.23}$$

Consider that we prepare the atom in the lower energy state $|E_s^-(\boldsymbol{x})\rangle$, and all subsequent operations on the state will be sufficiently slow such that the adiabatic approximation holds. In this case, we can assume that the spin of the atoms will always remain in the state $|E_s^-(\boldsymbol{x})\rangle$, and the wavefunction will be approximately

$$|\Psi(\boldsymbol{x},t)\rangle \approx \psi^-(\boldsymbol{x},t)|E_s^-(\boldsymbol{x})\rangle. \tag{9.24}$$

We now wish to obtain an equation of motion for the spatial wavefunction $\psi^-(\boldsymbol{x},t)$. First, rewrite the Hamiltonian (9.20) in the basis (9.21), yielding

$$\begin{aligned} H_{\text{Berry}}(\boldsymbol{x})|\Psi(\boldsymbol{x},t)\rangle = &\left(\frac{p^2}{2m} + V(\boldsymbol{x})\right)(|E_s^+(\boldsymbol{x})\rangle\langle E_s^+(\boldsymbol{x})| + |E_s^-(\boldsymbol{x})\rangle\langle E_s^-(\boldsymbol{x})|) \\ &+ E_s^+(\boldsymbol{x})|E_s^+(\boldsymbol{x})\rangle\langle E_s^+(\boldsymbol{x})| + E_s^-(\boldsymbol{x})|E_s^-(\boldsymbol{x})\rangle\langle E_s^-(\boldsymbol{x})|. \end{aligned} \tag{9.25}$$

Applying (9.25) to (9.24) and projecting the result onto $|E_s^-(\boldsymbol{x})\rangle$ under the assumption that the spin component will always remain in this state, we have the effective Hamiltonian,

$$\begin{aligned} H_{\text{eff}}^-(\boldsymbol{x}) &= \langle E_s^-(\boldsymbol{x})|H_{\text{Berry}}(\boldsymbol{x})|\Psi(\boldsymbol{x},t)\rangle \\ &= \langle E_s^-(\boldsymbol{x})|\frac{p^2}{2m}\psi^-(\boldsymbol{x},t)|E_s^-(\boldsymbol{x})\rangle + (V(\boldsymbol{x}) + E_s^-(\boldsymbol{x}))\,\psi^-(\boldsymbol{x},t). \end{aligned} \tag{9.26}$$

The first term above is the key difference from the more typical situation, where the spin eigenstates do not have a spatial dependence. Due to the spatial dependence of $|E_s^-(\boldsymbol{x})\rangle$, the derivative in the momentum $\boldsymbol{p} = -i\hbar\boldsymbol{\nabla}$ will act on the spin eigenstates as well as the spatial wavefunction. Evaluating this term, we have

$$H^-_{\text{eff}}(\boldsymbol{x}) = \left(\frac{p^2}{2m} + V(\boldsymbol{x}) + E^-_{\text{s}}(\boldsymbol{x})\right)\psi^-(\boldsymbol{x},t)$$
$$\frac{-i\hbar(\boldsymbol{p}\psi^-(\boldsymbol{x},t))\cdot\langle E^-_{\text{s}}(\boldsymbol{x})|\nabla|E^-_{\text{s}}(\boldsymbol{x})\rangle - \hbar^2\psi^-(\boldsymbol{x},t)\langle E^-_{\text{s}}(\boldsymbol{x})|\nabla^2|E^-_{\text{s}}(\boldsymbol{x})\rangle}{2m}. \tag{9.27}$$

The last term may be written

$$\begin{aligned}\langle E^-_{\text{s}}(\boldsymbol{x})|\nabla^2|E^-_{\text{s}}(\boldsymbol{x})\rangle &= \langle E^-_{\text{s}}(\boldsymbol{x})|\nabla\cdot(|E^+_{\text{s}}(\boldsymbol{x})\rangle\langle E^+_{\text{s}}(\boldsymbol{x})| + |E^-_{\text{s}}(\boldsymbol{x})\rangle\langle E^-_{\text{s}}(\boldsymbol{x})|)\,\nabla|E^-_{\text{s}}(\boldsymbol{x})\rangle \\ &= ||\langle E^-_{\text{s}}(\boldsymbol{x})|\nabla|E^-_{\text{s}}(\boldsymbol{x})\rangle||^2 + ||\langle E^-_{\text{s}}(\boldsymbol{x})|\nabla|E^+_{\text{s}}(\boldsymbol{x})\rangle||^2,\end{aligned} \tag{9.28}$$

where the $||\cdots||$ is the vector norm with respect to the spatial directions. Introducing the effective vector and scalar potentials

$$\begin{aligned}\boldsymbol{A}(\boldsymbol{x}) &= i\hbar\langle E^-_{\text{s}}(\boldsymbol{x})|\nabla|E^-_{\text{s}}(\boldsymbol{x})\rangle \\ W(\boldsymbol{x}) &= -\frac{\hbar^2}{2m}||\langle E^-_{\text{s}}(\boldsymbol{x})|\nabla|E^+_{\text{s}}(\boldsymbol{x})\rangle||^2,\end{aligned} \tag{9.29}$$

we obtain the effective Hamiltonian

$$H^-_{\text{eff}}(\boldsymbol{x}) = \left(\frac{(\boldsymbol{p}-\boldsymbol{A}(\boldsymbol{x}))^2}{2m} + V(\boldsymbol{x}) + E^-_{\text{s}}(\boldsymbol{x}) + W(\boldsymbol{x})\right)\psi^-(\boldsymbol{x},t). \tag{9.30}$$

The above has the form of (9.17) where the vector potential originates from the Berry phase due to the spatial dependence of the spin eigenstates. By suitably engineering the spatial dependence of the spin eigenstates (9.21), it is possible to produce a vector potential of the desired form. The effective magnetic field corresponding to the vector potential is then given by

$$\begin{aligned}\boldsymbol{B}(\boldsymbol{x}) &= \nabla\times\boldsymbol{A}(\boldsymbol{x}) \\ &= i\hbar\nabla\times\langle E^-_{\text{s}}(\boldsymbol{x})|\nabla|E^-_{\text{s}}(\boldsymbol{x})\rangle.\end{aligned} \tag{9.31}$$

9.4.4 Spin-Orbit Coupling

In some condensed matter systems, even more exotic Hamiltonians can arise beyond the relatively straightforward types of potentials and interactions that we have examined so far. One class of interactions that has gained a lot of interest in recent years is the spin-orbit-coupled Hamiltonians. Spin-orbit coupling can be one of the ingredients to form a topological insulator, where the bulk electronic band structure is that of a band insulator, but has topologically protected conductive states on the edge or surface. Due to the great interest of such materials in condensed matter physics, a natural aim of quantum simulation has been to produce novel types of interactions that may display exotic properties. The great advantage of such a quantum simulation approach – where interactions are produced from the "ground-up" – is that it is possible to engineer Hamiltonians that may be difficult, or even impossible, to produce in natural materials.

First, what is spin-orbit coupling? This is most familiar from traditional atomic physics, where the orbital angular momentum $\boldsymbol{L}$ of an electron in an atom couples to its spin $\boldsymbol{S}$, and the interaction takes the form $\propto \boldsymbol{L}\cdot\boldsymbol{S}$. The key point here is that the

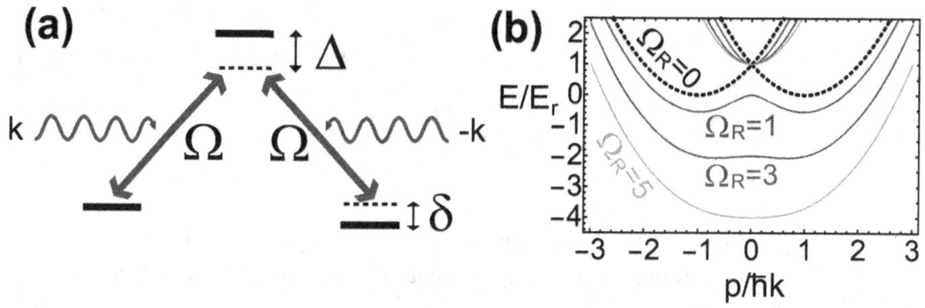

Fig. 9.3 Generating spin-orbit coupling in Bose–Einstein condensates. (a) Raman scheme for coupling two atomic spin states. The laser beams are counterpropagating with momenta $\boldsymbol{k}$, which transfers a momentum $2\boldsymbol{k}$ during the spin transition. (b) Energy spectrum of the spin-orbit coupled Hamiltonian (9.38). Here $\Omega_R = \frac{|\Omega|^2}{2\Delta}$ is the effective Raman coupling and $E_r = \frac{\hbar^2 k^2}{2m}$ is the recoil energy. Color version of this figure available at cambridge.org/quantumatomoptics.

motion of the particle affects the spin, such that there is momentum dependence to the energy of the spin. Recalling that the energy of a spin in a magnetic field takes the form $\propto \boldsymbol{B} \cdot \boldsymbol{S}$, we can view the spin-orbit coupling as an effective momentum-dependent Zeeman magnetic field.

For electrons within an atom, it is angular momentum that couples to spin, but in condensed matter physics, regular linear momentum can also couple to the spin. For example, a coupling of the form $\propto k_y S_x - k_x S_y$ is what is referred to as Rashba spin-orbit coupling and can arise in two-dimensional semiconductor heterostructure systems. Another type of spin-orbit coupling, referred to as the Dresselhaus effect, results in a coupling of the form $\propto k_x S_x - k_y S_y$ up to linear terms.

Such types of spin-orbit interaction can be produced in cold atoms by applying Raman fields between various internal states of atoms. The basic idea is that the Raman fields are applied in such way that spin transfer between two states also results in a net momentum transfer due to the absorption and emission of a photon. For example, suppose that in the Raman transition shown in Fig. 9.3(a) is produced by two counterpropagating laser beams each of wavenumber $\boldsymbol{k}$. On the first leg of the Raman transition a photon is absorbed, adding a momentum $\boldsymbol{k}$ to the atom. On the second leg of the Raman transition, a photon is emitted with momentum $-\boldsymbol{k}$, thus the atom loses momentum $-\boldsymbol{k}$. The total momentum change of the atom is $2\boldsymbol{k}$. For cold atoms, this momentum change is not negligible in comparison to their average momenta and causes a spin-dependent shift to the dispersion relation.

To show this in a concrete way, let us consider the situation of applying the two counterpropagating laser fields to a BEC. The single-particle Hamiltonian is

$$H_{\text{spin-orbit}}(\boldsymbol{x}) = \frac{p^2}{2m}\begin{pmatrix} 1 & 0 & 0 \\ 0 & 1 & 0 \\ 0 & 0 & 1 \end{pmatrix} + \hbar \begin{pmatrix} 0 & \Omega^* e^{-i\boldsymbol{k}\cdot\boldsymbol{x}}/2 & 0 \\ \Omega e^{i\boldsymbol{k}\cdot\boldsymbol{x}}/2 & \Delta & \Omega^* e^{-i\boldsymbol{k}\cdot\boldsymbol{x}}/2 \\ 0 & \Omega e^{i\boldsymbol{k}\cdot\boldsymbol{x}}/2 & \delta \end{pmatrix}, \quad (9.32)$$

where δ is a two-photon detuning. Here, we have ignored the trapping potential for simplicity and have included the spatial dependence of the Raman fields as in (4.47). Eliminating the intermediate level as in Section 4.5, we obtain the effective two-level Hamiltonian

$$H_{\text{spin-orbit}}(\boldsymbol{x}) \approx \frac{p^2}{2m}\begin{pmatrix}1 & 0\\ 0 & 1\end{pmatrix} + \hbar\begin{pmatrix}-|\Omega|^2/4\Delta & (\Omega^*)^2 e^{-2i\boldsymbol{k}\cdot\boldsymbol{x}}/4\Delta\\ \Omega^2 e^{2i\boldsymbol{k}\cdot\boldsymbol{x}}/4\Delta & \delta - |\Omega|^2/4\Delta\end{pmatrix}$$
$$= \frac{p^2}{2m}I - \frac{|\Omega|^2}{4\Delta}I + \frac{\delta}{2}(I-\sigma^z) + \frac{(\Omega^*)^2 e^{-2i\boldsymbol{k}\cdot\boldsymbol{x}}}{4\Delta}\sigma^+ + \frac{\Omega^2 e^{2i\boldsymbol{k}\cdot\boldsymbol{x}}}{4\Delta}\sigma^-, \tag{9.33}$$

where we wrote the various terms in terms of Pauli operators. We can rewrite this Hamiltonian to remove the spatially dependent Raman terms by applying a unitary transform:

$$U = e^{i\boldsymbol{k}\cdot\boldsymbol{x}\sigma^z}. \tag{9.34}$$

This removes the spatial dependence because the spin operators transform as

$$\begin{aligned} U\sigma^+U^\dagger &= e^{2i\boldsymbol{k}\cdot\boldsymbol{x}}\sigma^+\\ U\sigma^-U^\dagger &= e^{-2i\boldsymbol{k}\cdot\boldsymbol{x}}\sigma^-\\ U\sigma^zU^\dagger &= \sigma^z. \end{aligned} \tag{9.35}$$

Meanwhile, the momentum operator is shifted according to

$$U\boldsymbol{p}U^\dagger = \boldsymbol{p} - \hbar\boldsymbol{k}\sigma^z \tag{9.36}$$

due to the fact that $\boldsymbol{p} = -i\hbar\boldsymbol{\nabla}$. The transformed Hamiltonian then reads

$$UH_{\text{spin-orbit}}(\boldsymbol{x})U^\dagger = \frac{(\boldsymbol{p}-\hbar\boldsymbol{k}\sigma^z)^2}{2m}I - \frac{|\Omega|^2}{4\Delta}I + \frac{\delta}{2}(I-\sigma^z) + \frac{(\Omega^*)^2}{4\Delta}\sigma^+ + \frac{\Omega^2}{4\Delta}\sigma^- \tag{9.37}$$

$$= \begin{pmatrix}\frac{(\boldsymbol{p}-\hbar\boldsymbol{k})^2}{2m} - \frac{|\Omega|^2}{4\Delta} & \frac{(\Omega^*)^2}{4\Delta}\\ \frac{\Omega^2}{4\Delta} & \frac{(\boldsymbol{p}+\hbar\boldsymbol{k})^2}{2m} - \frac{|\Omega|^2}{4\Delta} + \delta\end{pmatrix}. \tag{9.38}$$

Expanding the first term in (9.37), we have

$$\frac{(\boldsymbol{p}-\hbar\boldsymbol{k}\sigma^z)^2}{2m} = \frac{p^2}{2m} - \frac{\hbar}{m}(\boldsymbol{k}\cdot\boldsymbol{p})\sigma^z + \frac{\hbar k^2}{2m}. \tag{9.39}$$

The component of the momentum parallel to the Raman fields is coupled to the spin of the particle, realizing the spin-orbit interaction.

Let us briefly analyze the Hamiltonian (9.38) to understand what the effect of the spin-orbit interaction is. First, consider the limit where $|\Omega| \to 0$. In this case, the two spin states are uncoupled and the dispersion corresponds to two parabola displaced by a momentum $\pm\hbar\boldsymbol{k}$. The two-photon detuning δ creates an energy offset between the two spin states. Turning on the Raman fields, the two spin states couple, producing an anticrossing of the levels, and a double-well structure forms in momentum space (see Fig. 9.3(b)). For larger values of $|\Omega|$, there is a point where the double wells merge into a single minimum. Such a phase transition was experimentally observed by [308], where spin-orbit coupling was produced in a BEC for the first time. We note that similar methods can be used to form artificial gauge fields, providing another method to that described in Section 9.4.3.

Exercise 9.4.1 Examining the $p = 0$ state, verify that the effective Hamiltonian (9.33) has the same spectrum as (9.32) for the lowest two levels, as long as $|\Omega|$, $\delta \ll \Delta$.

Exercise 9.4.2 Verify the unitary transforms (9.35) on the Pauli operators.

Exercise 9.4.3 Show that

$$e^{ikx} p_x e^{-ikx} \psi(x) = (p_x - \hbar k)\psi(x) \tag{9.40}$$

where $p_x = -i\hbar \frac{\partial}{\partial x}$. Use this to verify (9.36).

Exercise 9.4.4 Perform the unitary transformation to derive the spin-orbit Hamiltonian (9.38).

Exercise 9.4.5 Diagonalize the Hamiltonian (9.38) and show that the spectrum follows the form shown in Fig. 9.3(b).

9.4.5 Time-of-Flight Measurements

So far, we have discussed methods of creating various potentials and interactions for the trapped atoms. However, in order to understand the nature of the quantum many-body state of the system that has been created with the atoms, the atoms need to be measured in an appropriate way to extract information that characterizes the state. One of the most important measurement techniques in the context of quantum simulation is the *time-of-flight measurement*, which allows one to measure the momentum distribution of the atoms.

The way that this is performed is illustrated in Fig. 9.4(a). After the suitable quantum state has been prepared with the atoms (e.g., by applying a suitable Hamiltonian), they are released from the trap, and they fall under the influence of gravity. Typically, the trap size is much smaller than the distance that the atoms fall, so to first approximation we may neglect the size of the atom cloud when it is first released from the trap. A given state of the atoms will generally consist of a variety of different momenta, and hence velocities of the atoms. Below the trap, laser light illuminates the atoms and is tuned at a frequency such that the fluorescence of the atoms can be measured. Due to the distribution of velocities of the atoms, the atom cloud will expand while falling under the influence of gravity. Suppose an atom is initially moving with a velocity $\boldsymbol{v} = (v_x, v_y, v_z)$ when it is just released from the trap. Then, the coordinates of the atom after a time t are

$$\begin{aligned} x &= v_x t \\ y &= v_y t \\ z &= v_z t - \frac{1}{2} g t^2, \end{aligned} \tag{9.41}$$

where $g = 9.8$ m/s^2 is the acceleration due to gravity taken to be in the z-direction. A measurement of the positions of the atoms (x, y, z) and the time t can thus be used to infer the velocity $\boldsymbol{v}$ (and hence the momentum) of the atoms. Despite the classical nature of the argument above, even for a quantum state being released from the trap, the momentum distribution of the atoms can be measured. In this case, the atoms undergo interference while expanding, such that the fluorescence image corresponds to a measurement in momentum space. An example of this is discussed later in Fig. 9.5.

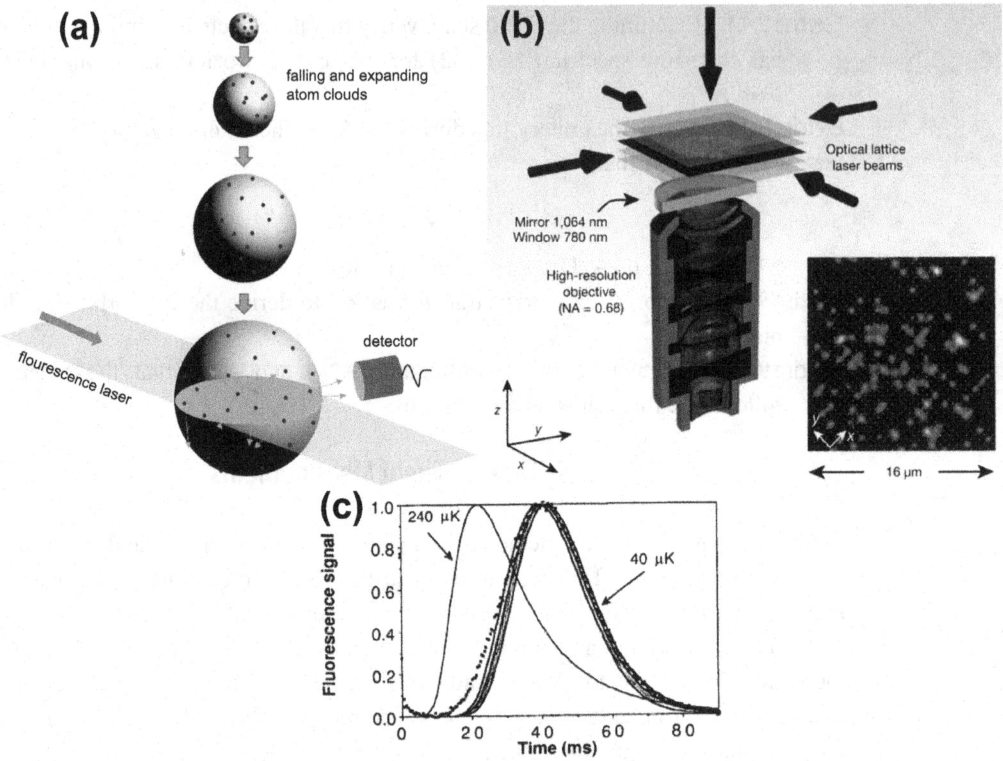

Fig. 9.4 Methods for measuring the quantum many-body state of cold atoms for quantum simulation. (a) Momentum space imaging using time-of-flight measurements (adapted from [372]). (b) Spatial imaging using the quantum gas microscope (reproduced from [426]). (c) Measurement data for cooled atoms above the BEC transition temperature showing a Maxwell–Boltzmann distribution at $T = 40\mu$K (reproduced from [299]). Color version of this figure available at cambridge.org/quantumatomoptics.

Time-of-flight measurements can also be used to infer the temperature of an atomic cloud, assuming a Maxwell–Boltzmann velocity distribution. An example of a time-of-flight measurement is shown in the inset of Fig. 9.4(c). Here, the fluorescence of the atoms is detected at a fixed distance $z = -l_0$ below the trap. The center of the fluorescence peak corresponds to atoms that had an initial velocity of $v_z = 0$ and arrive at a time $t = \sqrt{2l_0/g}$. The atoms that happened to be moving downwards, $v_z < 0$, will fluoresce earlier than the atoms that were moving upwards, $v_z > 0$, and thus we obtain a distribution to the fluorescence signal. By looking at the time distribution of the fluorescence, one can deduce the velocity distribution of the atoms, from which the temperature can be inferred.

9.4.6 Quantum Gas Microscope

Time-of-flight measurements allow one to examine the momentum space distribution of the atoms. It is then a natural question to ask whether one can also measure the spatial distribution of the atoms. Measuring the spatial distribution with high resolution is in fact a more technically demanding task than the momentum measurement. The reason is that in order to measure the spatial distribution, one requires

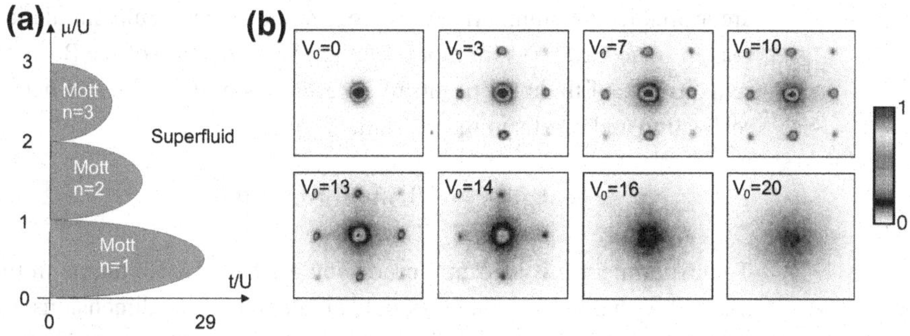

Fig. 9.5 Quantum simulation of the Bose–Hubbard model. (a) The theoretical zero-temperature phase diagram of the three-dimensional Bose–Hubbard model on a cubic lattice with $\epsilon = 0$. (b) Experimental results showing the time-of-flight measurements for the optical lattice potential amplitudes as indicated (reproduced from [177]). The potentials are measured in units of the recoil energy $\frac{\hbar^2 \kappa^2}{2m}$, where $\kappa = 2\pi/\lambda$ is the wavenumber of the optical lattice and m is the mass of an ^{87}Rb atom. Color version of this figure available at cambridge.org/quantumatomoptics.

the measurement of the BEC *in situ* (i.e., while the atoms are still in the trap). When the atoms are within the trap, they have a high optical density and the size of the condensate is very small, making absorption imaging (the most common method of measuring BECs) difficult to apply. In the first quantum simulation experiments with cold atoms, such as with the Bose–Hubbard model performed in 2002 (discussed in more detail in Section 9.5), no methods were available to image BECs at the resolution of the optical lattice period, and most of the results were inferred using time-of-flight measurements. However, from 2008 onward, several methods were developed to image BECs at the resolution of the optical lattice period.

The most well-known method of performing spatial measurements is fluorescence imaging. In fluorescence imaging, photons originating from spontaneous emission during the laser cooling cycle are collected and imaged. Using a high-resolution microscope (see Fig. 9.4), it is possible to detect the presence of atoms at the level of a single optical lattice site. In one of the first experiments to perform this (by Sherson *et al.*), the optical lattice period was 532 nm, while the wavelength of the detected light was 780 nm and the numerical aperture was NA = 0.68. This gives a optical resolution of about 700 nm, which is comparable to the optical lattice period. By collecting a sufficiently high number of photons (several thousand per pixel), one can accurately deduce the presence of an atom at a optical lattice site with fidelity over 99%.

Exercise 9.4.6 Derive the effective Hamiltonian (9.19).

9.5 Example: The Bose–Hubbard Model

One of the best-known examples of analogue quantum simulation was the realization of the Bose–Hubbard model using cold atoms realized by Greiner, Bloch, and coworkers ([177]). Since this experiment, there have been numerous examples of quantum simulation using a variety of different physical systems. Several reviews are available that summarize these realizations of quantum simulation; we refer the reader to these reviews for more details. The realization of the Bose–Hubbard model remains one of the most beautiful experiments in the field, and here we explain the realization and results in more detail.

9.5.1 The Experiment

The experiment involved first producing a BEC of ^{87}Rb atoms in the state $F = 2$, $m_F = 2$ with 2×10^5 atoms. An optical lattice in three dimensions was then applied to the BEC using three standing waves of light in orthogonal directions in a similar way as shown in Fig. 9.2. This produces a periodic potential of the form

$$V(\boldsymbol{x}) = \frac{V_0}{2}\left[\cos(2\kappa x) + \cos(2\kappa y) + \cos(2\kappa z)\right], \tag{9.42}$$

where we have generalized (9.16) to three dimensions. When the optical lattice was applied to the BEC, the desired potential V_0 was gradually increased such that the system remained in the quantum many-body ground state of the whole system. Once the lattice was applied, the atoms occupied approximately 65 lattice sites in a single direction, with approximately 2.5 atoms per site at the center. After the optical lattice was prepared, the momentum distribution of the atoms was measured using a time-of-flight measurement.

9.5.2 Effective Hamiltonian

In this section, we derive the effective Hamiltonian produced by the optical lattice. The derivation of an effective Hamiltonian is an essential step in many quantum simulation experiments, since several approximations are required to obtain the simple models that are often studied from a theoretical perspective. The same techniques can be used in this section applied to derive other related lattice models.

We start by substituting the potential (9.16) into (1.31), so that we have

$$H_0 = \int d\boldsymbol{x} a^\dagger(\boldsymbol{x}) \left(-\frac{\hbar^2}{2m}\nabla^2 + \frac{V_0}{2}\left[\cos(2\kappa x) + \cos(2\kappa y) + \cos(2\kappa z)\right] \right) a(\boldsymbol{x}), \tag{9.43}$$

where the potential is (9.42). Due to the rectangular symmetry, the equations for the x, y, and z directions are separable. First, examining the x-direction, the periodic potential implies that the eigenstates of the single-particle Hamiltonian obey Bloch's theorem and take a form

$$\psi_k^{(m)}(x) = u_k^{(m)}(x)e^{ikx}, \tag{9.44}$$

where $u_k(x) = u_k(x + \lambda/2)$ is a periodic function with periodicity of the optical lattice, and the m labels the mth band of the band structure that arises from the periodic potential. We take the lowest band to be the ground state and labeled as $m = 0$. In the y- and z-directions, the solutions also obey Bloch's theorem, and we can write the full eigenstate as

$$\Psi_{\boldsymbol{k}}^{(\boldsymbol{m})}(\boldsymbol{x}) = \psi_{k_x}^{(m_x)}(x)\psi_{k_y}^{(m_y)}(y)\psi_{k_z}^{(m_z)}(z), \tag{9.45}$$

where $\boldsymbol{m} = (m_x, m_y, m_z)$ is the band index.

The Bloch functions can be transformed in turn to give Wannier functions, defined as

$$w_m(\boldsymbol{x}-\boldsymbol{x}_j) = \int_{B_m} d\boldsymbol{k} e^{-i\boldsymbol{k}\boldsymbol{x}_j} \Psi_{\boldsymbol{k}}^{(m)}(\boldsymbol{x}), \tag{9.46}$$

where the integration is carried out over the momenta for the $\boldsymbol{m}$th band and $\boldsymbol{x}_j$ denotes the location of the jth lattice site. For example, for the lowest energy band, the range of $\boldsymbol{k}$ is $k_{x,y,z} \in [-\kappa, \kappa]$. The Wannier functions can be considered to be a localized version of Bloch functions, and they obey orthogonality properties:

$$\int d\boldsymbol{x} w_{\boldsymbol{m}}^*(\boldsymbol{x}-\boldsymbol{x}_j) w_{\boldsymbol{m}'}(\boldsymbol{x}-\boldsymbol{x}_{j'}) = \delta_{jj'}\delta_{\boldsymbol{m}\boldsymbol{m}'}. \tag{9.47}$$

The full set of Wannier functions form a complete set, which serves to expand the bosonic operators $a(x)$ in a similar way to (1.32). Expanding the operators according to

$$a(\boldsymbol{x}) = \sum_j \sum_{\boldsymbol{m}} w_{\boldsymbol{m}}(\boldsymbol{x}-\boldsymbol{x}_j) a_{j\boldsymbol{m}}, \tag{9.48}$$

we may substitute this into (9.43) such that

$$H_0 = \sum_{jj'} \sum_{\boldsymbol{m}\boldsymbol{m}'} T_{j,j'}^{\boldsymbol{m},\boldsymbol{m}'} a_{j\boldsymbol{m}}^\dagger a_{j'\boldsymbol{m}'}, \tag{9.49}$$

where the hopping matrix elements are defined as

$$T_{j,j'}^{\boldsymbol{m},\boldsymbol{m}'} = \int d\boldsymbol{x} w_{\boldsymbol{m}}^*(\boldsymbol{x}-\boldsymbol{x}_j) \left[-\frac{\hbar^2}{2m}\nabla^2 + V(\boldsymbol{x})\right] w_{\boldsymbol{m}'}(\boldsymbol{x}-\boldsymbol{x}_{j'}) \tag{9.50}$$

$$= \frac{\delta_{\boldsymbol{m}\boldsymbol{m}'}}{2\kappa} \int_{B_m} d\boldsymbol{k} E_{\boldsymbol{k}} e^{i\boldsymbol{k}(\boldsymbol{x}_j-\boldsymbol{x}_{j'})}. \tag{9.51}$$

Here, we substituted the definition of the Wannier functions in the first line and used the fact that the Bloch functions are eigenstates of the Hamiltonian $H_0\Psi_{\boldsymbol{k}}^{(m)}(\boldsymbol{x}) = E_{\boldsymbol{k}}\Psi_{\boldsymbol{k}}^{(m)}(\boldsymbol{x})$.

The expression (9.49) constitutes a change of basis from the position basis of the bosons to the Wannier basis, and no approximations have been made so far. Typically, one makes further assumptions to simplify the Hamiltonian. For example, for sufficiently low temperatures, the bosons will not occupy the higher energy bands, and one can consider the lowest band alone. Furthermore, since Wannier integrals (9.50) involve the overlap of two localized functions, we expect the magnitude of the integrals to decrease with $|\boldsymbol{x}_j - \boldsymbol{x}'_j|$. Taking only nearest neighbor (NN) terms, the Hamiltonian (9.49) is

$$H_0 \approx -t \sum_{j,j' \in \text{NN}} \left(a_j^\dagger a_{j'} + a_{j'}^\dagger a_j\right) + \epsilon \sum_j a_j^\dagger a_j, \tag{9.52}$$

where

$$\begin{aligned} t &\equiv -T_{j,j'}^{\boldsymbol{0},\boldsymbol{0}} \\ &= -\int d\boldsymbol{x} w_{\boldsymbol{m}}^*(\boldsymbol{x}-(\boldsymbol{x}_j-\boldsymbol{x}_{j'})) \left[-\frac{\hbar^2}{2m}\nabla^2 + V(\boldsymbol{x})\right] w_{\boldsymbol{m}'}(\boldsymbol{x}) \end{aligned} \tag{9.53}$$

is the nearest neighbor hopping and $\epsilon \equiv T_{j,j}^{\boldsymbol{0},\boldsymbol{0}}$. In (9.53), we use the fact that there is a translational symmetry such that t is the same for all lattice sites, which is also true of ϵ. Such a Hamiltonian takes the form of a typical lattice model that might be considered in many strongly correlated condensed matter problems. Although we have used bosons to derive the Hamiltonian (9.49), we note that the same steps can be used to derive a fermionic Hamiltonian, starting from fermion operators (1.5).

In the above, we have only considered the kinetic energy component of the Hamiltonian. We must also consider the interaction Hamiltonian, as defined by (4.10). In (1.44), a transformation was made into the basis of the single-particle Hamiltonian that traps the atoms. Due to the presence of the optical lattice, the relevant single-particle basis is now the Bloch wavefunction, or equivalently, the Wannier basis. As such, we substitute (9.48) into (4.10). In a similar way to (1.44), we obtain

$$H_I = \frac{1}{2} \sum_{j_1 j_2 j_3 j_4} \sum_{\boldsymbol{m}_1 \boldsymbol{m}_2 \boldsymbol{m}_3 \boldsymbol{m}_4} g_{j_1,j_2,j_3,j_4}^{\boldsymbol{m}_1,\boldsymbol{m}_2,\boldsymbol{m}_3,\boldsymbol{m}_4} a_{j_1 \boldsymbol{m}_1}^\dagger a_{j_2 \boldsymbol{m}_2}^\dagger a_{j_3 \boldsymbol{m}_3} a_{j_4 \boldsymbol{m}_4}, \tag{9.54}$$

where

$$g_{j_1,j_2,j_3,j_4}^{\boldsymbol{m}_1,\boldsymbol{m}_2,\boldsymbol{m}_3,\boldsymbol{m}_4} = \int d\boldsymbol{x} d\boldsymbol{x}' w_{\boldsymbol{m}_1}^*(\boldsymbol{x} - \boldsymbol{x}_{j_1}) w_{\boldsymbol{m}_2}^*(\boldsymbol{x}' - \boldsymbol{x}_{j_2}) \times U(\boldsymbol{x}, \boldsymbol{x}') w_{\boldsymbol{m}_3}(\boldsymbol{x} - \boldsymbol{x}_{j_3}) w_{\boldsymbol{m}_4}(\boldsymbol{x}' - \boldsymbol{x}_{j_4}). \tag{9.55}$$

For the case of the contact potential (4.13), we obtain

$$g_{j_1,j_2,j_3,j_4}^{\boldsymbol{m}_1,\boldsymbol{m}_2,\boldsymbol{m}_3,\boldsymbol{m}_4} = \frac{4\pi\hbar^2 a_s}{m} \int d\boldsymbol{x} w_{\boldsymbol{m}_1}^*(\boldsymbol{x} - \boldsymbol{x}_{j_1}) w_{\boldsymbol{m}_2}^*(\boldsymbol{x} - \boldsymbol{x}_{j_2}) w_{\boldsymbol{m}_3}(\boldsymbol{x} - \boldsymbol{x}_{j_3}) w_{\boldsymbol{m}_4}(\boldsymbol{x} - \boldsymbol{x}_{j_4}). \tag{9.56}$$

Again, due to similar assumptions as above, often one considers only the lowest band, $\boldsymbol{m}_1 = \boldsymbol{m}_2 = \boldsymbol{m}_3 = \boldsymbol{m}_4 = \boldsymbol{0}$. Due to the localized nature of the Wannier functions, the largest matrix element will correspond to the case when $j_1 = j_2 = j_3 = j_4$. This largest term is called the on-site interaction and can be defined as

$$U = g_{j,j,j,j}^{\boldsymbol{0},\boldsymbol{0},\boldsymbol{0},\boldsymbol{0}} = \frac{4\pi\hbar^2 a_s}{m} \int d\boldsymbol{x} |w_{\boldsymbol{0}}(\boldsymbol{x})|^4. \tag{9.57}$$

Since $|w_{\boldsymbol{0}}(\boldsymbol{x})|^2$ is a normalized probability function for the lowest energy band, it has dimensions of inverse length. This means that the smaller the spatial extent of the Wannier function, the larger the integral will be. A Wannier function with a small spatial extent can be achieved by increasing the laser intensity, thereby producing a larger potential V_0, which has a tighter potential for each site.

To lowest order, the interaction part of the Hamiltonian can be written as

$$H_I \approx \frac{U}{2} \sum_j a_j^\dagger a_j^\dagger a_j a_j. \tag{9.58}$$

Combined with (9.52), this Hamiltonian corresponds to a model of bosons that are hopping on a lattice. The total Hamiltonian

$$\begin{aligned} H &= H_0 + H_I \\ &= -t \sum_{j,j' \in \mathrm{NN}} \left(a_j^\dagger a_{j'} + a_{j'}^\dagger a_j \right) + \epsilon \sum_j a_j^\dagger a_j + \frac{U}{2} \sum_j a_j^\dagger a_j^\dagger a_j a_j \end{aligned} \tag{9.59}$$

is called the Bose–Hubbard model.

9.5.3 Experimental Observation of the Phase Transition

The Bose–Hubbard model has been well studied in condensed matter physics. The model has primarily two phases: a Mott insulating phase and a superfluid phase. The superfluid phase is characterized by long-range phase coherence, as is present in a BEC. The prototypical superfluid state occurs in the limit of $U = 0$ and takes the form

$$|\Phi(U = 0)\rangle = \frac{1}{\sqrt{N!}}\left(\frac{1}{\sqrt{M}}\sum_{j=1}^{M} a_j^\dagger\right)^N |0\rangle, \tag{9.60}$$

where N is the total number of bosons in the system. In this state, each boson is in a superposition across all lattice sites and all N bosons occupy this same state. In the reverse limit, where the interactions are large and the kinetic energy is small, the interactions cause the bosons to localize and there is a fixed number of atoms per lattice site. The prototypical state occurs when $t = 0$, and the Mott insulator state takes the form

$$|\Phi(t = 0)\rangle = \prod_j \frac{\left(a_j^\dagger\right)^n}{\sqrt{n!}}|0\rangle, \tag{9.61}$$

where we have assumed there are n bosons per site, which is also called the filling factor.

The schematic phase diagram of the model is shown in Fig. 9.5(a). The phase diagram consists of several separated regions (usually called "lobes") corresponding to different filling factors n. For $t = 0$, the ground state is always in a Mott insulating phase. The chemical potential μ controls the number of particles in the system. As the chemical potential increases, the lattice fills up in a stepwise fashion, where the number of bosons per site increases by one per lobe. To understand this, set $t = 0$ in (9.59) and add a chemical potential term,

$$H(t = 0) - \mu N = \epsilon \sum_j n_j + \frac{U}{2}\sum_j n_j(n_j - 1) - \mu \sum_j n_j, \tag{9.62}$$

where we have written pairs of bosonic operators as number operators $n_j = a_j^\dagger a_j$. Assuming M lattice sites, the energy can be written as

$$E_0(t = 0, n) = M\left(n(\epsilon - \mu) + \frac{U}{2}n(n-1)\right). \tag{9.63}$$

Since the chemical potential is the energy that is available per particle on a lattice site, let us calculate what additional energy is required to add an extra boson from n to $n + 1$ per lattice site. We can evaluate

$$\frac{E_0(t = 0, n + 1) - E_0(t = 0, n)}{M} = \epsilon - \mu + Un. \tag{9.64}$$

The filling corresponds to when (9.64) gives zero, hence we obtain the filling factor to particle number relation

$$\mu = \epsilon + Un, \tag{9.65}$$

which increases in units of U, as shown in the phase diagram Fig. 9.5(a).

For a particular filling factor n, starting from a Mott insulator phase with $t = 0$, as U is decreased and/or t is increased, at some point there is a phase transition to a superfluid state. The precise value of t/U depends upon the chemical potential, as shown in Fig. 9.5(a). For the $n = 1$ Mott lobe, the phase transition occurs at $U/t \approx 29$ for a three-dimensional cubic lattice. For larger filling factors, the phase transitions occur at smaller values of U/t.

In the experiment, the ratio of t/U can be changed by changing the lattice potential V_0. As the potential V_0 is increased, this has the effect of making the Wannier functions $w_m(\boldsymbol{x})$ more localized. Examining the formulas for t and U, (9.53) and (9.57), respectively, we can deduce that as V_0 is increased, t will decrease, while U will increase. The hopping energy t decreases because the overlap between the Wannier functions $w_m(\boldsymbol{x} - (\boldsymbol{x}_j - \boldsymbol{x}_{j'}))$ and $w_m(\boldsymbol{x})$ decreases due to the increased localization. Meanwhile, (9.57) increases because the square of a normalized probability distribution will increase with localization. For example, the square of a normalized Gaussian distribution is

$$\int p^2_{\text{Gauss}}(x) = \frac{1}{2\sigma\sqrt{\pi}}, \tag{9.66}$$

where

$$p_{\text{Gauss}}(x) = \frac{e^{-x^2/2\sigma^2}}{\sqrt{2\pi\sigma}}, \tag{9.67}$$

and thus is inversely proportional to the standard deviation. In this way, by increasing V_0, it has the effect of traversing the phase diagram of Fig. 9.5(a) from left to right.

Returning to the experiment, the time-of-flight images with various amplitudes of the potentials are shown in Fig. 9.5(b). For $V_0 = 0$, no optical lattice is applied, and the momentum distribution corresponds to a single peak at $\boldsymbol{k} = 0$. As the periodic potential amplitude is applied, the state corresponds to the lattice superfluid state (9.61), with a coherent superposition across lattice sites. This produces an interference pattern much like that observed from a diffraction grating, where the sources are the lattice sites and the interference occurs due to matter wave interference. Another way to view this is that the momentum representation of the Bloch states (9.45) contains components of integer multiples of the frequency κ of the lattice. As the potential is further increased, the characteristic interference pattern is lost, due to the Mott insulator transition. At this point, there is no longer any diffraction grating-like interference, since there is no phase coherence between sites in a state like (9.61). The pattern changes to the momentum distribution of atoms that are localized in position space with the dimensions of the optical lattice spacing. Taking the spatial distribution to be a Gaussian (9.67) with $\sigma = \lambda = 2\pi/\kappa$, the momentum space distribution should be another Gaussian with approximate standard deviation $\propto \pi^2/\sigma = \kappa/2\pi$, which is the observed result.

Exercise 9.5.1 Verify the orthogonality relations of the Wannier functions (9.47).

Exercise 9.5.2 Verify the expression for the hopping energies (9.51).

Exercise 9.5.3 Verify that the states (9.60) and (9.61) are the ground states of the Bose–Hubbard Hamiltonian (9.59) in their respective limits.

9.6 References and Further Reading

- Section 9.1: Review articles and commentaries on quantum simulation [165, 184, 65, 45, 44, 411, 91, 46, 358, 302, 82]. Feynman's conjecture, first introducing the idea of simulating quantum systems with quantum computers [146]. Review article on quantum Monte Carlo [150], DMRG [488, 416], density functional theory [249], dynamical mean field theory [164], series expansion methods [353], lattice QCD [390], and quantum chemistry [300, 325].
- Section 9.2: Original works showing the efficiency of quantum simulation [312, 146]. Original work exactly solving the one-dimensional transverse Ising model [371]. Finite size analysis of the transverse Ising model [198]. A textbook introduction to spin systems [83].
- Section 9.3: Original works introducing digital quantum simulation [312, 490]. Review articles discussing digital quantum simulation [165, 91]. Experimental demonstrations of digital quantum simulation [289, 28, 290, 413]. Theoretical work showing method of performing digital quantum simulation of lattice gauge theory [75]. Original works introducing the Suzuki–Trotter expansion [469, 447, 448].
- Section 9.4.1: Original works experimentally implementing optical lattices [473, 245, 212, 187, 213, 256]. Review articles and commentaries discussing optical lattices [182, 43, 44, 46, 45, 358]. Optical lattices have found a variety of different uses in addition to quantum simulation, including atom sorting and transport [336, 318], investigating physics of Anderson localization [41, 417] and Bose–Fermi mixtures [303], optical lattice clocks [450, 296, 471, 47], lattice solitons [132, 500], molecular formation [444], and producing three-dimensional atomic structures with tweezers [30].
- Section 9.4.2: For references relating to Feshbach resonances, see Section 4.6. Experimental realization of the BEC–BCS crossover with cold fermionic atoms [391, 53, 178, 392, 506]. Theoretical proposal and analysis of the atomic BEC–BCS crossover [351, 175]. Review articles on the BEC–BCS crossover [88, 507, 446]. Experimental observation of condensation of molecules using a Feshbach resonance [508].
- Section 9.4.3: Review articles discussing artificial gauge fields induced by rotation [144, 96, 94]. Review article discussing methods of producing artificial gauge fields for neutral atoms [105, 174]. Experimental works demonstrating synthetic fields for cold atoms [307, 306]. Original works introducing Berry's phase [359, 313, 39]. Further works discussing Berry phases in adiabatic evolution [40, 127, 326, 238].
- Section 9.4.4: Experiments realizing spin-orbit coupling in a BEC [308, 14, 495, 304]. Theoretical works analyzing spin-orbit coupling in BECs [356, 404, 125, 437, 232, 220, 478]. Review articles on spin-orbit coupling [204, 160, 504, 505].
- Section 9.4.5: Experiment demonstrating the time-of-flight method [299]. Theoretical calculations modeling the time-of-flight method [501, 64]. Review articles discussing time-of-flight measurements [372, 43].

- Section 9.4.6: Experiments demonstrating the quantum gas microscope [26, 426, 89, 196, 496]. Review articles and books on the quantum gas microscope [275, 358]. Experimental realization of scanning electron microscopy of a cold atomic gas [167].
- Section 9.5: Experimental realizations of the Bose–Hubbard model with cold atoms [177, 77, 487]. Theoretical proposals of the Bose–Hubbard model quantum simulator with cold atoms [239, 327]. Theoretical investigations of the physics of the Bose–Hubbard model [79, 328, 410, 158, 8]. Other examples of theoretical proposals for quantum simulation [400, 108, 107, 72]. Other examples of experimental realizations of quantum simulation [350, 360, 443, 498, 415, 87, 305, 288, 111, 244, 214, 292, 379, 283, 203, 113, 452, 38, 484]. Experimental realization of slow light [205, 309, 282, 55]. Review articles relating to cold atom quantum simulators [126, 453].

10 Entanglement Between Atom Ensembles

10.1 Introduction

Up to this point, we have primarily dealt with entanglement only within a single atomic ensemble or Bose–Einstein condensate (BEC). For example, for spin squeezed BECs, we introduced a criterion for detecting entanglement between the particles in Section 5.8. However, one of the distinctive features of entanglement is the nonlocal aspect, which Einstein famously called a "spooky action at a distance." In fact, the notion of entanglement introduced in Section 5.8 is a rather ambiguous one, since the bosons within a BEC are one composite system. A more conventional notion of entanglement is between the states of two spatially distinct particles. In this chapter, we will explore entanglement defined between two atomic ensembles. Due to the large number of degrees of freedom, we will see that the types of entanglement that can be created are much more complex than for qubits. We will also describe various ways that such entanglement can be quantified and detected.

10.2 Inseparability and Quantifying Entanglement

First, let us define precisely what it means for two systems to be entangled. Once it is defined, we will want to find measures of entanglement such that we can put a numerical value on the amount of entanglement present in a quantum state. We will treat pure states and mixed states differently, since each has a different approach to quantifying entanglement.

10.2.1 Pure States

Consider that the wavefunction consists of two subsystems, labeled by A and B, with basis states $|n\rangle_A$ and $|m\rangle_B$ in each. The Hilbert space dimension of each of the subsystems separately is $D_{A,B}$, such that the total dimension is $D = D_A D_B$. The most general pure state that can be written in this case is

$$|\Psi\rangle = \sum_{nm} \Psi_{nm} |n\rangle_A |m\rangle_B. \tag{10.1}$$

A subclass of the general pure state is the *product state*, which takes the form

$$|\psi\rangle_A \otimes |\phi\rangle_A = \sum_{nm} \psi_n \phi_m |n\rangle_A |m\rangle_B, \tag{10.2}$$

where we have expanded the product state wavefunction in terms of the basis states with coefficients ψ_n, ϕ_m. Clearly, it is possible to choose $\Psi_{nm} = \psi_n \phi_m$, but there are states of the form (10.1) that cannot be written in the form (10.2). Examples of such states are the Bell states:

$$\frac{1}{\sqrt{2}}(|0\rangle_A|0\rangle_B \pm |1\rangle_A|1\rangle_B)$$
$$\frac{1}{\sqrt{2}}(|0\rangle_A|1\rangle_B \pm |1\rangle_A|0\rangle_B). \tag{10.3}$$

The product states are a special class of states that by definition have no entanglement. Thus, a definition of an entangled pure state is any state that cannot be written in the form of (10.2).

For a pure bipartite system as we consider here, a convenient and complete way of quantifying the amount of entanglement is using the *von Neumann entropy*. This is defined by

$$E = -\mathrm{Tr}(\rho_A \log \rho_A), \tag{10.4}$$

where

$$\rho_A = \mathrm{Tr}_B(|\Psi\rangle\langle\Psi|) \tag{10.5}$$

is the reduced density matrix over subsystem B. The operation Tr_B denotes a partial trace, where the trace is only taken over the variable in subsystem B. For our state (10.1), the reduced density matrix is

$$\rho_A = \sum_{nn'm} \Psi_{nm}\Psi^*_{n'm}|n\rangle_A\langle n'|_A. \tag{10.6}$$

In (10.4), the logarithm denotes the matrix logarithm, and the base of the logarithm can be chosen by convenience. Common choices are base 2, which are used for qubit systems. To evaluate (10.4), it is most convenient to work in the diagonalized basis of the reduced density matrix ρ_A. Suppose (10.5) is diagonalized according to

$$\rho_A = \sum_n \lambda_n|\lambda_n\rangle\langle\lambda_n|. \tag{10.7}$$

Then, the von Neumann entropy can be written

$$E = -\sum_n \lambda_n \log \lambda_n. \tag{10.8}$$

This is a nonnegative quantity that takes a maximum value

$$E_{\max} = \log D_{\min}, \tag{10.9}$$

with $D_{\min} = \min(D_A, D_B)$. Then, the normalized von Neumann entropy is

$$\frac{E}{E_{\max}} = -\mathrm{Tr}(\rho_A \log_{D_{\min}} \rho_A). \tag{10.10}$$

Thus, taking the base of the logarithm to be the minimum dimension of A and B, we obtain a quantity that ranges from 0 to 1.

10.2.2 Mixed States

For pure states, defining and quantifying entanglement is relatively straightforward using product states and the von Neumann entropy. More generally, a quantum state is described by a mixed state. A mixed state consists of a statistical mixture of different quantum states and can arise in practice due to nonzero temperature or the presence of decoherence or noise. Thus, any quantum state in a realistic experimental system is usually described by a mixed state, since it is not possible to perfectly prepare a pure state.

The most general bipartite mixed state is written as

$$\rho = \sum_i P_i |\Psi_i\rangle\langle\Psi_i|, \tag{10.11}$$

where $|\Psi_i\rangle$ is a set of orthogonal bipartite states on A and B as before, and P_i are the probabilities of obtaining them. The state (10.11) could describe an ensemble of bipartite quantum states in several ways. The state could, for example, describe an ensemble of bipartite systems where there are numerous copies of the quantum system. The states $|\Psi_i\rangle$ occur with a relative population P_i. An example of this would be a nuclear magnetic resonance (NMR) experiment, where there are many molecules that all constitute the bipartite system. Another way that (10.11) could be realized is that there is only one quantum system, but the experiment is run numerous times. Each run of the experiment gives a different quantum state probabilistically, and P_i describes the statistics of obtaining each state.

A separable state is a mixed state consisting of the product states (10.2). The general form of a separable state is

$$\rho_{\text{sep}} = \sum_i p_i \rho_i^A \otimes \rho_i^B. \tag{10.12}$$

We may now give a more general definition of entanglement. Any state that cannot be written in the form (10.12) is then defined to be an entangled state. This is a more general definition than that of Section 10.2.1, because pure states are a specific example of the mixed states shown here: (10.1) is an instance of (10.11), and (10.2) is an instance of (10.12).

Unlike the pure state case, where there is a relatively simple formula (10.4) that quantifies the amount of entanglement, for mixed states it is more difficult even to give a test for finding whether a state is entangled or not. The most general way to do this is to search the full space of separable states and check whether it can be written in the form (10.12). This approach is quite numerically intensive, however, due to the large number of parameters that specify the general separable state. A more convenient method would be to have a formula, much like the von Neumann entropy, that depends only on the density matrix so that one may find the amount of entanglement.

One of the most popular methods to achieve this is the Peres–Horodecki criterion. To define this criterion, first write a general bipartite density matrix as

$$\rho = \sum_{n'm'nm} \rho_{n'm'nm} |n'\rangle_A |m'\rangle_B \langle n|_A \rangle m|_B, \tag{10.13}$$

where we have simply expanded all the matrix elements in terms of the basis states $|n\rangle_A$ and $|m\rangle_B$. One of the properties of a density matrix is that its eigenvalues are always positive, since it can always be written in the form (10.11). This property can be stated as the positive semidefiniteness of the density matrix:

$$\rho \geq 0. \tag{10.14}$$

Now define the partial transpose of the density matrix as

$$\begin{aligned}\rho^{T_B} &= \sum_{n'm'nm} \rho_{n'm'nm} |n'\rangle_A |m\rangle_B \langle n|_A \langle m'|_B \\ &= \sum_{n'm'nm} \rho_{n'mnm'} |n'\rangle_A |m'\rangle_B \langle n|_A \langle m|_B.\end{aligned}$$

This involves taking the transpose only on subsystem B. The Peres–Horodecki criterion then states that if ρ is separable, then its partial transpose is also a density matrix and is positive semidefinite:

$$\rho^{T_B} \geq 0. \qquad \text{(if } \rho \text{ is separable)} \tag{10.15}$$

Hence, if ρ^{T_B} has any negative eigenvalues, ρ must be entangled. Due to this test involving the positivity of the partial transpose (PPT), it is also often called the *PPT criterion*. We note that although we took the partial transpose with respect to B above, we could equally have done so with A instead, and the same results would be obtained.

The above procedure can be extended to quantify the amount of entanglement in a straightforward way. Since the signature of entanglement is the presence of negative eigenvalues of ρ^{T_B}, one would expect that the more entangled the state is, the more negative the eigenvalues are. A suitable measure of entanglement is the sum of only the negative eigenvalues, which is called the *negativity*, defined as

$$N = \frac{1}{2} \sum_i (|\lambda_i| - \lambda_i). \tag{10.16}$$

Here, λ_i are the eigenvalues of ρ^{T_B}. In matrix invariant form, this can be equivalently be written as

$$N = \frac{||\rho^{T_B}||_1 - 1}{2}, \tag{10.17}$$

where $||X||_1 = \text{Tr}\sqrt{X^\dagger X}$ is the trace norm of a matrix. We used the fact that the sum of the eigenvalues of the partial transposed density matrix still adds to 1:

$$\text{Tr}(\rho^{T_B}) = 1. \tag{10.18}$$

The negativity is a nonnegative number in the range

$$0 \leq N \leq \frac{D_{\min} - 1}{2}, \tag{10.19}$$

where $D_{\min} = \min(D_A, D_B)$.

It is sometimes more natural to use a slightly different form of the negativity with the same basic idea, so that it has a form more similar to the von Neumann entropy (10.4). In this case, we define the *logarithmic negativity* as

$$E_N = \log ||\rho^{T_B}||_1. \tag{10.20}$$

The logarithmic negativity is also a nonnegative number in the range

$$0 \leq E_N \leq \log D_{\text{min}}, \tag{10.21}$$

which has the same range as the von Neumann entropy (10.9).

The above shows that the PPT criterion is simple to calculate and gives a quantification of entanglement for mixed states in the form of negativity and logarithmic negativity. It has unfortunately one problem, that it is only a *necessary* condition for separability for systems larger than 2×2 or 2×3. This means that there are states that are in fact entangled but do not violate the PPT criterion. In other words, there are states that $N = E_N = 0$, despite the fact that they are in fact entangled. However, if the PPT criterion is violated and $N > 0$ or $E_N > 0$, one can know for sure that the state is entangled. For system dimensions that are 2×2 or 2×3, the PPT criterion is necessary and sufficient, meaning that there is no ambiguity for states that do not violate the PPT criterion. In practice, the PPT criterion can detect many entangled states, hence it provides a powerful and convenient way of detecting entanglement for mixed states.

Exercise 10.2.1 Why doesn't the von Neumann entropy (10.4) work as an entanglement measure for a mixed state? Show that it cannot work by contradiction by evaluating the entropy for a separable state (10.12).

Exercise 10.2.2 Evaluate the von Neumann entropy and logarithmic negativity for the state

$$|\Psi\rangle = \cos\theta|0\rangle_A|0\rangle_B + \sin\theta|1\rangle_A|1\rangle_B. \tag{10.22}$$

Does the von Neumann entropy and logarithmic negativity have the same value for all θ?

10.3 Correlation-Based Entanglement Criteria

The entanglement measures of Section 10.2.2 give a straightforward way of evaluating the amount of entanglement between two systems. For systems with relatively small Hilbert space dimensions (e.g. two qubits), it is within the capabilities of modern experiments to perform measurements on a given quantum state such that the density matrix is reconstructed. But when the system dimension is much larger, such as with an atomic gas or BEC, this becomes impractical very quickly. To see this, recall that for a D-dimensional density matrix, there are $D^2 - 1$ free parameters. Thus, to tomographically reconstruct a two-qubit density matrix, one requires $4^2 - 1 = 15$ independent measurements. On the other hand, to fully reconstruct the density matrix of 10 qubits, we require 1,048,575 independent measurements!

Therefore, it is more practical for larger systems to find other ways of detecting entanglement that do not require a complete set of measurements. Correlation-based entanglement criteria provide exactly such a way of detecting entanglement. Typically, the way that these are formulated is in the following way. First, a mathematical statement about any separable state is made, often in the form of an inequality. Then, this expression is evaluated for an entangled state, and if it is violated, it is concluded that the state is entangled. Several such criteria have been formulated in the literature; here we provide some of the more well-known methods.

10.3.1 Variance-Based Criteria

The first criterion we introduce was derived first for quantum optical states, where the natural observables are position and momentum operators, as seen in (5.13). Entanglement occurs in such systems between two modes labeled by $i \in \{A, B\}$, for which we define the observables as

$$\begin{aligned} x_i &= \frac{1}{\sqrt{2}}(a_i + a_i^\dagger) \\ p_i &= -\frac{i}{\sqrt{2}}(a_i - a_i^\dagger). \end{aligned} \tag{10.23}$$

The *Duan–Giedke–Cirac–Zoller criterion* then states that

$$\sigma_u^2 \sigma_v^2 \geq \xi^2 |\langle [x_A, p_A] \rangle| + \frac{|\langle [x_B, p_B] \rangle|}{\xi^2}. \qquad \text{(for separable states)} \tag{10.24}$$

In the case of the position and momentum operators, we can immediately evaluate the right-hand side to give

$$\sigma_u^2 \sigma_v^2 \geq \xi^2 + \frac{1}{\xi^2}, \qquad \text{(for separable states)} \tag{10.25}$$

where σ_u^2, σ_v^2 are the variances of the operators

$$\begin{aligned} u &= |\xi| x_A + \frac{x_B}{\xi} \\ v &= |\xi| p_A - \frac{p_B}{\xi} \end{aligned} \tag{10.26}$$

and ξ is an arbitrary nonzero parameter. If it is found that (10.25) is violated, then the state is entangled. The criterion is derived using a combination of the uncertainty relations between x_i and p_i and the Cauchy–Schwarz inequality.

The above is defined in terms of observables of quantum optical states. How can we apply it to spin systems? As we saw in Section 5.9, it is possible to transform spin variables into bosonic operators using the Holstein–Primakoff transformation. Due to the square root factors in (5.118), this cannot be done perfectly. In the case that $N \gg 1$, we can approximate the square root

$$\sqrt{N - a^\dagger a} \approx \sqrt{N} - \frac{a^\dagger a}{2} \dots \tag{10.27}$$

using a Taylor expansion. Taking just the leading order term, we have

$$\begin{aligned} S_+ &\approx \sqrt{N} a^\dagger \\ S_- &\approx \sqrt{N} a. \end{aligned} \tag{10.28}$$

To ensure that this is a reasonable approximation, we should assume that $\langle a^\dagger a \rangle$ is small, which implies that

$$\langle S_z \rangle \approx -N \tag{10.29}$$

from (5.116). We can then approximate the position and momentum operators as

$$x \approx \frac{S_- + S_+}{\sqrt{2N}} = \frac{S_x}{\sqrt{2N}}$$
$$p \approx -i\frac{(S_- - S_+)}{\sqrt{2N}} = -\frac{S_y}{\sqrt{2N}} \tag{10.30}$$

for the ensembles A and B. These variables can be used in (10.26) to evaluate the criterion. To evaluate (10.26), we only require second-order correlations in the spin operators – a much easier task than tomographically reproducing the full density matrix.

In the above, we obtained minus signs in the polarization direction and the momentum. We can equally polarize the spins in the reverse direction and redefine the operators in different way. For S_z operators polarized in the positive direction, we have

$$[S_x, S_y] = 2iS_z \approx 2iN. \tag{10.31}$$

Defining the position and momentum variables as in (10.30) without the minus sign, we have variables that obey canonical commutation relations $[x, p] = i$. Similar types of inequalities to (10.25) were derived with tighter bounds, which can be used as more sensitive detectors of entanglement [171, 222].

10.3.2 Hillery–Zubairy Criteria

Another correlation-based entanglement criterion was originally discovered in the context of quantum optics. The first *Hillery–Zubairy criterion* states that

$$|\langle a_A a_B \rangle|^2 \leq \langle a_A^\dagger a_A \rangle \langle a_B^\dagger a_B \rangle. \qquad \text{(for separable states)} \tag{10.32}$$

In order to conclude that entanglement is present, one requires a violation of the inequality, which requires that the left-hand side is as large as possible. Since the expectation value on the left-hand side destroys two bosons simultaneously, one typically requires a state with quantum correlations that are a superposition of states that differ by an a_A and a a_B photon. A two-mode squeezed state takes exactly the form

$$\sqrt{1 - \tanh^2 r} \sum_{n=0}^{\infty} \tanh r |n\rangle_A |n\rangle_B, \tag{10.33}$$

where r is the squeezing parameter and the states $|n\rangle$ are photonic Fock states (1.17).

A similar type of inequality can be written where the correlations on the left-hand side are of the form where the total boson number is conserved. The second Hillery–Zubairy criterion states that

$$|\langle a_A a_B^\dagger \rangle|^2 \leq \langle a_A^\dagger a_A a_B^\dagger a_B \rangle. \qquad \text{(for separable states)} \tag{10.34}$$

Again, to conclude that entanglement is present, one must have a larger correlation on the left-hand side of the inequality than on the right. In order to apply this to spin systems, one may apply the Holstein–Primakoff approximation, as in the last section. Alternatively, it can be directly applied to systems involving two spatial modes of a BEC – for example, in a double-well potential.

10.3.3 Entanglement Witness

In the previous sections, several methods were given that relied upon calculating particular correlations of specific operators. In a general experiment, the specific operators and correlations may not be available, in which case they cannot be used. One may require the construction of an entanglement witness based on some arbitrary correlations of observables. The *entanglement witness* approach provides such a general approach of determining whether entanglement is present from a set of observables of the form

$$\langle \xi_i^A \otimes \xi_j^B \rangle, \tag{10.35}$$

where the ξ_i^A and ξ_j^B label a set of operators on subsystems A and B, respectively. In the approach, one constructs a witness operator,

$$W = \sum_{i=1}^{M_A} \sum_{j=1}^{M_B} c_{ij} \xi_i^A \otimes \xi_j^B, \tag{10.36}$$

where c_{ij} are real coefficients to be determined and M_A and M_B are the number of operators in A and B, respectively. The procedure is then to optimize the coefficients such that

$$\begin{aligned} &\text{Minimize } \langle W \rangle \\ &\text{Subject to: (1) } W = P + Q^{T_A} \\ &\text{(2) } P \geq 0 \\ &\text{(3) } Q \geq 0 \\ &\text{(4) } \mathrm{Tr}(W) = 1. \end{aligned} \tag{10.37}$$

If it is found that $\langle W \rangle < 0$, then the state is entangled.

Generally, the optimization is performed numerically such that the coefficients are obtained iteratively. Given a set of measurement operators ξ_i^A and ξ_j^B (which may involve the identity), one randomly chooses a set of coefficients c_{ij}. One then normalizes the coefficients to satisfy condition (4). If it is found that W satisfies condition (1), then one evaluates $\langle W \rangle$ to see if it is negative. If it is negative, then the state is entangled; if it is positive, another trial version of W is generated and the process is repeated.

10.3.4 Covariance Matrix

Another useful correlation-based method is based on constructing a covariance matrix. A general covariance matrix is defined with respect to a set of Hermitian operators ξ_i, where $i \in [1, M]$. The $M \times M$ covariance matrix is defined as

$$V_{jk} \equiv \frac{1}{2}\langle \{\xi_j, \xi_k\} \rangle - \langle \xi_j \rangle \langle \xi_k \rangle, \tag{10.38}$$

which is a real symmetric matrix. We may also define the commutation relations between these operators by the $M \times M$ commutation matrix, defined as

$$\Omega_{jk} \equiv -i \langle [\xi_j, \xi_k] \rangle, \tag{10.39}$$

which is a real antisymmetric matrix.

The covariance matrix and commutation matrix can be put together to give a generalized statement of the uncertainty relation. The matrix inequation

$$V + \frac{i}{2}\Omega \geq 0 \tag{10.40}$$

succinctly summarizes the uncertainty relation between the operators ξ_j. The meaning of (10.40) is in terms of the semipositive nature of the matrix (i.e., that it has no negative eigenvalues). Equation (10.40) is true for any set of operators and is never violated for any kind of quantum state, whether separable or entangled.

Covariance Matrix Formulation of the Uncertainty Relation

Equation (10.40) summarizes the generalized uncertainty relations between an arbitrary number of operators. In Section 5.6, we encountered the Schrodinger uncertainty relation, which gave an inequality between two operators. This same idea can be generalized to any number of operators, forming increasingly complex inequalities.

For example, the Schrodinger uncertainty relation gives the relationship bounding the product of the variances between two operators ξ_1, ξ_2,

$$I_{12} \equiv \sigma^2_{\xi_1}\sigma^2_{\xi_2} - \left|\frac{\langle\{\xi_1, \xi_2\}\rangle}{2} - \langle\xi_1\rangle\langle\xi_2\rangle\right|^2 - \left|\frac{\langle[\xi_1, \xi_2]\rangle}{2i}\right|^2 \geq 0,$$

where $\sigma^2_\xi \equiv \langle\xi^2\rangle - \langle\xi\rangle^2$. Following the same procedure, one can straightforwardly derive the Schrodinger uncertainty relations for M operators $\xi_1, \ldots, \xi_M$. For example, for $M = 1$, we obtain

$$I_1 \equiv \sigma^2_{\xi_1} \geq 0, \tag{10.41}$$

and for $M = 3$, we obtain

$$\begin{aligned} I_{123} \equiv &\sigma^2_{\xi_1}\sigma^2_{\xi_2}\sigma^2_{\xi_3} - \langle f_1|f_1\rangle|\langle f_2|f_3\rangle|^2 - \langle f_2|f_2\rangle|\langle f_3|f_1\rangle|^2 \\ &- \langle f_1|f_2\rangle\langle f_2|f_3\rangle\langle f_3|f_1\rangle - \langle f_2|f_1\rangle\langle f_3|f_2\rangle\langle f_1|f_3\rangle \\ &- \langle f_3|f_3\rangle|\langle f_1|f_2\rangle|^2 \geq 0, \end{aligned} \tag{10.42}$$

where $|f_i\rangle = (\xi_i - \langle\xi_i\rangle)|\Psi\rangle$.

The remarkable feature of (10.40) is that it contains information about all the Schrodinger uncertainty relations, as described above. Taking the $M = 3$ case, for example, the first-order invariant of (10.40) yields $I_1 + I_2 + I_3 \geq 0$; that is, the sum of the variances of the operators is nonnegative. The second-order invariant (sum of principle minors) yields $I_{12} + I_{23} + I_{13} \geq 0$, which is the sum of the standard Schrodinger uncertainty relations between all operator pairs. Finally, the third-order invariant (i.e., the determinant) yields $I_{123} \geq 0$, which is the three-operator Schrodinger uncertainty relation. In the M-operator case, (10.40) summarizes the $1, 2, \ldots, M$–operator Schrodinger uncertainty relation via the matrix invariants in a highly succinct way.

The covariance matrix criterion then says that

$$\tilde{V} + \frac{i}{2}\tilde{\Omega} \geq 0, \qquad \text{(for separable states)} \tag{10.43}$$

where

$$\tilde{V}_{jk} = \frac{1}{2}\langle\{\xi_j, \xi_k\}^{T_B}\rangle - \langle\xi_j^{T_B}\rangle\langle\xi_k^{T_B}\rangle$$
$$\tilde{\Omega}_{jk} = -i\langle[\xi_j, \xi_k]^{T_B}\rangle. \tag{10.44}$$

Again, if (10.43) is violated, then the state is entangled; otherwise, the test is inconclusive. The meaning of the inequality is again in relation to the positivity of the $M \times M$ matrix. For separable states, (10.43) possesses only positive or zero eigenvalues; for entangled states, (10.43) can possess negative eigenvalues. The matrix elements of (10.43) can be written equally as

$$\left[\tilde{V} + \frac{i}{2}\tilde{\Omega}\right]_{jk} = \langle(\xi_j\xi_k)^{T_B}\rangle - \langle\xi_j^{T_B}\rangle\langle\xi_k^{T_B}\rangle. \tag{10.45}$$

The most well-known example of the covariance matrix entanglement criterion is again in the context of quantum optical systems. The operators of the covariance matrix are taken to be

$$\xi = (x_A, p_A, x_B, p_B). \tag{10.46}$$

The commutation matrix can be written in this case as

$$\Omega = \begin{pmatrix} 0 & -1 & 0 & 0 \\ 1 & 0 & 0 & 0 \\ 0 & 0 & 0 & -1 \\ 0 & 0 & 1 & 0 \end{pmatrix} = \begin{pmatrix} -J & 0 \\ 0 & -J \end{pmatrix}, \tag{10.47}$$

where we defined

$$J = \begin{pmatrix} 0 & 1 \\ -1 & 0 \end{pmatrix}. \tag{10.48}$$

Evaluating the entanglement criterion requires evaluating the covariance and commutation matrices involving the partial transposed operators (10.43). For the choice (10.46), the partial transposed operators have the simple relation

$$\xi^{T_B} = (x_A, p_A, x_B, -p_B), \tag{10.49}$$

when working in the x-basis. The matrix elements of the partial transposed operators are

$$\tilde{V} = \begin{pmatrix} V_{11} & V_{12} & V_{13} & -V_{14} \\ V_{12} & V_{22} & V_{23} & -V_{24} \\ V_{13} & V_{23} & V_{33} & -V_{34} \\ -V_{14} & -V_{24} & -V_{34} & V_{44} \end{pmatrix}. \tag{10.50}$$

That is, the matrix elements involving p_B are inverted in sign. For the commutation matrix, all elements are unchanged,

$$\tilde{\Omega}_{jk} = \Omega_{jk}, \tag{10.51}$$

where we used the fact that $(x_B p_B)^{T_B} = p_B^{T_B} x_B^{T_B} = -p_B x_B$. One can then substitute (10.50) and (10.51) into (10.43) and check for positivity of the matrix.

Checking the positivity of (10.43) typically involves finding the eigenvalues of the matrix $\tilde{V} + \frac{i}{2}\tilde{\Omega}$. One can bypass the diagonalization by evaluating the determinant,

$$\det\left(\tilde{V} + \frac{i}{2}\tilde{\Omega}\right) \geq 0, \qquad \text{(for separable states)} \tag{10.52}$$

since the determinant is the product of all the eigenvalues. As long as the number of negative eigenvalues is an odd number, entanglement will still be detected using the determinant approach. For the example of the operators (10.46), this can be written in the form

$$\begin{aligned} &\det A \det B + \left(\frac{1}{4} - |\det C|\right)^2 - \mathrm{Tr}\left(AJCJBJC^T J\right) \\ &\geq \frac{\det A + \det B}{4}, \qquad \text{(for separable states)} \end{aligned} \tag{10.53}$$

where we have defined the covariance submatrix operators

$$\begin{aligned} V &= \begin{pmatrix} A & C \\ C^T & B \end{pmatrix} \\ A &= \begin{pmatrix} V_{11} & V_{12} \\ V_{12} & V_{22} \end{pmatrix} \\ B &= \begin{pmatrix} V_{33} & V_{34} \\ V_{34} & V_{44} \end{pmatrix} \\ C &= \begin{pmatrix} V_{13} & V_{14} \\ V_{23} & V_{24} \end{pmatrix}. \end{aligned} \tag{10.54}$$

In the case of total spin operators, the procedure is similar. For example, one can instead use the operators

$$\xi = (S_x^A, S_y^A, S_z^A, S_x^B, S_y^B, S_z^B). \tag{10.55}$$

The partial transpose of the operators gives, in the S_z-basis,

$$\xi^{T_B} = (S_x^A, S_y^A, S_z^A, S_x^B, -S_y^B, S_z^B). \tag{10.56}$$

In the previous case, the commutation matrix was independent of the state because the commutator $[x, p] = i$ evaluates to a constant. Here, the commutation matrix will depend upon the expectation value with respect to the state in question. When evaluating (10.44) or (10.45), one must be careful to interchange the order of the operators if both operators are on subsystem B; for example, $(S_x^B S_y^B)^{T_B} = (S_y^B)^{T_B}(S_x^B)^{T_B} = -S_y^B S_x^B$.

Exercise 10.3.1 Explain why there are $D^2 - 1$ free parameters in a D-dimensional density matrix. Hint: An arbitrary density matrix can be written

$$\begin{aligned} \rho_{nn} &= p_n \qquad (n < D) \\ \rho_{DD} &= 1 - \sum_{n=1}^{D-1} \\ \rho_{n'n} &= \rho_{nn'}^* = a_{n'n}, \end{aligned} \tag{10.57}$$

where p_n are real numbers and $a_{n'n}$ are complex numbers.

Exercise 10.3.2 Evaluate the Hillery–Zubairy criterion (10.32) for the two mode squeezed state (10.33). What values of r does the criterion detect entanglement for?

Exercise 10.3.3 Verify that the covariance and commutation matrices for the transposed operators satisfy (10.50) and (10.51) for (10.46).

Exercise 10.3.4 Verify that the partial transpose of the operators (10.55) gives (10.56). Hint: Use the definition of the spin operators in the form (5.20) and directly take the transpose.

10.4 One-Axis Two-Spin Squeezed States

Up to this point, we have discussed entanglement from a rather general perspective, without considering specific examples. Due to the large Hilbert space available to spin ensembles, the types of entangled states that can be generated between spins are rather complex, much more so than for qubits. In this section, we discuss an example of a type of entangled state that can be generated between two atomic ensembles.

In (5.85), we saw an example of a squeezing Hamiltonian that can be produced by interactions between the bosons. This was an interaction on the same ensemble, but it is also possible to have interactions of the same form between two ensembles. The Hamiltonian in this case is

$$H_{\mathrm{1A2S}} = \hbar\kappa S_z^A S_z^B, \tag{10.58}$$

where the total S_z spin for the two ensembles is labeled by A and B. We call this the one-axis two-spin (1A2S) squeezed state, since it is a two-spin generalization of the one-axis squeezed state on a single ensemble. Applying such a Hamiltonian on two ensembles that are in maximal S_x-eigenstates, we obtain (see Fig. 10.1(a))

$$\begin{aligned} & e^{-iS_z^A S_z^B \tau} \left| \frac{1}{\sqrt{2}}, \frac{1}{\sqrt{2}} \right\rangle\!\!\right\rangle_A \left| \frac{1}{\sqrt{2}}, \frac{1}{\sqrt{2}} \right\rangle\!\!\right\rangle_B \\ & = \frac{1}{\sqrt{2^N}} \sum_k \sqrt{\binom{N}{k}} \left| \frac{e^{i(N-2k)\tau}}{\sqrt{2}}, \frac{e^{-i(N-2k)\tau}}{\sqrt{2}} \right\rangle\!\!\right\rangle_A |k\rangle_B, \end{aligned} \tag{10.59}$$

where the subscripts A and B on the states label the two ensembles and $\tau = \kappa t$ is the dimensionless time.

Using the techniques shown in the previous sections, we can directly quantify that entanglement is present. Since (10.59) is a pure state and the full quantum state is available, it is straightforward to evaluate the entanglement using the von Neumann entropy (10.4). We compare the entanglement generated between the two ensembles by comparing to the standard qubit case ($N = 1$) in Fig. 10.1(b). The entanglement E is normalized to $E_{\mathrm{max}} = \log_2(N+1)$, which is the maximum entanglement possible between two N particle systems. For $N = 1$, we see the expected behavior, where a maximally entangled state is reached at $\tau = \pi/4$. For $N = 100$, the entanglement shows a repeating structure with period $\pi/2$ but otherwise no apparent periodicity in between. As N grows, the fluctuations increase at finer timescales, on the order

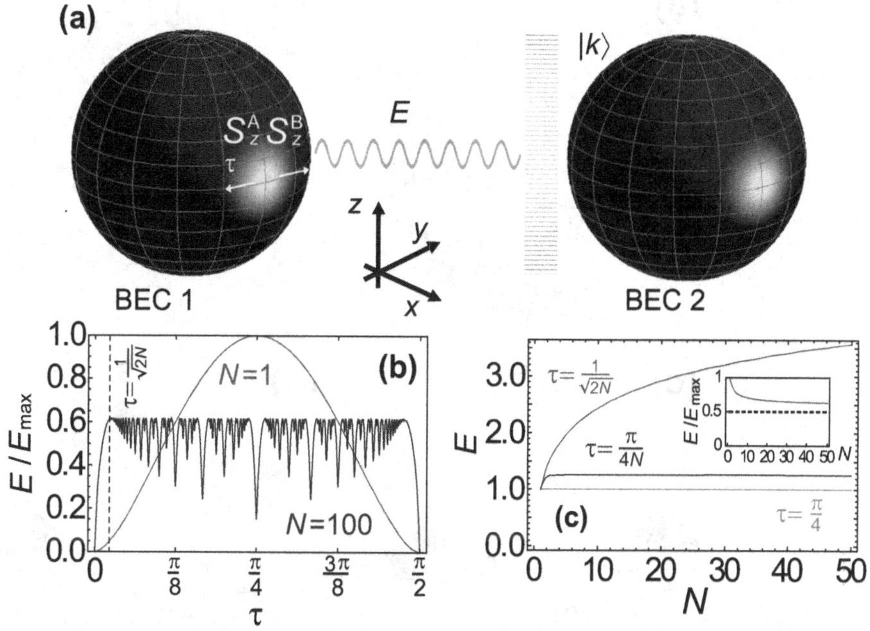

Fig. 10.1 Entangling operation between two atomic ensembles. (a) The schematic operation considered in this paper. The Q-functions for the initial spin coherent state $|\frac{1}{\sqrt{2}}, \frac{1}{\sqrt{2}}\rangle\rangle$ on each BEC are shown. The $S_z^A S_z^B$ interactions induces a fanning of coherent states in the directions shown by the arrows. (b) The von Neumann entropy normalized to the maximum entanglement ($E_{\text{max}} = \log_2(N+1)$) between two BEC qubits for $N = 1$ and $N = 100$ after operation of $S_z^A S_z^B$ for a time τ. (c) Entanglement at times $\tau = \frac{\pi}{4N}, \frac{1}{\sqrt{2N}}, \frac{\pi}{4}$ for various boson numbers N. Inset: The same data for $\tau = \frac{1}{\sqrt{2N}}$ but normalized to E_{max}. Color version of this figure available at cambridge.org/quantumatomoptics.

of $\sim 1/N$. The basic behavior can be summarized as follows. The entanglement increases monotonically until a characteristic time $\tau = 1/\sqrt{2N}$, after which very fast fluctuations occur, bounded from above by a "ceiling" in the entanglement.

What is the origin of the complex behavior of the entanglement? Figure 10.1(a) shows the Q-function for the initial state $\tau = 0$ of one of the BECs on the surface of the normalized Bloch sphere. Evolving the system now in τ, a visualization of the states is shown in Fig. 10.2. Keeping in mind that the spin coherent states form a quasi-orthogonal set of states (5.153), we represent each of the various spin coherent states in the sum (10.59) as a circle of radius $r = \sqrt{\frac{2}{N}}$, which corresponds to the distance where the overlap between two spin coherent states starts to diminish exponentially. As all spin coherent states in (10.59) are along the equator of the Bloch sphere $S_z = 0$, we flatten the Bloch sphere along the S_z-direction, such that each spin coherent state is located at an angle ϕ in the S_x–S_y plane. Each circle is marked to show the correlation between a particular $|k\rangle$ state and the spin coherent state that it is entangled with (see Fig. 10.1(a)).

There are several characteristic times that we consider separately. As τ is increased, the circles fan out in both clockwise and counterclockwise directions. At $\tau = \frac{\pi}{4N}$, the extremal states $k = 0, N$ reach the $\pm S_y$ directions. For $N = 1$ (the standard qubit case), this gives a maximally entangled state. Figure 10.1(c)

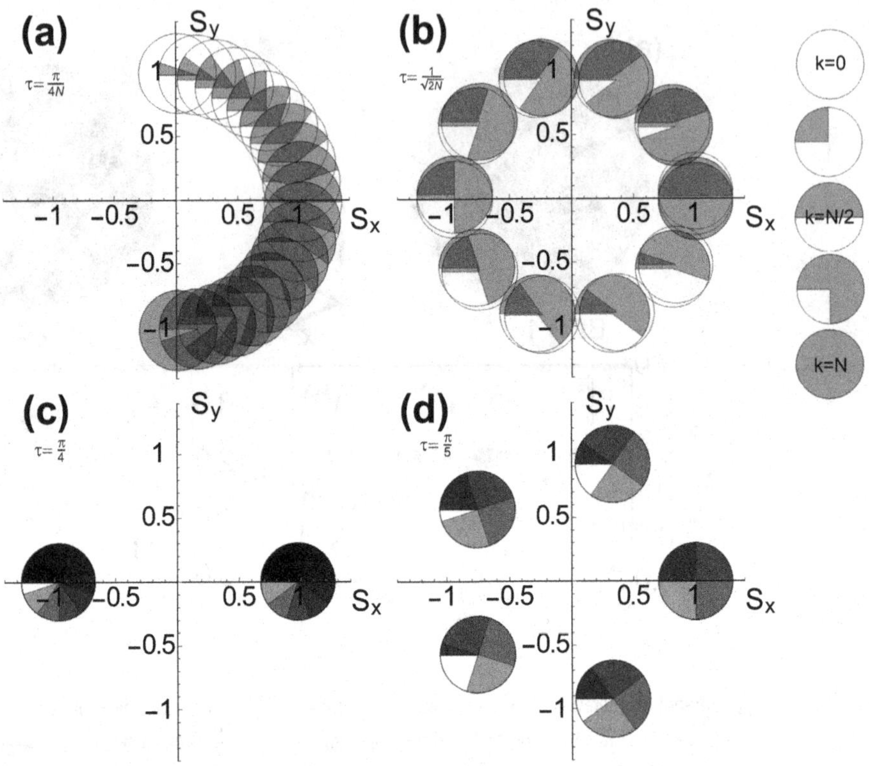

Fig. 10.2 Visualization of the entangled state (10.59) for gate times (a) $\tau = \frac{\pi}{4N}$ (b) $\tau = \frac{1}{\sqrt{2N}}$ (c) $\tau = \frac{\pi}{4}$ (d) $\tau = \frac{\pi}{5}$ (adapted from Ref. [71]). Eigenstates of the S_z operator, $|k\rangle$, are entangled with spin coherent states, as marked in the key (far right). The radii of the circles correspond to approximate distances for diminishing overlap of the spin coherent states. All plots correspond to $N = 20$. Color version of this figure available at cambridge.org/quantumatomoptics.

shows the amount of entanglement for times $\tau = \frac{\pi}{4N}$, which shows that it quickly approaches the asymptotic value of $E \approx 1.26$. A maximally entangled state would have entanglement $E_{\max} = \log_2(N+1)$, and thus gate times of $\tau = \frac{\pi}{4N}$ correspond to relatively small amounts of entanglement, equivalent to approximately one pair of entangled qubits.

The next characteristic time is $\tau = \frac{1}{\sqrt{2N}}$. This is the time when the spin coherent states are separated far enough such that adjacent circles have no overlap (see Fig. 10.2(b)). The entanglement, as can be seen in Fig. 10.1(b), reaches its maximal value at this time, beyond which it executes fast fluctuations briefly returning to this value. It is also a characteristic time due to the weight factors in (10.59). Approximating

$$\sqrt{\frac{1}{2^N}\binom{N}{k}} \approx \left(\frac{2}{\pi N}\right)^{1/4} \exp\left[-\frac{1}{N}(k - N/2)^2\right], \tag{10.60}$$

we see that the only the terms between $k = N/2 \pm \sqrt{N}$ have a significant weight in the summation and the other terms only contribute an exponentially small amount. At time $\tau = \frac{1}{\sqrt{2N}}$, these states are spread out over the unit circle, since the angular positions of the spin coherent states in (10.59) reach order unity for the first time.

These two observations give us a simple way of estimating the asymptotic value of the entanglement in the limit $N \to \infty$. Starting from (10.59) and discarding small weight-contributing states, we have

$$\frac{1}{\sqrt{2N}} \sum_{k=N/2-\sqrt{N}}^{k=N/2+\sqrt{N}} \left| \frac{e^{i(N-2k)\tau}}{\sqrt{2}}, \frac{e^{-i(N-2k)\tau}}{\sqrt{2}} \right\rangle\!\!\Big\rangle_A |k\rangle_B, \tag{10.61}$$

where we have approximated the binomial factor within the range to be constant. Assuming that the spin coherent states are orthogonal, this gives an entanglement of $E \approx \log_2 \sqrt{N}$. Relative to the maximal value, this gives

$$\lim_{N\to\infty} E/E_{\max} = 1/2. \tag{10.62}$$

This shows that even in the limit of $N \to \infty$, entanglement survives. Thus, although the limit $N \to \infty$ is considered to be a classical limit under certain situations, in this case it is clear that quantum effects are present at all N. The inset of Fig. 10.1(c) shows results consistent with the asymptotic result (10.62).

We now consider the origin of the fast fluctuations in the entanglement in Fig. 10.1(b). Due to the periodicity of the spin coherent states, the circles align with high symmetry at particular τ times, as seen in Fig. 10.2(c) and 10.2(d). These points correspond to sharp dips in the entanglement, as seen in Fig. 10.1(b). For example, at $\tau = \pi/4$, (10.59) can be written for even N:

$$\begin{aligned}&\frac{1}{2}\left((-1)^{N/2}\left|\frac{1}{\sqrt{2}},\frac{1}{\sqrt{2}}\right\rangle\!\!\Big\rangle_A + \left|\frac{1}{\sqrt{2}},-\frac{1}{\sqrt{2}}\right\rangle\!\!\Big\rangle_A\right)\left|\frac{1}{\sqrt{2}},\frac{1}{\sqrt{2}}\right\rangle\!\!\Big\rangle_B \\ &+\frac{1}{2}\left(\left|\frac{1}{\sqrt{2}},\frac{1}{\sqrt{2}}\right\rangle\!\!\Big\rangle_A - (-1)^{N/2}\left|\frac{1}{\sqrt{2}},-\frac{1}{\sqrt{2}}\right\rangle\!\!\Big\rangle_A\right)\left|\frac{1}{\sqrt{2}},-\frac{1}{\sqrt{2}}\right\rangle\!\!\Big\rangle_B.\end{aligned} \tag{10.63}$$

For large N, the states $|\frac{1}{\sqrt{2}}, \frac{1}{\sqrt{2}}\rangle\rangle$ and $|\frac{1}{\sqrt{2}}, -\frac{1}{\sqrt{2}}\rangle\rangle$ require flipping N bosons from $+S_x$ to $-S_x$ eigenstates. Thus, the terms in the brackets are Schrodinger cat states. The positive-parity Schrodinger cat state contains only even-numbered Fock states, while the negative-parity Schrodinger cat state contains only odd-numbered Fock states. Taking into account that

$$\Big\langle\!\!\Big\langle \frac{1}{\sqrt{2}}, -\frac{1}{\sqrt{2}} \Big| \frac{1}{\sqrt{2}}, \frac{1}{\sqrt{2}} \Big\rangle\!\!\Big\rangle = 0, \tag{10.64}$$

(10.63) is precisely equivalent to a Bell state, which has an entanglement $E = \log_2 2 = 1$. For odd N and $\tau = \pi/4$, the states can be written in terms of the eigenstates of S_y:

$$\begin{aligned}&\frac{1}{2}\left((-1)^{(N+1)/2}\left|\frac{e^{-i\pi/4}}{\sqrt{2}},\frac{e^{i\pi/4}}{\sqrt{2}}\right\rangle\!\!\Big\rangle_A + \left|\frac{e^{-i\pi/4}}{\sqrt{2}},-\frac{e^{i\pi/4}}{\sqrt{2}}\right\rangle\!\!\Big\rangle_A\right)\left|\frac{e^{-i\pi/4}}{\sqrt{2}},\frac{e^{i\pi/4}}{\sqrt{2}}\right\rangle\!\!\Big\rangle_B \\ &+\frac{1}{2}\left(\left|\frac{e^{-i\pi/4}}{\sqrt{2}},\frac{e^{i\pi/4}}{\sqrt{2}}\right\rangle\!\!\Big\rangle_A - (-1)^{(N+1)/2}\left|\frac{e^{-i\pi/4}}{\sqrt{2}},-\frac{e^{i\pi/4}}{\sqrt{2}}\right\rangle\!\!\Big\rangle_A\right)\left|\frac{e^{-i\pi/4}}{\sqrt{2}},-\frac{e^{i\pi/4}}{\sqrt{2}}\right\rangle\!\!\Big\rangle_B.\end{aligned} \tag{10.65}$$

This state also has entanglement $E = 1$, hence for all N, the entanglement is unity, in agreement with the numerical results of Figure 10.1(c).

Similar arguments may be made for any τ that is a rational multiple of $\pi/4$. Consider a general time $\tau = \frac{m\pi}{4d}$, where m, d are integers. The angular difference

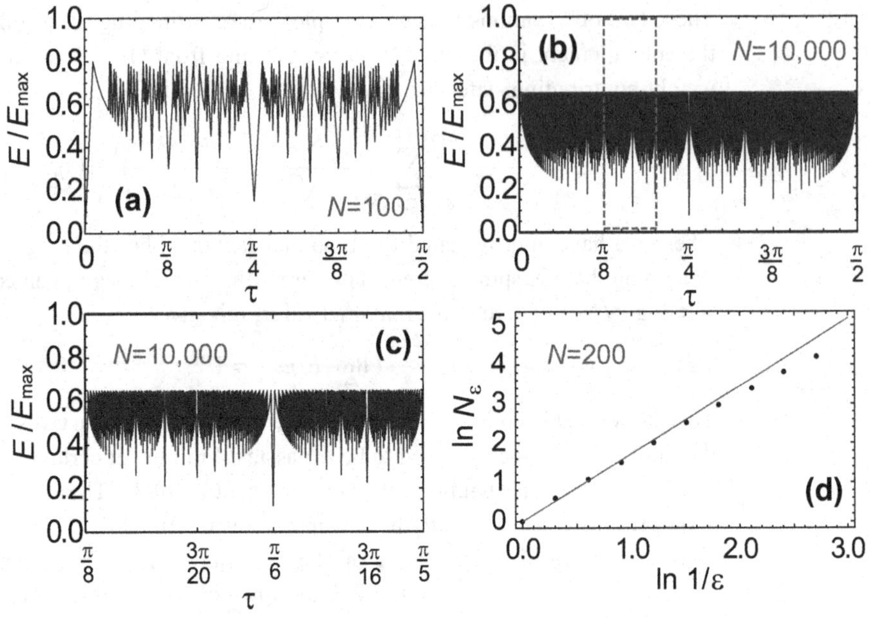

Fig. 10.3 The entanglement as given by the approximate formula (10.66) (valid for times τ that are a rational multiple of $\pi/4$) for (a) $N = 100$ and (b) $N = 10{,}000$. (c) The self-similar behavior of the entanglement can be observed in a zoomed in view of (b). (d) Logarithmic plot of the number of boxes N_ε versus the size of the box ε in the box-counting method to determine the Haussdorff (fractal) dimension of the exact entanglement curve for $N = 200$. Color version of this figure available at cambridge.org/quantumatomoptics.

between adjacent circles in Figure 10.2 may be deduced to be $\Delta\phi = 4\tau = \frac{m\pi}{d}$. The number of circles must then be the first integer multiple of $\Delta\phi$ that gives a multiple of 2π. This is LCM$(m/d, 2)d/m$, where LCM is the least common multiple. The amount of entanglement, valid for $d \lesssim \sqrt{N}$, can then be written as

$$E = \log_2 \left[\frac{\text{LCM}(m/d, 2)d}{m} \right]. \tag{10.66}$$

The entanglement at times that is an irrational multiple of $\pi/4$ do not have any particular alignment of the circles and thus can be considered to be approximately randomly arranged over ϕ. Similar arguments to (10.62) may then be applied, and at these points the entanglement jumps back up to $E/E_{\text{max}} = 1/2$ for $N \to \infty$. We see that (10.66) correctly reproduces the result for $\tau = \pi/4$ in Fig. 10.1(c), with $E = 1$ for all N.

The formula (10.66) is directly plotted in Fig. 10.3(a). We see that the dips in the entanglement are well reproduced in comparison to the exact result of Fig. 10.1(b). The ceiling value of the entanglement does not agree exactly since this depends upon the cutoff imposed on d, a quantity that does not have a precise cutoff. Neglecting the ceiling value, this allows us to obtain the structure for large N, as seen in Fig. 10.3(b) and (c). We now see more clearly the self-similar repeating structure characteristic of a fractal. Since the set of rational numbers is everywhere dense, in the limit of $N \to \infty$, the entanglement has a pathological behavior where there are an infinite number of dips between any two values of τ. Such a dependence of whether a parameter is either rational or irrational is familiar in physics contexts from models such as the long-range antiferromagnetic Ising model, where the filling factors equal to a rational number occupy a finite extent of the chemical potential, forming the so-called "devil's staircase." As an analogy, the behavior of the entanglement here is described by a *devil's crevasse*, where every rational multiple of $\pi/4$ in the gate time gives a sharp dip in the entanglement. The fractal dimension of the curve may be estimated using the box-counting method (Figure 10.3d), applied to the exact entanglement curve (i.e., not (10.66)). Using several different N values and extrapolating to $N \to \infty$, the Haussdorff dimension is found to be $d \approx 1.7$, showing clearly the fractal behavior of the entanglement.

10.5 Two-Axis Two-Spin Squeezed States

In Section 5.7, we saw the one-axis (1A1S) and two-axis (2A1S) squeezed states on single ensembles. In Section 10.4, we generalized the one-axis squeezed state to its two-spin version, the one-axis two-spin (1A2S) squeezed state. It is thus only natural to ask what is the two-spin version of the two-axis squeezed state. Accordingly, we define the two-axis two-spin (2A2S) Hamiltonian according to

$$H_{2A2S} = \frac{\hbar\kappa}{2}(S_x^A S_x^B - S_y^A S_y^B) = \hbar\kappa(S_+^A S_+^B + S_-^A S_-^B), \tag{10.67}$$

where κ is an interaction constant. The 2A2S squeezed states are produced by a unitary evolution according to the Hamiltonian (10.67) for a time t,

$$\begin{aligned} |\psi_{2A2S}(t)\rangle\rangle &= e^{-iH_{2A2S}t/\hbar}|1,0\rangle\rangle_A|1,0\rangle\rangle_2 \\ &= e^{-i(S_+^A S_+^B + S_-^A S_-^B)\tau}|1,0\rangle\rangle_A|1,0\rangle\rangle_2, \end{aligned} \tag{10.68}$$

where we have defined a dimensionless time $\tau = \kappa t$. The initial states are maximally polarized states in the S_z-direction, the same as for the 2A1S Hamiltonian. In a similar way to the 2A1S Hamiltonian, the 2A2S Hamiltonian cannot be diagonalized using a linear transformation of the bosonic operators. Thus, generally, it must be studied using numerical methods. For small times, we may expand the exponential in (10.68) to second order, where the first terms in the expansion are

$$\begin{aligned} |\psi(t)\rangle\rangle \approx &(1 - \tau^2 N^2)|N\rangle_1|N\rangle_2 - i\tau N|N-1\rangle_1|N-1\rangle_2 \\ &- \tau^2 N(N-1)|N-2\rangle_1|N-2\rangle_2 + \dots, \end{aligned} \tag{10.69}$$

where the states we have used are Fock states (5.1). All terms in the superposition have the same Fock numbers for the two BECs, hence the first thing to notice is that the state is perfectly correlated in S_j^z.

To obtain some intuition about the state (10.68), let us first examine the state for small evolution times. For a system with fixed particle number N, the Holstein–Primakoff approximation (5.119) can be used, where we approximate the 2A2S Hamiltonian by

$$H_{2M} \approx \hbar\kappa N(a_A a_B + a_A^\dagger a_B^\dagger). \tag{10.70}$$

The Hamiltonian (10.70) is exactly the two-mode squeezing Hamiltonian considered in quantum optics, with an multiplicative factor of N. The transformation of the mode operators is

$$\begin{aligned} e^{iH_{2M}t/\hbar} a_A e^{-iH_{2M}t/\hbar} &= a_A \cosh N\tau - i a_B^\dagger \sinh N\tau \\ e^{iH_{2M}t/\hbar} a_B e^{-iH_{2M}t/\hbar} &= a_B \cosh N\tau - i a_A^\dagger \sinh N\tau. \end{aligned} \tag{10.71}$$

We can deduce the time for which the Holstein–Primakoff approximation is valid by evaluating the population of the a_1 and a_2 states. The populations of the two states are always equal $\langle a_1^\dagger a_1 \rangle = \langle a_2^\dagger a_2 \rangle$, and we obtain

$$\begin{aligned} \langle a_1^\dagger(t) a_1(t) \rangle &= \langle 0 | (a_1^\dagger \cosh N\tau + i a_2 \sinh N\tau)(a_1 \cosh N\tau - i a_2^\dagger \sinh N\tau) | 0 \rangle \\ &= \sinh^2 N\tau \approx e^{2N\tau}/4, \end{aligned} \tag{10.72}$$

where in the last step we assumed $N\tau \gg 1$. Demanding that $\langle a_1^\dagger a_1 \rangle \ll N$, we have the criterion for the validity of the Holstein–Primakoff approximation:

$$\tau \ll \frac{\ln(4N)}{2N}. \tag{10.73}$$

Next, let us define the canonical position and momentum operators as

$$\begin{aligned} x_j &= \frac{a_j + a_j^\dagger}{\sqrt{2}} \approx \frac{S_+^j + S_-^j}{\sqrt{2N}} = \frac{S_x^j}{\sqrt{2N}} \\ p_j &= \frac{-i a_j + i a_j^\dagger}{\sqrt{2}} \approx \frac{-i S_+^j + i S_-^j}{\sqrt{2N}} = \frac{S_y^j}{\sqrt{2N}}, \end{aligned} \tag{10.74}$$

where $j \in \{A, B\}$. For the choice of phase between the two terms in (10.70), the relevant operators are those that are rotated by 45° with respect to the quadrature axes,

$$\begin{aligned} \tilde{x}_j &= \frac{x_j + p_j}{\sqrt{2}} \approx \frac{\tilde{S}_x^j}{\sqrt{2N}} \\ \tilde{p}_j &= \frac{p_j - x_j}{\sqrt{2}} \approx \frac{\tilde{S}_y^j}{\sqrt{2N}}, \end{aligned} \tag{10.75}$$

where we used the definitions (5.107). The correlations for which the quantum noise is suppressed are then $\tilde{x}_A + \tilde{x}_B$ and $\tilde{p}_A - \tilde{p}_B$. This can be seen by evaluating

$$\begin{aligned} e^{iH_{2M}t/\hbar}(\tilde{x}_A + \tilde{x}_B) e^{-iH_{2M}t/\hbar} &= e^{-N\tau}(\tilde{x}_A + \tilde{x}_B) \\ e^{iH_{2M}t/\hbar}(\tilde{p}_A - \tilde{p}_B) e^{-iH_{2M}t/\hbar} &= e^{-N\tau}(\tilde{p}_A - \tilde{p}_B), \end{aligned} \tag{10.76}$$

which become suppressed for large squeezing times. The corresponding anti-squeezed variables are

$$\begin{aligned} e^{iH_{2M}t/\hbar}(\tilde{x}_A - \tilde{x}_B) e^{-iH_{2M}t/\hbar} &= e^{N\tau}(\tilde{x}_A - \tilde{x}_B) \\ e^{iH_{2M}t/\hbar}(\tilde{p}_A + \tilde{p}_B) e^{-iH_{2M}t/\hbar} &= e^{N\tau}(\tilde{p}_A + \tilde{p}_B). \end{aligned} \tag{10.77}$$

We now directly evaluate the correlations produced by the 2A2S Hamiltonian numerically, without applying the Holstein–Primakoff approximation. From (10.76), we expect that the variances of the observables

$$O_{\text{sq}} \in \{\tilde{S}_x^A + \tilde{S}_x^B, \tilde{S}_y^A - \tilde{S}_y^B\} \tag{10.78}$$

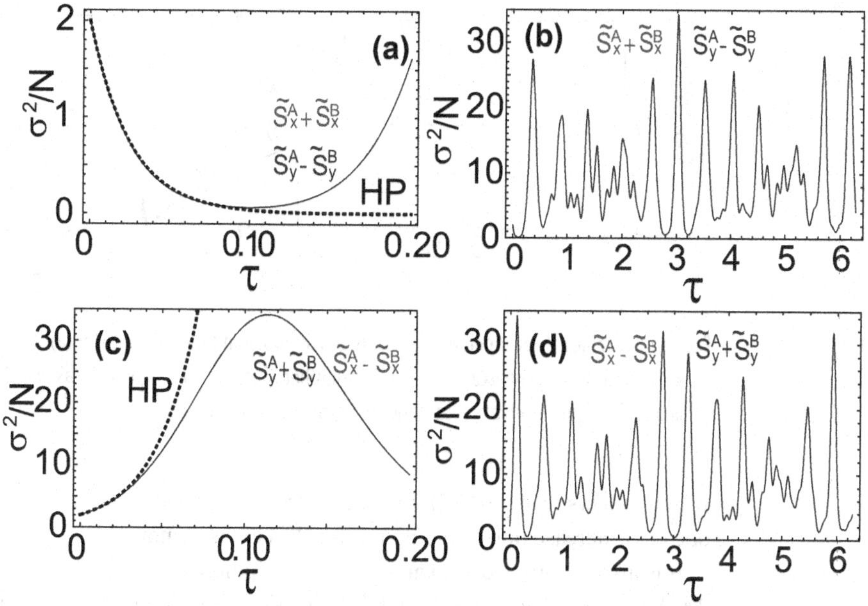

Fig. 10.4 Variances of EPR-like observables in the two-axis two-spin squeezed state. The variances of the (a)–(b) squeezed variables $\tilde{S}_x^A + \tilde{S}_x^B$ and $\tilde{S}_y^A - \tilde{S}_y^B$, and (c)–(d) antisqueezed variables $\tilde{S}_x^A - \tilde{S}_x^B$ and $\tilde{S}_y^A + \tilde{S}_y^B$ are plotted as a function of the dimensionless interaction time τ. The Holstein–Primakoff (HP) approximated variances are shown by the dotted lines. (a)–(c) show timescales in the range $\tau \sim 1/N$ and (b)–(d) show longer timescales $\tau \sim 1$. The number of atoms per ensemble is taken as $N = 20$. Color version of this figure available at cambridge.org/quantumatomoptics.

become suppressed, for short times when the Holstein–Primakoff approximation holds. The observables in the perpendicular directions (10.77),

$$O_{\text{asq}} \in \{\tilde{S}_x^A - \tilde{S}_x^B, \tilde{S}_y^A + \tilde{S}_y^B\}, \tag{10.79}$$

are the antisqueezed variables.

In Fig. 10.4(a), the variances of the observables (10.78) are plotted for short timescales $\tau \sim 1/N$. We see that the two variances have exactly the same time dependence and take a minimum at a time that we define as the optimal squeezing time τ_{opt}. In the time region $0 \le \tau \le \tau_{\text{opt}}$, the variance agrees well with Holstein–Primakoff approximation, giving

$$\sigma^2_{\tilde{S}_x^A + \tilde{S}_x^B} = \sigma^2_{\tilde{S}_y^A - \tilde{S}_y^B} \approx 2N e^{-2N\tau}, \tag{10.80}$$

which follows from the relations (10.76) and the fact that $\text{Var}(O_{\text{sq}}, \tau = 0) = 2N$. Beyond these times, the variance increases and no longer follows (10.80). For longer timescales, as shown in Fig. 10.4(b), the variance follows aperiodic oscillations between low and high variance states. Some relatively low variance states are achieved (e.g., particularly around $\tau \approx 3$), although the minimum variance at the times τ_{opt} is not attained again. The antisqueezed variables (10.79) are shown in Fig. 10.4(c) for short timescales $\tau \sim 1/N$. Again, the two variables (10.79) have exactly the same time dependence and initially increase according to

$$\sigma^2_{\tilde{S}_x^A - \tilde{S}_x^B} = \sigma^2_{\tilde{S}_y^A + \tilde{S}_y^B} \approx 2N e^{2N\tau}, \tag{10.81}$$

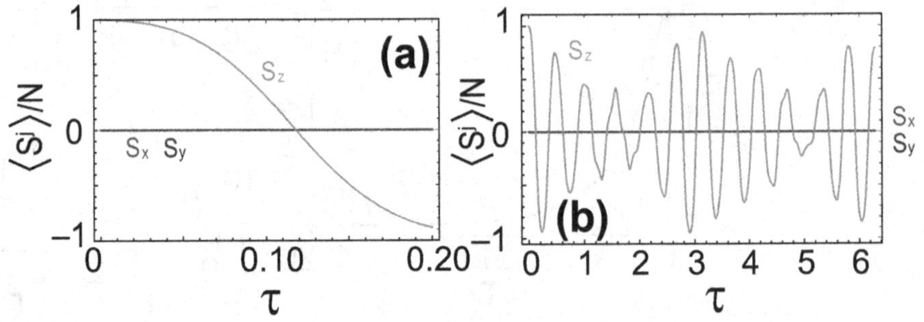

Fig. 10.5 Expectation values of spin operators for the two-axis two-spin squeezed state for (a) short timescales $\tau \sim 1/N$ and (b) long timescales $\tau \sim 1$. The number of atoms per ensemble is taken as $N = 20$. Color version of this figure available at cambridge.org/quantumatomoptics.

which follows from (10.77). In contrast to genuine two-mode squeezing, the variance does not increase unboundedly but reaches a maximum.

It is also instructive to examine the expectation values of the spin operators in the 2A2S squeezed state. Figure 10.5 shows the expectation values of the operators S_x^j, S_y^j, S_z^j. Due to the symmetry between the initial state of the two ensembles and the 2A2S Hamiltonian, identical values are obtained for the two ensembles $j \in \{A, B\}$. Furthermore, the expectation values of two of the operators are always zero:

$$\langle S_x^j(\tau)\rangle = \langle S_y^j(\tau)\rangle = 0. \tag{10.82}$$

This can be seen from (10.69), where the Hamiltonian creates pairs of equal-number Fock states. Since the S_x and S_y operators shift the Fock states by one unit, the expectation values of S_x and S_y are zero for all time. Meanwhile, the expectation value of S_z undergoes aperiodic oscillations and flips sign numerous times during the evolution. In particular, a sign change is observed in the vicinity of τ_{opt}, although it is not exactly at this time. This can be understood from (10.69), where at $\tau \sim 1/N$ the sum contains all terms with a similar magnitude.

The optimal squeezing time, obtained by minimizing the variance of the squeezed variable $\tilde{S}_x^A + \tilde{S}_x^B$, has a dependence approximated by

$$\tau_{\text{opt}} \approx \frac{p_0 + p_1 \ln N}{N}, \tag{10.83}$$

where the parameters are $p_0 = 0.467, p_1 = 0.508$ for large N. Using the optimal squeezing times, the maximal squeezing that can be attained can be estimated from the Holstein–Primakoff relation according to

$$\begin{aligned} \min_\tau \text{Var}(\tilde{S}_1^x + \tilde{S}_2^x, \tau) &= \min_\tau \text{Var}(\tilde{S}_1^y - \tilde{S}_2^y, \tau) \\ &\approx 2Ne^{-2N\tau_{\text{opt}}^{(\text{sq})}} \\ &\approx \frac{2N}{e^{2p_0}N^{2p_1}}, \end{aligned} \tag{10.84}$$

where we substituted the expression (10.83) in the last line. The minimal squeezing level tends to improve approximately as $1/N$ for larger ensemble sizes due to the factor of N^{2p_1} in the denominator, relative to the spin coherent state variance $2N$.

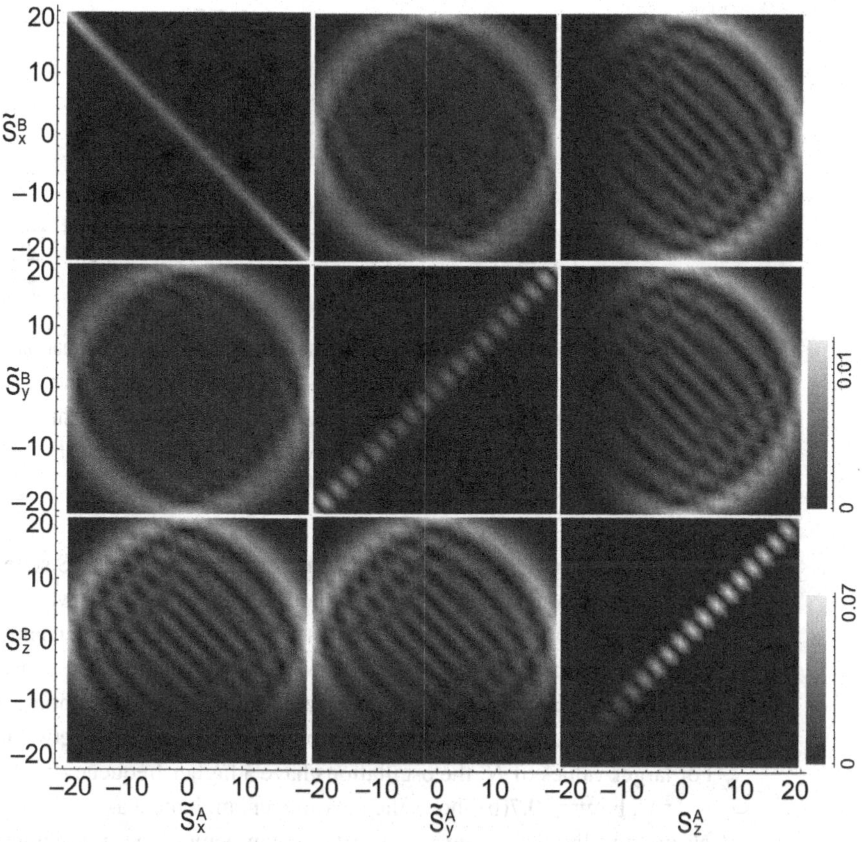

Fig. 10.6 Probability distributions of the two-axis two-spin state measured in various bases for $\tau = 0.1 \approx \tau_{\text{opt}}$ and $N = 20$. The density plot legends for similar-shaped distributions are the same. The legend for the $(\tilde{S}_y^A, \tilde{S}_x^B)$ and $(S_z^A, \tilde{S}_x^B)$ is the same as $(S_z^A, \tilde{S}_y^B)$. Color version of this figure available at cambridge.org/quantumatomoptics.

Another way to visualize the correlations is to plot the probability distributions when the state (10.68) is measured in various bases. Specifically, the probability of a measurement outcome k_A, k_B for various Fock states is then

$$p_{l_A l_B}(k_A, k_B) = |\langle\psi(t)| \left(|k_A\rangle_{l_A} \otimes |k_B\rangle_{l_B}\right)|^2, \tag{10.85}$$

where $l_A, l_B \in \{x, y, z\}$ and are defined in Section 5.12.3. The probabilities for two evolution times near τ_{opt} are shown in Fig. 10.6. The effect of the correlations is seen in the $(\tilde{S}_x^A, \tilde{S}_x^B)$ and $(\tilde{S}_y^A, \tilde{S}_y^B)$ measurement combinations, where the most likely probabilities occur when $\tilde{S}_x^A = -\tilde{S}_x^B$ and $\tilde{S}_y^A = \tilde{S}_y^B$, respectively. This means that the quantities $\tilde{S}_x^A + \tilde{S}_x^B$ and $\tilde{S}_y^A - \tilde{S}_y^B$ always take small values and hence are squeezed. The probability distribution for the measurements for the four cases $(\tilde{S}_{x,y}^A, \tilde{S}_{x,y}^B)$ initially starts as a Gaussian centered around $\tilde{S}_{x,y} = 0$ and becomes increasingly squeezed. For the (S_z^A, S_z^B) measurement, we see Fock state correlations arising from the fact that the 2A2S Hamiltonian always produces Fock states in pairs, as shown in (10.69). For the remaining correlation pairs, the distributions are always symmetrical in the variables $\tilde{S}_x$ and $\tilde{S}_y$, hence they give zero when averaged. Thus there is no correlation between the remaining variables. The lack of correlations in the off-diagonal combinations in Fig. 10.6 is also a feature of standard Bell states.

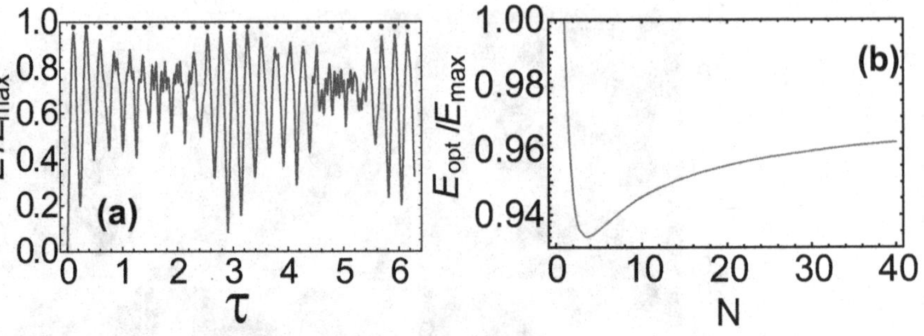

Fig. 10.7 Entanglement of the two-axis two-spin squeezed state, as measured by the von Neumann entropy E normalized to the maximum value $E_{max} = \log_2(N+1)$. (a) As a function of the interaction time τ for $N = 20$. The dots in each figure denote the times when $\langle S_z^A \rangle = 0$ for each N. (b) The maximum value of the von Neumann entropy as a function of N. Color version of this figure available at cambridge.org/quantumatomoptics.

We now turn to the entanglement that is generated in a 2A2S state. The entanglement that we consider is that present between the two BECs, which forms a natural bipartition in the system. Figure 10.7(a) shows the von Neumann entropy normalized to the maximum value $E_{max} = \log_2(N+1)$ for two $N+1$ level systems. We see that the entanglement first reaches a maximum at a similar time to the optimal squeezing time τ_{opt}, and reaches nearly the maximum possible entanglement between the two BECs. For larger values of N, the oscillations have a higher frequency, with a period that is $\sim 2\tau_{opt}$. Figure 10.7(b) shows the maximal entanglement as a function of N. For each N, we find the maximum value of the entanglement by optimizing the time in the vicinity of the first maximum. We see that the optimized entanglement approaches the maximum possible entanglement E_{max} for large N. The entanglement oscillates between large and small values and tends to occur at the values corresponding to $\langle S_z \rangle = 0$. This is true not only in the vicinity of the first maximum in the entanglement but for all τ. Figure 10.7(a) marks all the times (with a dot in the figure) where $\langle S_z \rangle = 0$. We see that each peak in the entanglement occurs when $\langle S_z \rangle = 0$.

We have seen in Fig. 10.7(b) that near-maximal entanglement can be obtained at optimized evolution times of the 2A2S Hamiltonian. We have also seen in Fig. 10.6 that at the optimized squeezing times, very flat distributions of the correlations can be obtained. These facts suggest that a good approximation for the state in the large N regime is

$$|\psi(\tau_{opt})\rangle \approx |\text{EPR}_-\rangle, \tag{10.86}$$

where we defined the state

$$|\text{EPR}_-\rangle = \frac{1}{\sqrt{N+1}} \sum_{k=0}^{N} |k\rangle_{\tilde{x}}^{A} |N-k\rangle_{\tilde{x}}^{B}. \tag{10.87}$$

This state has the maximum possible entanglement E_{max} between the two BECs and exhibits squeezing in the variable $\tilde{S}_x^A + \tilde{S}_x^B$. Algebraic manipulation allows one to rewrite this state equally as

$$|\text{EPR}_-\rangle = \frac{1}{\sqrt{N+1}} \sum_{k=0}^{N} (-1)^k |k\rangle_{\tilde{y}}^{A} |k\rangle_{\tilde{y}}^{B} \tag{10.88}$$

$$= \frac{1}{\sqrt{N+1}} \sum_{k=0}^{N} (-1)^k |k\rangle_z^A |k\rangle_z^B, \tag{10.89}$$

which have the correct $\tilde{S}_y^A - \tilde{S}_y^B$ and $\tilde{S}_z^A - \tilde{S}_z^B$ correlations, in agreement with Fig. 10.6. Such a state is a type of spin-EPR state that exhibits correlations in a similar way to Bell states and continuous variable two-mode squeezed states, in all possible bases. In fact, the correlations for any choice of basis are such that

$$|\text{EPR}_-\rangle = \frac{1}{\sqrt{N+1}} \sum_{k=0}^{N} |k\rangle_{\boldsymbol{n}}^A |k\rangle_{\tilde{\boldsymbol{n}}}^B, \tag{10.90}$$

where $\tilde{\boldsymbol{n}} = (-\sin\theta\cos\phi, \sin\theta\sin\phi, \cos\theta)$.

In Fig. 10.8, we show the fidelity of the 2A2S squeezed state with reference to the spin-EPR state, defined as

$$F_- = |\langle \text{EPR}_- | \psi(\tau)\rangle|^2. \tag{10.91}$$

We also plot the fidelity with respect to another spin-EPR state defined without phases,

$$F_+ = |\langle \text{EPR}_+ | \psi(\tau)\rangle|^2, \tag{10.92}$$

where

$$|\text{EPR}_+\rangle = \frac{1}{\sqrt{N+1}} \sum_{k=0}^{N} |k\rangle_z^A |k\rangle_z^B. \tag{10.93}$$

We see that the state attains high overlap with the $|\text{EPR}_-\rangle$ at a time τ_{opt} as expected and oscillates with peaks at similar times as the peaks in the antisqueezing parameters seen in Fig. 10.4(d). On comparison with Fig. 10.7(b), we see that every second peak in the entanglement corresponds to the peaks for the fidelity F_-. The remaining peaks occur for the fidelity F_+. The timing of the peaks in F_+ also match the peaks in squeezing parameters in Fig. 10.4(b). This makes it clear that the effect of the 2A2S Hamiltonian is to first generate a state closely approximating $|\text{EPR}_-\rangle$, which then subsequently evolves to $|\text{EPR}_+\rangle$, and this cycle repeats itself in an aperiodic fashion. In Fig. 10.8(b), we examine the scaling of the fidelity with N. We optimize

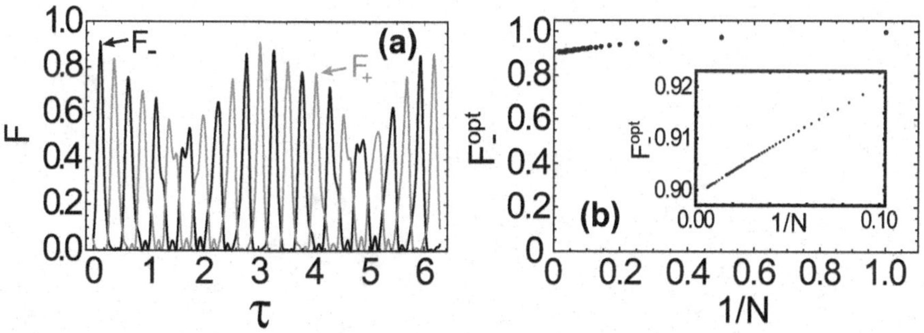

Fig. 10.8 (a) Fidelities of the two-axis two-spin squeezed state (10.68) with respect to the spin-EPR states (10.89) and (10.93). Time dependence of the fidelity for $N = 20$. (b) Optimized fidelity (10.91) as a function of $1/N$. Inset shows zoomed in region for large N. Color version of this figure available at cambridge.org/quantumatomoptics.

the interaction time τ such as to maximize F_- in the region of the optimal squeezing time. The optimal time in terms of fidelity is again found to be most similar to times when $\langle S_z \rangle = 0$, but not precisely the same. The state approaches $F_- \approx 0.9$, showing that the 2A2S squeezed state has a high overlap with the $|\text{EPR}_-\rangle$ state.

Exercise 10.5.1 Show that the form of the spin-EPR state (10.90) is algebraically equivalent to (10.87). Hint: Use the definition (5.162) of the rotated Fock states and show that the rotation factors can be eliminated.

10.6 References and Further Reading

- Section 10.2: General reviews on entanglement [188, 231, 378, 11, 154].
- Section 10.2.1: Original reference on von Neumann entropy [475]. For further reviews and books on entropy [345, 482].
- Section 10.2.2: Original works for the PPT criterion [364, 229, 230]. Negativity entanglement measure for mixed states [376, 474].
- Section 10.3.1: The Duan–Giedke–Cirac Zoller criterion [124]. Other variance-based entanglement criteria [171, 222].
- Section 10.3.2: The Hillery–Zubairy criterion [215].
- Section 10.3.3: Methods of constructing entanglement witnesses [226, 301, 454, 449].
- Section 10.3.4: The connection between covariance matrices and uncertainty relations [398]. The covariance matrix entanglement criterion for optical quadratures was introduced in [429]. This was then generalized to arbitrary operators [468].
- Section 10.4: In an atomic context, the one-axis two-spin squeezed states were theoretically investigated in [71, 277]. Various proposals for producing this in cold atoms [381, 466, 246, 402].
- Section 10.5: Theoretical investigation of the two-axis two-spin squeezed states [264].

11 Quantum Information Processing with Atomic Ensembles

11.1 Introduction

In this chapter, we will give an overview of how atomic ensembles can be used for quantum information processing. What we mean here by "quantum information processing" is in a different sense from that used to describe the whole field of quantum information. We have seen up to this point that atomic ensembles can be used for a variety of different applications, ranging from matter wave and spin interferometry, which can be applied to precision measurements (the field of "quantum metrology"), and quantum simulations of various many-body systems. The field of quantum information encompasses these types of applications in the sense that they are all applications of quantum mechanics. On the other hand, "quantum information processing" refers to tasks that are more connected to information and how quantum operations can be performed on them. The tasks that we will discuss are simple protocols such as quantum teleportation, which has a fundamental place in the field as a simple yet nontrivial operation. Other tasks include quantum computation, where the aim is to perform algorithmic tasks that cannot be performed using a conventional classical computer. In particular, we will examine Deutsch's algorithm as a simple example showing quantum mechanical speedup, and how it is implemented with atomic ensembles. Another example shows how adiabatic quantum computing can be performed with atomic ensembles.

11.2 Continuous Variables Quantum Information Processing

11.2.1 Mapping between Spin and Photonic Variables

The first and simplest way that atomic ensembles can be used for quantum information processing is to use the Holstein–Primakoff transformation (see Section 5.9) to approximate spin variables as quadrature operators. Consider an atomic ensemble that contains a large number of atoms $N \gg 1$ and is strongly polarized in the $+S_x$-direction (e.g., the maximally polarized state $|1/\sqrt{2}, 1/\sqrt{2}\rangle\rangle$). Writing the spin operators in the x basis using the transformation (5.79), we have

$$
\begin{aligned}
S_x &= a_x^\dagger a_x - b_x^\dagger b_x \\
S_y &= -i b_x^\dagger a_x + i a_x^\dagger b_x \\
S_z &= b_x^\dagger a_x + a_x^\dagger b_x.
\end{aligned}
\tag{11.1}
$$

Since the state is polarized in the $+S_x$-direction, we expect that the number of atoms in the b_x state will be small, $\langle b_x^\dagger b_x \rangle \approx 0$. Applying the Holstein–Primakoff transformation (5.118), we may write the spin operators as

$$\begin{aligned} S_x &\approx N \\ S_y &\approx \sqrt{N}(-ib_x + ib_x^\dagger) = \sqrt{2N}p \\ S_z &\approx \sqrt{N}(b_x + b_x^\dagger) = \sqrt{2N}x, \end{aligned} \tag{11.2}$$

where we used the definitions of the quadrature operators (5.13).

One of the main techniques to visualize photonic states is to plot the (optical) Wigner function of a quantum optical state in phase space (x, p). We can visualize the state in a similar way using the (spin) Wigner functions, as discussed in Section 5.11.2. For example, the maximally polarized state $|1/\sqrt{2}, 1/\sqrt{2}\rangle\rangle$, which corresponds to the photonic vacuum state, has a similar Wigner function to the optical case, as shown in Fig. 5.5(a). Performing a one-axis twisting squeezing operation on such a state produces a squeezed state, as shown in Fig. 5.5(e) and (f). In addition to squeezing operations, one can perform displacement operations that rotate the maximally polarized state to a spin coherent state centered at a different point using (5.59). This would be performed by applying a Hamiltonian $\cos\theta S_z + \sin\theta S_y$, which rotates the state away from the maximally S_x-polarized state. As long as the displacement is not too large, so that $\langle b_x^\dagger b_x \rangle$ is small, this corresponds to an optical coherent state, as discussed in (5.10). Such states are all dominantly polarized in the $+S_x$-direction, which is why they have an equivalence to the photonic case.

States that deviate significantly from the maximally $+S_x$-polarized state cannot be mapped exactly to photonic states since the Holstein–Primakoff approximation breaks down. As a simple example, a spin coherent state that is rotated too far from the $+S_x$-polarized state will have different properties from a genuine optical state. An optical coherent state can be displaced in any direction by an arbitrary amount,

$$e^{\beta a^\dagger - \beta^* a}|\alpha\rangle = |\alpha + \beta\rangle, \tag{11.3}$$

where the states are optical coherent states (5.10). In contrast, a displacement of a spin coherent state (5.59) will eventually result in the coherent state returning to the original position after a rotation angle of 2π (see Fig. 5.4). One can visualize that the Holstein–Primakoff approximation is performing quantum operations such that it is on a locally flat region of the Bloch sphere.

For measurements, two common types of optical measurements that are performed are photon counting and homodyne measurements. For the former, according to the transformation (11.2), it is evident that counting the number of atoms in the b_x state corresponds to a projection to the photon Fock basis. For homodyne measurements, one would like to measure the expectation value of the generalized quadrature operator:

$$x(\theta) = \frac{1}{\sqrt{2}}(e^{-i\theta}a + e^{i\theta}a^\dagger) \leftrightarrow \frac{\cos\theta S_z + \sin\theta S_z}{\sqrt{2N}}. \tag{11.4}$$

The measurement of such an operator can be made by measurement of spin operators in the basis $\cos\theta S_z + \sin\theta S_z$.

The above shows that one atomic spin ensemble can be used as an effective photonic mode. This can be extended to the multimode case by simply having more atomic ensembles. One important operation that is required between modes is to perform entangling operations between them. We have discussed in Section 10.4 one type of entangling operation that can be produced between atomic ensembles. In the Holstein–Primakoff approximation, this corresponds to a $x_A x_B$ type of Hamiltonian. Other types of quantum states can also be produced that are analogous to two-mode squeezing and have been produced by shining a common light beam through two atomic ensembles (see [252]).

Thus, we see that many of the operations that are possible in the photonic case have exact analogues to the atomic case. One operation that is somewhat less natural in the atomic case in comparison to the photonic case is the beam-splitter operation, which mixes two photonic modes,

$$
\begin{aligned}
a &\to \cos\theta a + \sin\theta a' \leftrightarrow \frac{\cos\theta}{2\sqrt{N}}(S_z + iS_y) + \frac{\sin\theta}{2\sqrt{N'}}(S_z' + iS_y') \\
a' &\to \sin\theta a - \cos\theta a' \leftrightarrow \frac{\sin\theta}{2\sqrt{N}}(S_z + iS_y) - \frac{\cos\theta}{2\sqrt{N'}}(S_z' + iS_y'),
\end{aligned}
\tag{11.5}
$$

where the dashes denote the second photonic mode or second atomic ensemble. Physically, this would correspond to mixing two atomic ensembles and interfering the spins with one another. While it is not impossible to perform such an operation using matter interferometry techniques, this is more experimentally challenging than performing rotations and squeezing operations between atomic levels on the same ensemble.

11.2.2 Example: Quantum Teleportation

We now give an example of a quantum information protocol that can be implemented using the Holstein–Primakoff approximation. In quantum teleportation, an unknown quantum state is transferred from one party (which we refer to as Alice) to another (referred to as Bob) with the use of entanglement and classical communication. The teleportation protocol is most easily represented as a optical quantum circuit diagram, as shown in Fig. 11.1. The sequence of operations consists of the following steps:

1. Produce a two-mode squeezed state on modes a and b.
2. Prepare a state in mode a' to be teleported.
3. Interfere modes a and a' with a 50/50 beam splitter.
4. Measure the output modes c and c' with respect to the p and x quadratures respectively and send the results to Bob.
5. Perform a conditional displacement on mode b based on the measurement outcomes.

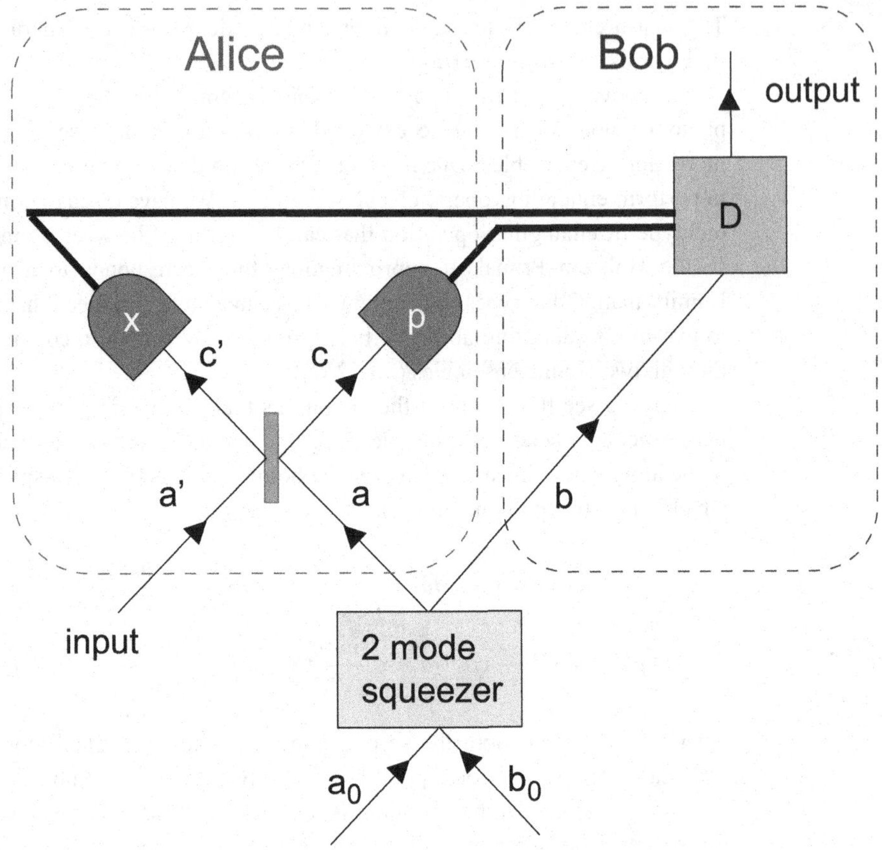

Fig. 11.1 The quantum teleportation protocol for continuous variable optics. Thin lines denote the optical modes and thick lines denote classical communication. A two-mode squeezed state is produced in modes *a* and *b*. The state to be teleported in mode *a′* enters a beam splitter interfering with mode *a*. The modes exiting the beam splitter are measured using a homodyne detectors, measuring the *x* and *p* quadratures. The results of the measurements are sent to a displacement operator *D*, which makes a displacement of the mode *b*. Color version of this figure available at cambridge.org/quantumatomoptics.

The simplest way that the protocol can be described is in the Heisenberg picture. Let us work through each of the steps in the protocol above and obtain the transformations in the mode operators.

The two-mode squeezing operation corresponds to applying the operator $U = e^{-r(a_0 b_0 - a_0^\dagger b_0^\dagger)}$. This produces the transformation

$$\begin{aligned} a &= a_0 \cosh r + b_0^\dagger \sinh r \\ b &= b_0 \cosh r + a_0^\dagger \sinh r, \end{aligned} \tag{11.6}$$

where r is the squeezing parameter. In terms of quadrature operators, using (5.13) we have

$$x_a = \frac{1}{\sqrt{2}}(e^r x_{a_0} + e^{-r} x_{b_0})$$

$$p_a = \frac{1}{\sqrt{2}}(e^{-r} p_{a_0} + e^r p_{b_0})$$

$$x_b = \frac{1}{\sqrt{2}}(e^r x_{a_0} - e^{-r} x_{b_0})$$
$$p_b = \frac{1}{\sqrt{2}}(e^{-r} p_{a_0} - e^r p_{b_0}), \tag{11.7}$$

where $x_a = (a + a^\dagger)/\sqrt{2}$ and $p_a = -i(a - a^\dagger)/\sqrt{2}$ and similarly for the other modes. This state has correlations in the variables $x_a - x_b$ and $p_a + p_b$ because these are exponentially suppressed according to

$$x_a - x_b = \sqrt{2}e^{-r} x_{b_0}$$
$$p_a + p_b = \sqrt{2}e^{-r} p_{b_0}. \tag{11.8}$$

The next step is to prepare a quantum state in mode a. In the Heisenberg picture, we evolve the operators, hence no calculation is required since the state remains time invariant. Next, we interfere the modes a and a', giving the transformation

$$c' = \frac{1}{\sqrt{2}}(a' - a)$$
$$c = \frac{1}{\sqrt{2}}(a' + a). \tag{11.9}$$

In terms of quadrature operators, this is

$$x_{c'} = \frac{1}{\sqrt{2}}(x_{a'} - x_a) \tag{11.10}$$

$$p_{c'} = \frac{1}{\sqrt{2}}(p_{a'} - p_a) \tag{11.11}$$

$$x_c = \frac{1}{\sqrt{2}}(x_{a'} + x_a) \tag{11.12}$$

$$p_c = \frac{1}{\sqrt{2}}(p_{a'} + p_a). \tag{11.13}$$

Next, the modes c and c' are measured, and the $x_{c'}$ and p_c quantum variables collapse to classical random variables $x_{c'}^M$ and p_c^M, respectively. These two variables are at this point classical c-numbers and are classically communicated to Bob.

Now let us look at the quadratures of Bob's mode before it enters the displacement operation D. We can rearrange the correlations (11.8) to write

$$x_b = x_a - \sqrt{2}e^{-r} x_{b_0} = x_{a'} - \sqrt{2}x_{c'}^M - \sqrt{2}e^{-r} x_{b_0}$$
$$p_b = -p_a + \sqrt{2}e^{-r} p_{b_0} = p_{a'} - \sqrt{2}p_c^M + \sqrt{2}e^{-r} p_{b_0}, \tag{11.14}$$

where we used (11.10) and (11.13) to rewrite x_a and p_a. For an infinitely squeezed state $r \to \infty$, the last term in the above quadratures becomes zero. This shows that the quadratures of the mode b are the same as for the mode a', up to a constant offset. Since the offsets $x_{c'}^M$ and p_c^M are classical numbers, these can be canceled by a displacement operation:

$$D(x_{c'}^M, p_c^M) = e^{-\sqrt{2}(x_{c'}^M + ip_c^M)b^\dagger + \sqrt{2}(x_{c'}^M - ip_c^M)b}. \tag{11.15}$$

This completes the teleportation operation, and the mode emerging from the displacement operation has the same state as the input mode a'.

The above protocol was first performed experimentally in 1998 with optical continuous variable modes, following the above optical circuit. In the context of atomic ensembles, the above protocol was first performed by Krauter, Polzik, and coworkers in 2013 using two atomic ensembles and optical modes. In their experiment, the atomic ensembles were trapped in paraffin-coated glass cells, where the effect of the paraffin was to prolong the spin coherence time despite the numerous collisions the atoms undergo with the glass walls. Polarized atomic ensembles were used to realize the modes a' and b in the Holstein–Primakoff approximation. Meanwhile, mode a was implemented by an optical mode. A light mode was sent through to the atomic ensemble corresponding to mode b, producing a two-mode squeezed state between the ensemble and the light. The emerging light mode a then interacted with the second atomic ensemble corresponding to the mode a', this time with a beam-splitter type of interaction. The light mode was then measured, and the classical results were fed back to the atomic ensemble corresponding to mode b. For further details, we refer the reader to the original experiment [274].

The above example shows the way in which atomic ensembles can be used to perform an experiment that is initially designed for optical modes. Numerous other examples of continuous variables quantum information processing are possible, which we do not reproduce here. We refer the reader to the excellent review for other protocols that can be implemented [59].

11.3 Spinor Quantum Computing

In Section 11.2.2, we saw that it is possible to use spin ensembles as effective bosonic modes that can be used for continuous variables quantum information processing. In order to perform this mapping, the Holstein–Primakoff approximation was used, which requires that the spin ensembles are dominantly polarized in the S_x direction. This means that one must always work with states that are close to the fully polarized S_x eigenstate, or in terms of the Wigner or Q-distribution, it must be primarily centered around $\theta = \phi = 0$ (see Figs. 5.5(a) and 5.6(a)). This is a restriction in terms of the types of states that can be used, and experimentally there is no reason to look only at such a small class of states. We may ask whether it is possible to use more of the Hilbert space available to the ensembles. Due to the strong analogy between quantum optical states and spin ensemble states, which we have seen in Chapter 5, this suggests that it might be possible to find a way of performing quantum information processing that is more naturally suited to the structure of states that one encounters with spin ensembles.

The most natural way to encode a single qubit with ensembles is to use the spin coherent states that we encountered in Section 5.2. Suppose that one wishes to encode the quantum information associated with a single qubit. This could be part of a quantum memory in a quantum computer, for example, that could be used as part of an quantum algorithm. Physically, this could be implemented using thermal atomic ensembles in glass cells, as described in Section 11.2.2, or with atom chips, which can be patterned to have multiple traps. The same quantum information could be stored in a qubit or a spin coherent state, according to the mapping

$$\alpha|0\rangle + \beta|1\rangle \to |\alpha, \beta\rangle\rangle = \frac{1}{\sqrt{N!}}(\alpha a^\dagger + \beta b^\dagger)^N|0\rangle. \tag{11.16}$$

Each of these states contains exactly the same amount of quantum information, since they involve the same parameters α, β. Of course, the spin ensemble could potentially have many other types of states, not only spin coherent states, since the Hilbert space dimension is in fact much larger. However, for the set of all spin coherent states, there is a one-to-one mapping between the qubit states and spin coherent states.

Single qubit gates also have a perfect analogy with spin coherent states. As we saw in (5.59), applying a Hamiltonian $H = \boldsymbol{n} \cdot \boldsymbol{S}$ on a spin coherent state has exactly the corresponding effect for qubits, in terms of the mapping of the parameters $(\alpha, \beta) \to (\alpha', \beta')$. For example, common gates such as the Z phase flip gate, X bit flip gate, and Hadamard gate are performed by the operations

$$e^{-i\pi S_z/2}|\alpha, \beta\rangle\rangle = e^{-i\pi N/2}|\alpha, -\beta\rangle\rangle \qquad (Z\text{-gate}) \tag{11.17}$$

$$e^{-i\pi S_x/2}|\alpha, \beta\rangle\rangle = (-i)^N|\beta, \alpha\rangle\rangle \qquad (X\text{-gate}) \tag{11.18}$$

$$e^{-i3\pi S_y/4}|\alpha, \beta\rangle\rangle = (-1)^N\left|\frac{\alpha+\beta}{\sqrt{2}}, \frac{\alpha-\beta}{\sqrt{2}}\right\rangle\rangle. \qquad (\text{Hadamard gate}) \tag{11.19}$$

We have written the global phase factors to match the expressions in (5.62), but they are physically irrelevant.

For two qubit gates, the types of states that are produced with ensembles are analogous, but do not have the perfect equivalence as single qubits. A typical type of interaction that might be present between ensembles is the $H = S_z^A S_z^B$ interaction, where A and B label the two ensembles. For the qubit case, the analogous Hamiltonian is $H = \sigma_z^A \sigma_z^B$ and can produce an entangled state:

$$e^{-i\sigma_z^A\sigma_z^B\tau}\left(\frac{|0\rangle+|1\rangle}{\sqrt{2}}\right)\left(\frac{|0\rangle+|1\rangle}{\sqrt{2}}\right) = \left(\frac{e^{i\tau}|0\rangle + e^{-i\tau}|1\rangle}{\sqrt{2}}\right)|0\rangle + \left(\frac{e^{-i\tau}|0\rangle + e^{i\tau}|1\rangle}{\sqrt{2}}\right)|1\rangle. \tag{11.20}$$

For a time $\tau = \pi/4$, the above state becomes a maximally entangled state,

$$e^{-i\sigma_z^A\sigma_z^B\pi/4}\left(\frac{|0\rangle+|1\rangle}{\sqrt{2}}\right)\left(\frac{|0\rangle+|1\rangle}{\sqrt{2}}\right) = \frac{1}{\sqrt{2}}(|-y\rangle|0\rangle + |+y\rangle|1\rangle), \tag{11.21}$$

where the eigenstates of the σ^y Pauli operator are

$$|+y\rangle = \frac{|0\rangle + i|1\rangle}{\sqrt{2}}$$
$$|-y\rangle = \frac{i|0\rangle + |1\rangle}{\sqrt{2}}. \tag{11.22}$$

Such a Hamiltonian can be used as the basis of a controlled-NOT (CNOT) gate, which produces a maximally entangled state.

For ensembles, the $H = S_z^A S_z^B$ interaction produces the state (10.59):

$$e^{-iS_z^A S_z^B \tau} \left| \frac{1}{\sqrt{2}}, \frac{1}{\sqrt{2}} \right\rangle\!\!\right\rangle_A \left| \frac{1}{\sqrt{2}}, \frac{1}{\sqrt{2}} \right\rangle\!\!\right\rangle_B$$

$$= \frac{1}{\sqrt{2^N}} \sum_k \sqrt{\binom{N}{k}} \left| \frac{e^{i(N-2k)\tau}}{\sqrt{2}}, \frac{e^{-i(N-2k)\tau}}{\sqrt{2}} \right\rangle\!\!\right\rangle_A |k\rangle_B$$

$$= \frac{1}{\sqrt{2^N}} \left[\left| \frac{e^{iN\tau}}{\sqrt{2}}, \frac{e^{-iN\tau}}{\sqrt{2}} \right\rangle\!\!\right\rangle_A |0\rangle_B + \sqrt{N} \left| \frac{e^{i(N-2)\tau}}{\sqrt{2}}, \frac{e^{-i(N-2)\tau}}{\sqrt{2}} \right\rangle\!\!\right\rangle_A |1\rangle_B + \cdots \right.$$

$$\left. + \sqrt{\binom{N}{N/2}} \left| \frac{1}{\sqrt{2}}, \frac{1}{\sqrt{2}} \right\rangle\!\!\right\rangle_A |N/2\rangle_B + \cdots + \left| \frac{e^{-iN\tau}}{\sqrt{2}}, \frac{e^{iN\tau}}{\sqrt{2}} \right\rangle\!\!\right\rangle_A |N\rangle_B \right]. \tag{11.23}$$

Comparing (11.20) and (11.23), we see that a similar type of state is produced where the first ensemble is rotated by an angle $2(N - 2k)\tau$ around the S_z axis, for a Fock state $|k\rangle$ on the second ensemble (see Fig. 10.2). The types of correlations are similar for both the qubit and ensemble cases, but there are also differences. The first, most obvious difference is that the ensemble consists of $N + 1$ terms, whereas the qubit version only has two terms. Thus, although the same type of correlation is present to the qubit entangled state, it is a higher dimensional generalization. The other difference is the presence of the binomial factor, which weights the terms. The binomial function approximately follows a Gaussian distribution with a standard deviation $\sim\sqrt{N}$. Thus, the most important terms are in the range $k \in [N/2 - \sqrt{N}, N/2 + \sqrt{N}]$.

Using a sequence of single and two-ensemble Hamiltonians, it is possible to produce a large range of effective Hamiltonians. A well-known quantum information theorem states that if it is possible to perform an operation with Hamiltonians H_A and H_B, then it is also possible to perform the operation corresponding to $H_C = i[H_A, H_B]$ [311]. Therefore, the combination of single and two-ensemble Hamiltonians may be combined to form more complex effective Hamiltonians. Using this theorem, one can show that for an M ensemble system, it is possible to produce any Hamiltonian of the form

$$H_{\text{eff}} \propto \prod_{n=1}^{M} S_{j(n)}^{(n)}, \tag{11.24}$$

where $j(n) \in \{0, x, y, z\}$ and $S_0 = I$. An arbitrary sum of such a Hamiltonian may also be produced by performing a Trotter expansion [311]. For Bose–Einstein condensate (BEC) qubits in general, higher-order operators can also be constructed (e.g., $(S_j)^l$ with $l \geq 2$).

The above results suggest that it should be possible to use the single and two-ensemble Hamiltonians $H_n = \boldsymbol{n} \cdot \boldsymbol{S}^{(n)}$ and $H_{nm} = S_z^{(n)} S_z^{(m)}$ together to perform various quantum information processing tasks. We show several examples that demonstrate this explicitly in the following sections. When constructing a quantum algorithm using ensembles, often there is no unique mapping since the Hilbert space of the ensemble case is much larger. Thus, there is a lot of freedom in implementing a given quantum algorithm using ensembles, with some mappings being more favorable than others. We now discuss what factors should be included in a good mapping from qubits to ensembles.

Equivalence of the Circuit

The first most obvious requirement is that the ensemble version of the algorithm does in fact perform the same quantum computation as its qubit counterpart. Once the quantum algorithm is complete, one should be able to view the result of quantum computation on a readout of the BEC qubits, which may involve some simple encoding rule to obtain the standard qubit version. As we show in the example for Deutsch's algorithm below, this may also include the requirement that certain circuit elements, such as the oracle, behave in the same way as the qubit counterparts when operated outside of the quantum circuit.

Elementary Gates

Generally, it is assumed that the only operations that are available are the single and two-ensemble Hamiltonians $H_n = \boldsymbol{n} \cdot \boldsymbol{S}^{(n)}$ and $H_{nm} = S_z^{(n)} S_z^{(m)}$. This is because these operations are readily producible experimentally. The same goes for the measurements that are performed; these should be in a basis that can be implemented in a realistic way in the lab. This might involve measuring ensembles in the S_z basis, for example.

Decoherence

The algorithm should be robust against decoherence. This is a particularly important issue, since the spin ensembles may use a macroscopic number of atoms. Encoding of the quantum information using particularly sensitive states – such as Schrodinger cat states – will generally not be practical, because in the presence of decoherence, the quantum algorithm will not complete with high fidelity.

Algorithmic Complexity

The complexity of the mapped algorithm should still retain the quantum speedup of the original qubit version of the algorithm. Here, the way complexity is calculated is with respect to the number of gate operations of the Hamiltonians H_n and H_{nm} are executed. Thus, gates such as (11.17)–(11.19) would count as 1 gate. Although each ensemble may include a large number of atoms N, there is little dependence to the experimental complexity with N, justifying this way of counting resources.

In Fig. 11.2, we summarize the differences between qubit, continuous variables, and spinor quantum computing. We can see that spinor quantum computing is in many ways a hybrid approach of the qubit and continuous variables approaches. Spin ensembles have a similar structure of states in terms of the Bloch sphere, and

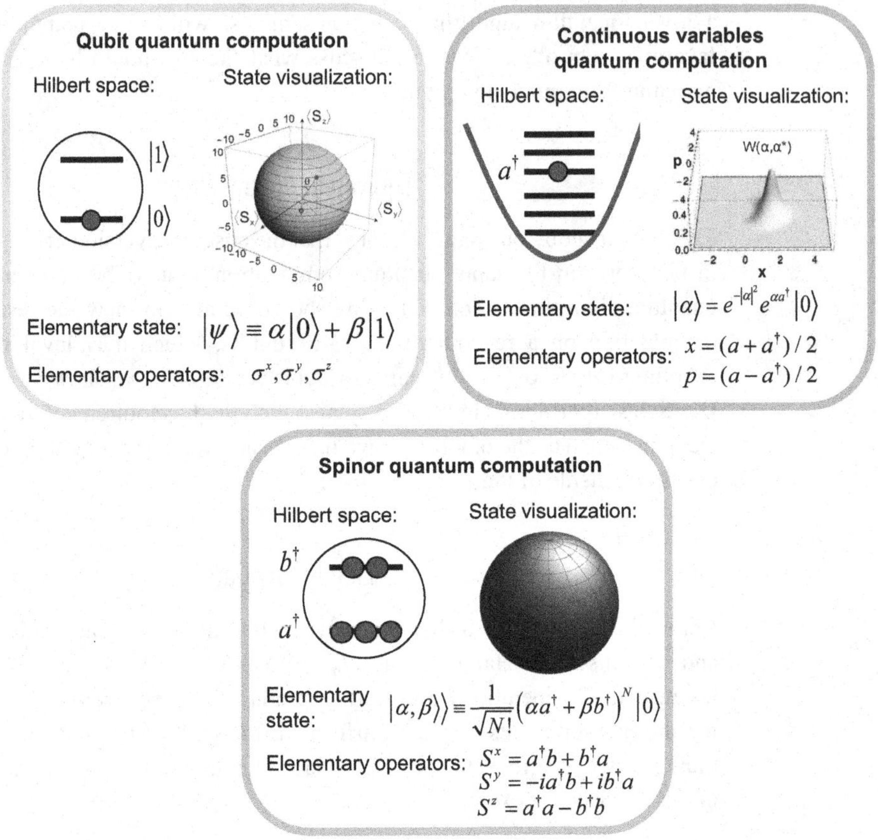

Fig. 11.2 Relationship between qubit, continuous variables, and spinor quantum computation. For each case, the Fock states spanning the Hilbert space, a typical state visualization using Wigner or Q-functions, an elementary quantum state, and the Hamiltonian operators used to manipulate them are shown. Color version of this figure available at cambridge.org/quantumatomoptics.

their operators have similar commutation relations. Meanwhile, spin ensembles also are similar to continuous variables in that large Hilbert spaces are used to store the quantum information.

11.4 Deutsch's Algorithm

An example of a simple quantum algorithm where we see a quantum advantage is Deutsch's algorithm. This is a straightforward example where it is possible to implement the same algorithm on ensembles, in a regime beyond the Holstein–Primakoff approximation. We first briefly explain how the algorithm works for standard qubits and then show an explicit translation to spin ensembles.

11.4.1 Standard Qubit Version

Consider the binary function $f(x)$, which takes a binary variable x as its argument and returns another binary number. There are in fact only four types of such

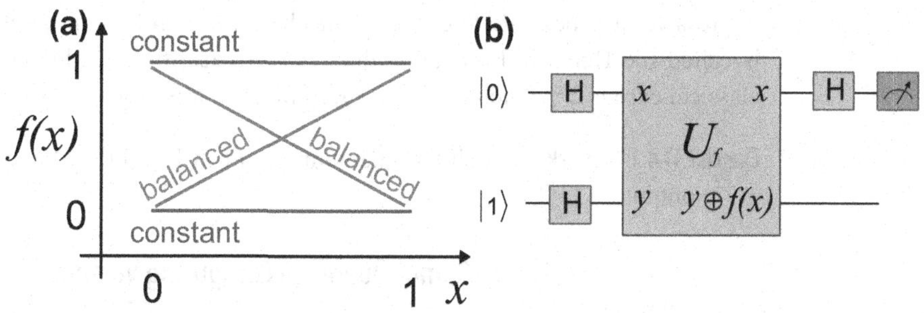

Fig. 11.3 Deutsch's algorithm. (a) The four constant and balanced algorithms as implemented by the oracle. (b) The quantum circuit implementing Deutsch's algorithm. Color version of this figure available at cambridge.org/quantumatomoptics.

functions; they are shown in Fig. 11.3(a). Two of these functions always give the same output regardless of the input x. These are called *constant* functions. The other two functions, which do have a dependence on x, are called *balanced* functions. Now, say that someone gives you a black box that gives the output $f(x)$ given the input x. In order to identify whether a particular black box was in the constant or balanced categories, classically, at least two evaluations would be necessary. For instance, if it is found that $f(0) = 0$, this rules out two of the four functions, but one cannot tell which of the remaining two it is, of which there is one constant and one balanced.

Now let us see if we can achieve the same task in a quantum mechanical setting. The quantum version of the black box is the oracle, which performs the same operation as the black box when given classical inputs. The main difference with a quantum mechanical operation is that we must make sure that any operation we perform is a unitary operation, which is always a reversible operation. At the moment, the black box is not necessarily reversible. For example, for the constant function $f(x) = 1$, you would not be able to work out what x was from the output, since in both cases the same result of 1 is the result. The way to make the oracle reversible is to have an oracle with two inputs and outputs, following the operation

$$U_f|x\rangle|y\rangle = |x\rangle|y \oplus f(x)\rangle, \tag{11.25}$$

where $x, y \in \{0, 1\}$ and $\oplus$ is addition modulo 2. It is clear that this is a reversible operation, since one can perform the same gate twice and recover the original state $U_f^2 = I$. The oracle works in essentially the same way as the black box mentioned above. For example, if one sets $y = 0$ and chooses an input x, the output of the second register is the same as the output of the black box.

Quantum mechanically, it is possible to work out whether a given oracle is constant or balanced with only one evaluation. The quantum circuit for this is shown in Fig. 11.3(b). Following the circuit, we can show that the output of the circuit is

$$\begin{cases} |0\rangle|-\rangle & \text{if} f(0) = f(1) \\ |1\rangle|-\rangle & \text{if} f(0) \neq f(1) \end{cases}, \tag{11.26}$$

where we have discarded irrelevant global phases. Therefore, the output of the first qubit is dependent only upon whether the oracle is balanced or constant. Since in the quantum circuit the oracle is only called once, this gives a speedup of a factor of 2 compared to the classical case. While a factor of 2 speedup does not sound too impressive, it is possible to extend Deutsch's algorithm to an n-qubit oracle, which is called the Deutsch–Jozsa algorithm. In this case, the speedup compared to the classical case is 2^n, which is a huge speedup when n is large.

Exercise 11.4.1 Work through the quantum circuit in Fig. 11.3(b) and show that the output is (11.26).

11.4.2 Spinor Quantum Computing Version

We now show that the above logical operations can also be performed using atomic ensembles. Following Section 11.3, we can map the input states, such that the input state is

$$|0,1\rangle\rangle|1,0\rangle\rangle. \tag{11.27}$$

The Hadamard gates then operate on this state, such that the state before the oracle is

$$\left|\frac{1}{\sqrt{2}},\frac{1}{\sqrt{2}}\right\rangle\!\!\right\rangle\left|\frac{1}{\sqrt{2}},-\frac{1}{\sqrt{2}}\right\rangle\!\!\right\rangle. \tag{11.28}$$

Now we must make sure that the oracle can be mapped in a way that has an analogous operation to the qubit example. To show that the ensemble implementation of the oracle has the same effect as the qubit implementation, we distinguish two cases when the input states $|x\rangle$ and $|y\rangle$ are (1) in the computational basis $\{|0\rangle,|1\rangle\}$ and (2) involve superpositions such as the $\{|+\rangle,|-\rangle\}$ states. For the input states (1), we will say that the oracle is working in "classical mode," and for (2) we will say it is working in "quantum mode." Of course, for genuine qubits the quantum mode is simply the superposition of the classical cases, so this never needs to be distinguished. However, for the ensemble case, since a state

$$\left|\frac{1}{\sqrt{2}},\frac{1}{\sqrt{2}}\right\rangle\!\!\right\rangle \neq \frac{|1,0\rangle\rangle+|0,1\rangle\rangle}{\sqrt{2}}, \tag{11.29}$$

we must ensure that the oracle works in the desired way for both cases.

It is possible to implement the oracle in many different ways, but a particularly simple choice corresponds to taking the Hamiltonians for the four cases

$$H_{f=0} = 0 \tag{11.30}$$

$$H_{f=1} = S_2^x - N \tag{11.31}$$

$$H_{f=\{1,0\}} = \frac{1}{2}\left(1 - S_1^z/N\right)(S_2^x - N) \tag{11.32}$$

$$H_{f=\{0,1\}} = \frac{1}{2}\left(1 + S_1^z/N\right)(S_2^x - N), \tag{11.33}$$

and it is implied that we evolve these for a time $t = \pi/2$. First, let us verify that in classical mode, the Hamiltonians produce the correct outputs,

$$e^{-iH_{f=0}t}|x,1-x\rangle\rangle|y,1-y\rangle\rangle = |x,1-x\rangle\rangle|y,1-y\rangle\rangle$$
$$e^{-iH_{f=1}t}|x,1-x\rangle\rangle|y,1-y\rangle\rangle = |x,1-x\rangle\rangle|1-y,y\rangle\rangle$$
$$e^{-iH_{f=\{1,0\}}t}|0,1\rangle\rangle|y,1-y\rangle\rangle = |0,1\rangle\rangle|1-y,y\rangle\rangle$$

$$e^{-iH_{f=\{1,0\}}t}|1,0\rangle\rangle|y,1-y\rangle\rangle = |1,0\rangle\rangle|y,1-y\rangle\rangle$$
$$e^{-iH_{f=\{0,1\}}t}|0,1\rangle\rangle|y,1-y\rangle\rangle = |0,1\rangle\rangle|y,1-y\rangle\rangle$$
$$e^{-iH_{f=\{0,1\}}t}|1,0\rangle\rangle|y,1-y\rangle\rangle = |1,0\rangle\rangle|1-y,y\rangle\rangle, \quad (11.34)$$

where $x, y \in \{0, 1\}$ and we have discarded any irrelevant global phase factors. The above shows that the Hamiltonians for the oracles (11.33) give the same results as for qubits when operated in classical mode.

Operating in quantum mode, after the initial Hadamard gates, the Hamiltonian is applied on the state (11.28). For the constant cases, the resulting state is

$$e^{-iH_{f=0}t}\left|\frac{1}{\sqrt{2}},\frac{1}{\sqrt{2}}\right\rangle\!\!\right\rangle\left|\frac{1}{\sqrt{2}},-\frac{1}{\sqrt{2}}\right\rangle\!\!\right\rangle = \left|\frac{1}{\sqrt{2}},\frac{1}{\sqrt{2}}\right\rangle\!\!\right\rangle\left|\frac{1}{\sqrt{2}},-\frac{1}{\sqrt{2}}\right\rangle\!\!\right\rangle \quad (11.35)$$

$$e^{-iH_{f=1}t}\left|\frac{1}{\sqrt{2}},\frac{1}{\sqrt{2}}\right\rangle\!\!\right\rangle\left|\frac{1}{\sqrt{2}},-\frac{1}{\sqrt{2}}\right\rangle\!\!\right\rangle = \left|\frac{1}{\sqrt{2}},\frac{1}{\sqrt{2}}\right\rangle\!\!\right\rangle\left|\frac{1}{\sqrt{2}},-\frac{1}{\sqrt{2}}\right\rangle\!\!\right\rangle, \quad (11.36)$$

which is the same state up to a global phase. For the balanced cases, the resulting state is

$$e^{-iH_{f=\{1,0\}}t}\left|\frac{1}{\sqrt{2}},\frac{1}{\sqrt{2}}\right\rangle\!\!\right\rangle\left|\frac{1}{\sqrt{2}},-\frac{1}{\sqrt{2}}\right\rangle\!\!\right\rangle = \left|\frac{1}{\sqrt{2}},-\frac{1}{\sqrt{2}}\right\rangle\!\!\right\rangle\left|\frac{1}{\sqrt{2}},-\frac{1}{\sqrt{2}}\right\rangle\!\!\right\rangle \quad (11.37)$$

$$e^{-iH_{f=\{1,0\}}t}\left|\frac{1}{\sqrt{2}},\frac{1}{\sqrt{2}}\right\rangle\!\!\right\rangle\left|\frac{1}{\sqrt{2}},-\frac{1}{\sqrt{2}}\right\rangle\!\!\right\rangle = \left|\frac{1}{\sqrt{2}},-\frac{1}{\sqrt{2}}\right\rangle\!\!\right\rangle\left|\frac{1}{\sqrt{2}},-\frac{1}{\sqrt{2}}\right\rangle\!\!\right\rangle, \quad (11.38)$$

which is again the same state up to a global phase.

Finally, after the Hadamard gate, the state for constant cases become

$$|0,1\rangle\rangle\left|\frac{1}{\sqrt{2}},-\frac{1}{\sqrt{2}}\right\rangle\!\!\right\rangle. \quad (11.39)$$

For the balanced cases, the state is

$$|1,0\rangle\rangle\left|\frac{1}{\sqrt{2}},-\frac{1}{\sqrt{2}}\right\rangle\!\!\right\rangle. \quad (11.40)$$

This is the equivalent result as the qubit case (11.26). Since the quantum circuit for the ensemble case is the same as the qubit case, and the oracle is called only once, this has the same quantum speedup over the classical case by a factor of two.

The Hamiltonians (11.33) involve either single ensemble rotations (5.62) or two-ensemble interactions of the form (10.1). Furthermore, the time evolution of the interaction term is $\tau = \pi/4N$ in dimensionless units. From the discussion in Section 10.4, we know that the types of states that are generated with this interaction do not involve fragile Schrodinger cat–like states. This means that the ensemble version of the quantum algorithm satisfies the requirements discussed in Section 11.3 and is a satisfactory mapping to spinor quantum computing.

11.5 Adiabatic Quantum Computing

Adiabatic quantum computing (AQC) is an alternative approach to traditional gate-based quantum computing, where quantum adiabatic evolution is performed in order to achieve a computation. In the scheme, the aim is to find the ground state of a Hamiltonian H_Z that encodes the problem to be solved. In addition, an initial Hamiltonian H_X that does not commute with the problem Hamiltonian is prepared, where the ground state is known. For example, in the qubit formulation, common choices for these Hamiltonians are

$$H_Z = \sum_{i=1}^{M}\sum_{j=1}^{M} J_{ij}\sigma_z^{(i)}\sigma_z^{(j)} + \sum_{i=1}^{M} K_i\sigma_z^{(i)} \tag{11.41}$$

$$H_X = -\sum_{i=1}^{M} \sigma_x^{(i)}, \tag{11.42}$$

where $\sigma_{x,z}^{(i)}$ are Pauli matrices on site i, J_{ij} and K_i are coefficients that determine the problem to be solved, and there are M qubits. We take $J_{ij} = J_{ji}$ and $J_{ii} = 0$. The form of (11.41) as chosen is rather general and can encode a wide variety of optimization problems. For example, MAX-2-SAT and MAXCUT can be directly encoded in (11.41), which is an NP-complete problem, meaning that any other NP-complete problem can be mapped to it in polynomial time. AQC then proceeds by preparing the initial state of the quantum computer in the ground state of H_X (which in the case of the above is a superposition of all states), then applying the time-varying Hamiltonian,

$$H = (1-\lambda)H_X + \lambda H_Z \tag{11.43}$$

is prepared, where λ is a time-varying parameter that is swept from 0 to 1.

In the AQC framework, the speed of the computation is given by how fast the adiabatic sweep is performed. To maintain adiabaticity, one must perform the sweep sufficiently slowly, such that the system remains in the ground state throughout the evolution. The sweep time is known to be proportional to the inverse square of the minimum energy gap of the Hamiltonian (11.43). One of the issues with AQC is that the minimum gap energy is typically unknown prior to a computation, and it is difficult to make general statements of the size of this gap. One of the attractive features of AQC is that time-sequenced gates do not need to be applied, but it is nevertheless known to be equivalent to gate-based quantum computation.

The qubit Hamiltonians (11.41) and (11.42) can be mapped onto spin ensembles in a straightforward way. The equivalent Hamiltonian is

$$H_Z = \frac{1}{N}\sum_{i=1}^{M}\sum_{j=1}^{M} J_{ij}S_z^{(i)}S_z^{(j)} + \sum_{i=1}^{M} K_iS_z^{(i)} \tag{11.44}$$

$$H_X = -\sum_{i=1}^{M} S_x^{(i)}, \tag{11.45}$$

where the $S_{x,z}^{(i)}$ are the total spin operators as usual. Each of the ensembles is first prepared in a fully polarized state of S_i^x, according to

$$|\psi(0)\rangle = \prod_{j=1}^{M}\left|\frac{1}{\sqrt{2}}, \frac{1}{\sqrt{2}}\right\rangle\!\!\Big\rangle_j, \tag{11.46}$$

then adiabatically evolved to the ground state of H_Z. After the adiabatic evolution, measurements of the spin ensembles are made in the S^z basis. If the evolution is successful, one should find that

$$\text{sgn}(\langle S_z^{(j)} \rangle_{\text{ens}}) = \text{sgn}(\langle \sigma_z^{(j)} \rangle_{\text{qubit}}), \tag{11.47}$$

where the right-hand side of the expression is the spins for the original qubit operators, as in (11.41), and the expectation value is taken with respect to an adiabatic evolution of the qubit problem.

The ground states of (11.44) and (11.41) are equivalent in the sense of (11.47). This is easy to see for states such that $\langle S_z^{(j)} \rangle = \pm N$, since substitution of this into (11.44) gives exactly the same energy up to a factor of N as (11.41). The state of the ground state can be written as

$$\langle S_z^{(j)} \rangle_0 = \sigma_0^{(j)} N, \tag{11.48}$$

where $\sigma_0^{(j)} \in \{-1, 1\}$ is the ground state configuration. The energetic ordering of the states does not change for these states, hence the lowest energy states are equivalent in terms of (11.47). The ensemble Hamiltonian, however, also possesses many more states that lie in the range $\langle S_z^{(j)} \rangle \in [-N, N]$. Is it possible that such states have a lower energy than those with (11.48)? We do not show the arguments here, but it can be shown that these intermediate states always have an energy exceeding the ground state (see [338] for further details). This means that as long as it is possible to maintain adiabaticity throughout the evolution, evolving (11.43) with (11.44) and (11.45) gives an equivalent way of solving the qubit problem.

While the above shows it is possible to use the ensemble version of the Hamiltonian in place of the qubits, an important question is whether the performance of the adiabatic procedure changes. One crucial question is whether the minimum gap increases or decreases due to the replacement with ensembles. Here, we can distinguish two cases depending upon the specific instance of the problem Hamiltonian (11.44). The first is when the first excited state is a perturbation of the original ground state. For a specific ground state configuration (11.48), the first excited state involves a single spin flip on one of the ensembles k such that one of the ensembles is

$$\langle S_z^{(j)} \rangle_1 = \begin{cases} \sigma_0^{(j)} N & j \neq k \\ \sigma_0^{(j)} (N-2) & j = k \end{cases}. \tag{11.49}$$

The second case is when the first excited state has a completely different character to the ground state. In this case, the first excited state is

$$\langle S_z^{(j)} \rangle_1 = \sigma_1^{(j)} N, \tag{11.50}$$

where $\sigma_1^{(j)} \in \{-1, 1\}$ and denotes the excited qubit state configuration. It can be easily shown that the energy difference between the ground and excited state (11.50) is proportional to N, while for (11.49) it is independent of N. Thus, for large N, as with ensembles, the state (11.49) is the more typical case.

Figure 11.4 shows energy spectrum, energy gap, and error probabilities for randomly generated problem instances. In Fig. 11.4(a)–(b), the energy spectrum of the Hamiltonian (11.43) is plotted for the same problem instance. The use of ensembles gives many more states, but the energy scale of the Hamiltonian is also increased by a factor of N. The combination of these effects means that the gap energy stays approximately the same, as can be seen in Fig. 11.4(c). In fact, for the case where the first excited state takes the form (11.49), the gap energy slightly increases. More importantly, due to the ensemble mapping (11.45), the

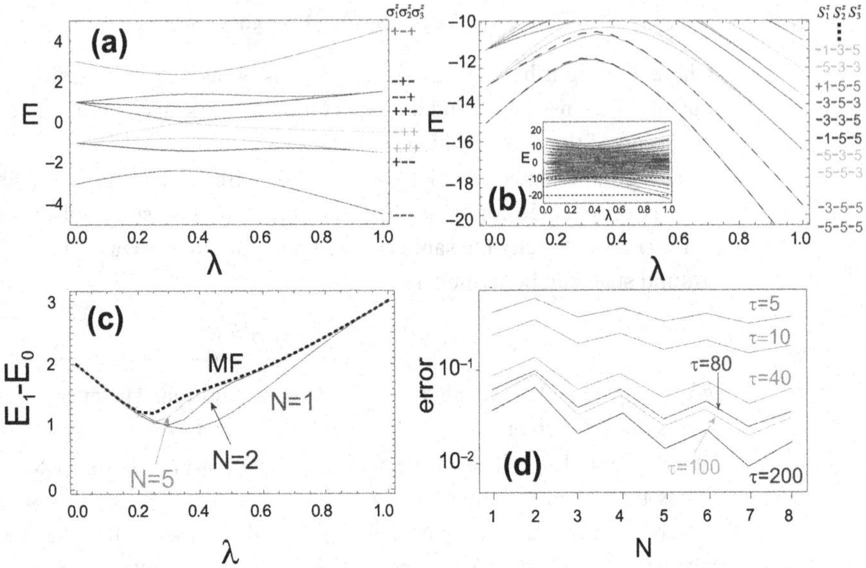

Fig. 11.4 Energy spectrum and gap energies of the adiabatic quantum computing Hamiltonian. Spectrum of (11.43) with $M = 3$ with (a) $N = 1$ and (b) $N = 5$ for parameters $J_{12} = -0.5, J_{13} = 0, J_{23} = -1, K_1 = 0.5, K_2 = 0, K_3 = 1$. The mean field approximation for the $N = 5$ is shown as the dashed lines for the ground and first excited state. (c) The gap energy for the ensemble qubit numbers as shown. (d) The final error probability for 60 instances averaged versus N for various τ and $M = 3$. Here, τ is the time for changing the adiabatic parameter $\lambda = t/\tau$, and the Hamiltonian (11.43) is evolved for a time τ. Color version of this figure available at cambridge.org/quantumatomoptics.

low energy states are all logically equivalent states according to (11.47), as can be seen in Fig. 11.4(b). This means that even if the system is diabatically excited, or decoherence produces some excitations, this does not contribute to a logical error, as long as the number of excitations is small. This translates to a reduced error for the adiabatic evolution, as shown in Fig. 11.4(d), which improves with the ensemble size N. We note that the improvement with N comes only when the first excited state is of the form (11.49), which occurs at a critical ensemble size N_c, and below this the performance can initially degrade with N. In this way, the ensemble encoding achieves quantum error suppression of logical errors in adiabatic quantum computing.

11.6 References and Further Reading

- Section 11.2: Reviews of articles and books on continuous variables quantum information processing [59, 166, 481, 60, 4, 157].
- Section 11.2.1: Original work on Holstein–Primakoff transformation [224]. Review article on atomic ensemble continuous variables using the mapping [199].
- Section 11.2.2: Experimental demonstrations of continuous variable quantum teleportation [155, 274, 425]. Theoretical proposals of discrete [34, 472] and

continuous variable quantum teleportation [58]. Review of continuous variable quantum teleportation [156]. Entanglement between atomic ensembles: [252, 273].

- Section 11.3: Original theoretical proposals of spinor quantum computing [74, 73, 3]. Experimental demonstrations of atom chips [147, 200, 50, 394]. Review articles and books reviewing atom chips [393, 259]. Proof of universality of gate decompositions [311]. Other theoretical proposals for quantum information schemes using atomic ensembles [314, 383, 62, 131, 76].
- Section 11.4: Original theoretical work introducing Deutsch's algorithm [115]. Reviews and books further discussing the algorithm [92, 345]. Spinor quantum computing mapping of Deutsch's algorithm [424].
- Section 11.5: Original theoretical work introducing adiabatic quantum computing [142]. Review articles on adiabatic quantum computing [109, 6]. Equivalence between adiabatic quantum computing and standard quantum computing [337, 5]. Spinor quantum computing mapping of adiabatic quantum computing [338].

References

[1] Abo-Shaeer, J. R., Raman, C., Vogels, J. M., and Ketterle, Wolfgang. 2001. Observation of vortex lattices in Bose–Einstein condensates. *Science*, **292**(5516), 476–479.

[2] Adams, C. S., and Riis, E. 1997. Laser cooling and trapping of neutral atoms. *Progress in Quantum Electronics*, **21**(1), 1–79.

[3] Adcock, Mark R. A., Høyer, Peter, and Sanders, Barry C. 2016. Quantum computation with coherent spin states and the close Hadamard problem. *Quantum Information Processing*, **15**(4), 1361–1386.

[4] Adesso, Gerardo, Ragy, Sammy, and Lee, Antony R. 2014. Continuous variable quantum information: Gaussian states and beyond. *Open Systems and Information Dynamics*, **21**(01n02), 1440001.

[5] Aharonov, Dorit, Van Dam, Wim, Kempe, Julia, Landau, Zeph, Lloyd, Seth, and Regev, Oded. 2008. Adiabatic quantum computation is equivalent to standard quantum computation. *SIAM Review*, **50**(4), 755–787.

[6] Albash, Tameem, and Lidar, Daniel A. 2018. Adiabatic quantum computation. *Reviews of Modern Physics*, **90**(Jan), 015002.

[7] Allen, L., and Eberly, J. H. 1975. *Optical Resonance and Two-Level Atoms*. John Wiley.

[8] Alon, Ofir E., Streltsov, Alexej I., and Cederbaum, Lorenz S. 2005. Zoo of quantum phases and excitations of cold bosonic atoms in optical lattices. *Physical Review Letters*, **95**(3), 030405.

[9] Altland, Alexander, and Simons, Ben D. 2010. *Condensed Matter Field Theory*. Cambridge University Press.

[10] Amico, Luigi, Birkl, Gerhard, Boshier, Malcolm, and Kwek, Leong-Chuan. 2017. Focus on atomtronics-enabled quantum technologies. *New Journal of Physics*, **19**(2), 020201.

[11] Amico, Luigi, Fazio, Rosario, Osterloh, Andreas, and Vedral, Vlatko. 2008. Entanglement in many-body systems. *Reviews of Modern Physics*, **80**(2), 517.

[12] Amo, Alberto, Pigeon, S., Sanvitto, D., Sala, V. G., Hivet, R., Carusotto, Iacopo, Pisanello, F., Leménager, G., Houdré, R., Giacobino, E., et al. 2011. Polariton superfluids reveal quantum hydrodynamic solitons. *Science*, **332**(6034), 1167–1170.

[13] Anderson, B. P., Haljan, P. C., Regal, C. A., Feder, D. L., Collins, L. A., Clark, Charles W., and Cornell, Eric A. 2001. Watching dark solitons decay into vortex rings in a Bose–Einstein condensate. *Physical Review Letters*, **86**(14), 2926.

[14] Anderson, Brandon M., Juzeliūnas, Gediminas, Galitski, Victor M., and Spielman, Ian B. 2012. Synthetic 3D spin-orbit coupling. *Physical Review Letters*, **108**(23), 235301.

[15] Anderson, Mike H., Ensher, Jason R., Matthews, Michael R., Wieman, Carl E., and Cornell, Eric A. 1995. Observation of Bose–Einstein condensation in a dilute atomic vapor. *Science*, 198–201.

[16] Andrews, M. R., Townsend, C. G., Miesner, H.-J., Durfee, D. S., Kurn, D. M., and Ketterle, W. 1997. Observation of Interference Between Two Bose Condensates. *Science*, **275**, 637–641.

[17] Anglin, James R., and Ketterle, Wolfgang. 2002. Bose–Einstein condensation of atomic gases. *Nature*, **416**(6877), 211.

[18] Appel, Jürgen, Windpassinger, Patrick Joachim, Oblak, Daniel, Hoff, U. Busk, Kjærgaard, Niels, and Polzik, Eugene Simon. 2009. Mesoscopic atomic entanglement for precision measurements beyond the standard quantum limit. *Proceedings of the National Academy of Sciences*, **106**(27), 10960–10965.

[19] Arecchi, F. T., Courtens, E., Gilmore, R., and Thomas, H. 1972. Atomic coherent states in quantum optics. *Physical Review A*, **6**(6), 2211–2237.

[20] Arfken, G. B., and Weber, H. J. 2005. *Mathematical Methods for Physicists*. 6th ed. Elsevier Academic Press.

[21] Ashkin, Arthur. 1997. Optical trapping and manipulation of neutral particles using lasers. *Proceedings of the National Academy of Sciences*, **94**(10), 4853–4860.

[22] Ashkin, Arthur. 2006. *Optical Trapping and Manipulation of Neutral Particles Using Lasers: A Reprint Volume with Commentaries*. World Scientific.

[23] Auerbach, Assa. 2012. *Interacting Electrons and Quantum Magnetism*. Springer Science and Business Media.

[24] Autler, Stanley H., and Townes, Charles H. 1955. Stark effect in rapidly varying fields. *Physical Review*, **100**(2), 703.

[25] Bacry, H. 1978. Physical significance of minimum uncertainty states of an angular momentum system. *Physical Review A*, **18**(2), 617.

[26] Bakr, Waseem S., Gillen, Jonathon I., Peng, Amy, Fölling, Simon, and Greiner, Markus. 2009. A quantum gas microscope for detecting single atoms in a Hubbard-regime optical lattice. *Nature*, **462**(7269), 74.

[27] Bao, Weizhu, Jaksch, Dieter, and Markowich, Peter A. 2003. Numerical solution of the Gross–Pitaevskii equation for Bose–Einstein condensation. *Journal of Computational Physics*, **187**(1), 318–342.

[28] Barends, R., Lamata, L., Kelly, J., García-Álvarez, L., Fowler, A. G., Megrant, A., Jeffrey, E., White, T. C., Sank, D., Mutus, J. Y., et al. 2015. Digital quantum simulation of fermionic models with a superconducting circuit. *Nature Communications*, **6**, 7654.

[29] Barnett, S. M., and Radmore, P. M. 1997. *Methods in Theoretical Quantum Optics*. 2nd ed. Oxford University Press.

[30] Barredo, Daniel, Lienhard, Vincent, De Leseleuc, Sylvain, Lahaye, Thierry, and Browaeys, Antoine. 2018. Synthetic three-dimensional atomic structures assembled atom by atom. *Nature*, **561**(7721), 79.

[31] Barrett, M. D., Sauer, J. A., and Chapman, M. S. 2001. All-optical formation of an atomic Bose–Einstein condensate. *Physical Review Letters*, **87**, 010404.

[32] Baym, G., and Pethick, C. J. 1996. Ground-state properties of magnetically trapped Bose-condensed rubidium gas. *Physical Review Letters*, **76**, 6–9.

[33] Bender, C. M., and Orszag, S. A. 1978. *Advanced Mathematical Methods for Scientists and Engineers*. McGraw-Hill.

[34] Bennett, C. H., Brassard, G., Crépeau, C., Jozsa, R., Peres, A., and Wootters, W. K. 1993. Teleporting an unknown quantum state via dual classical and Einstein–Podolsky–Rosen channels. *Physical Review Letters*, **70**, 1895–1899.

[35] Benson, O., Raithel, G., and Walther, Herbert. 1994. Quantum jumps of the micromaser field: dynamic behavior close to phase transition points. *Physical Review Letters*, **72**(22), 3506.

[36] Bergquist, J. C., Hulet, Randall G., Itano, Wayne M., and Wineland, D. J. 1986. Observation of quantum jumps in a single atom. *Physical Review Letters*, **57**(14), 1699.

[37] Berman, P. R. 1997. *Atom Interferometry*. Academic Press.

[38] Bernien, Hannes, Schwartz, Sylvain, Keesling, Alexander, Levine, Harry, Omran, Ahmed, Pichler, Hannes, Choi, Soonwon, Zibrov, Alexander S., Endres, Manuel, Greiner, Markus, et al. 2017. Probing many-body dynamics on a 51-atom quantum simulator. *Nature*, **551**(7682), 579.

[39] Berry, M. V. 1984. Quantal phase factors accompanying adiabatic changes. *Proceedings of the Royal Society of London A: Mathematical and Physical Sciences*, **392**(1802), 45–57.

[40] Berry, M. V. 1989. The quantum phase, five years after. Page 7 of: Wilczek, Frank, and Shapere, Alfred (eds.), *Geometric Phases in Physics*. Vol. 5. World Scientific.

[41] Billy, Juliette, Josse, Vincent, Zuo, Zhanchun, Bernard, Alain, Hambrecht, Ben, Lugan, Pierre, Clément, David, Sanchez-Palencia, Laurent, Bouyer, Philippe, and Aspect, Alain. 2008. Direct observation of Anderson localization of matter waves in a controlled disorder. *Nature*, **453**(7197), 891.

[42] Blatt, S., Ludlow, A. D., Campbell, G. K., Thomsen, Jan Westenkær, Zelevinsky, T., Boyd, M. M., Ye, J., Baillard, X., Fouché, M., Le Targat, R., et al. 2008. New limits on coupling of fundamental constants to gravity using ^{87}Sr optical lattice clocks. *Physical Review Letters*, **100**(14), 140801.

[43] Bloch, Immanuel. 2005. Ultracold quantum gases in optical lattices. *Nature Physics*, **1**(1), 23.

[44] Bloch, Immanuel. 2008. Quantum coherence and entanglement with ultracold atoms in optical lattices. *Nature*, **453**(7198), 1016.

[45] Bloch, Immanuel, Dalibard, Jean, and Nascimbene, Sylvain. 2012. Quantum simulations with ultracold quantum gases. *Nature Physics*, **8**(4), 267.

[46] Bloch, Immanuel, Dalibard, Jean, and Zwerger, Wilhelm. 2008. Many-body physics with ultracold gases. *Reviews of Modern Physics*, **80**(3), 885.

[47] Bloom, B. J., Nicholson, T. L., Williams, J. R., Campbell, S. L., Bishof, M., Zhang, X., Zhang, W., Bromley, S. L., and Ye, J. 2014. An optical lattice clock with accuracy and stability at the 10-18 level. *Nature*, **506**(7486), 71–75.

[48] Bo-sture, Skagerstam, et al. 1985. *Coherent States: Applications in Physics and Mathematical Physics*. World Scientific.

[49] Bogolyubov, Nikolay Nikolaevich. 1947. On the theory of superfluidity. *Izv. Akad. Nauk Ser. Fiz.*, **11**, 23–32.

[50] Böhi, P., Riedel, M. F., Hoffrogge, J., Reichel, J., Hänsch, T. W., and Treutlein, P. 2009. Coherent manipulation of Bose–Einstein condensates with state-dependent microwave potentials on an atom chip. *Nature Physics*, **5**, 592.

[51] Bose, Satyendra Nath. 1924. Plancks Gesetz und Lichtquantenhypothese. *Zeitschrift für Physik*, **26**(1).

[52] Bouchoule, Isabelle, and Mølmer, Klaus. 2002. Spin squeezing of atoms by the dipole interaction in virtually excited Rydberg states. *Physical Review A*, **65**(4), 041803.

[53] Bourdel, Thomas, Khaykovich, Lev, Cubizolles, Julien, Zhang, Jun, Chevy, Frédéric, Teichmann, M., Tarruell, L., Kokkelmans, SJJMF, and Salomon, Christophe. 2004. Experimental study of the BEC–BCS crossover region in lithium 6. *Physical Review Letters*, **93**(5), 050401.

[54] Bouyer, P., and Kasevich, M. A. 1997. Heisenberg-limited spectroscopy with degenerate Bose–Einstein gases. *Physical Review A*, **56**, R1083– R1086.

[55] Boyer, V., McCormick, C. F., Arimondo, Ennio, and Lett, Paul D. 2007. Ultraslow propagation of matched pulses by four-wave mixing in an atomic vapor. *Physical Review Letters*, **99**(14), 143601.

[56] Bransden, Brian Harold, Joachain, Charles Jean, and Plivier, Theodor J. 2003. *Physics of Atoms and Molecules*. Pearson Education India.

[57] Braunstein, S. L., and Caves, C. M. 1994. Statistical distance and the geometry of quantum states. *Physical Review Letters*, **72**, 3439–3443.

[58] Braunstein, S. L., and Kimble, H. Jeff. 1998. Teleportation of continuous quantum variables. *Physical Review Letters*, **80**(4), 869.

[59] Braunstein, S. L., and van Loock, P. 2005. Quantum information with continuous variables. *Reviews of Modern Physics*, **77**(April), 513.

[60] Braunstein, S. L., and Pati, Arun K. 2012. *Quantum Information with Continuous Variables*. Springer Science and Business Media.

[61] Brink, David Maurice, and Satchler, George Raymond. 1968. *Angular Momentum*. Clarendon Press.

[62] Brion, E., Mølmer, K., and Saffman, M. 2007a. Quantum computing with collective ensembles of multilevel systems. *Physical Review Letters*, **99**(26), 260501.

[63] Brion, E., Pedersen, L. H., and Mølmer, K. 2007b. Adiabatic elimination in a lambda system. *Journal of Physics A: Mathematic and Theoretical*, **40**(5), 1033.

[64] Brzozowski, Tomasz M., Maczynska, Maria, Zawada, Michal, Zachorowski, Jerzy, and Gawlik, Wojciech. 2002. Time-of-flight measurement of the temperature of cold atoms for short trap-probe beam distances. *Journal of Optics B: Quantum and Semiclassical Optics*, **4**(1), 62.

[65] Buluta, Iulia, and Nori, Franco. 2009. Quantum simulators. *Science*, **326**(5949), 108–111.

[66] Burger, Stefan, Bongs, Kai, Dettmer, Stefanie, Ertmer, Wolfgang, Sengstock, Klaus, Sanpera, Anna, Shlyapnikov, Gora V., and Lewenstein, Maciej. 1999. Dark solitons in Bose–Einstein condensates. *Physical Review Letters*, **83**(25), 5198.

[67] Burke, J. H. T., Deissler, B., Hughes, K. J., and Sackett, C. A. 2008. Confinement effects in a guided-wave atom interferometer with milimeter-scale arm separation. *Physical Review A*, **78**, 023619.

[68] Burke, J. H. T., and Sackett, C. A. 2009. Scalable Bose–Einstein condensate Sagnac interferometer in a linear trap. *Physical Review A*, **80**, 061603(R).

[69] Busch, T., and Anglin, J. R. 2000. Motion of dark solitons in trapped Bose–Einstein condensates. *Physical Review Letters*, **84**(11), 2298.

[70] Busch, T., and Anglin, J. R. 2001. Dark-bright solitons in inhomogeneous Bose–Einstein condensates. *Physical Review Letters*, **87**(1), 010401.

[71] Byrnes, Tim. 2013. Fractality and macroscopic entanglement in two-component Bose–Einstein condensates. *Physical Review A*, **88**, 023609.

[72] Byrnes, Tim, Recher, Patrik, and Yamamoto, Yoshihisa. 2010. Mott transitions of exciton polaritons and indirect excitons in a periodic potential. *Physical Review B*, **81**(20), 205312.

[73] Byrnes, Tim, Rosseau, Daniel, Khosla, Megha, Pyrkov, Alexey, Thomasen, Andreas, Mukai, Tetsuya, Koyama, Shinsuke, Abdelrahman, Ahmed, and Ilo-Okeke, Ebubechukwu O. 2015. Macroscopic quantum information processing using spin coherent states. *Optics Communications*, **337**, 102–109.

[74] Byrnes, Tim, Wen, Kai, and Yamamoto, Yoshihisa. 2012. Macroscopic quantum computation using Bose–Einstein condensates. *Physical Review A*, **85**(4), 040306.

[75] Byrnes, Tim, and Yamamoto, Yoshihisa. 2006. Simulating lattice gauge theories on a quantum computer. *Physical Review A*, **73**(2), 022328.

[76] Calarco, Tommaso, Hinds, E. A., Jaksch, D., Schmiedmayer, J., Cirac, J. I., and Zoller, P. 2000. Quantum gates with neutral atoms: controlling collisional interactions in time-dependent traps. *Physical Review A*, **61**(2), 022304.

[77] Campbell, Gretchen K., Mun, Jongchul, Boyd, Micah, Medley, Patrick, Leanhardt, Aaron E., Marcassa, Luis G., Pritchard, David E., and Ketterle, Wolfgang. 2006. Imaging the Mott insulator shells by using atomic clock shifts. *Science*, **313**(5787), 649–652.

[78] Campos, R. A., Gerry, C. C., and Benmoussa, A. 2003. Optical interferometry at the Heisenberg limit with twin Fock states and parity measurements. *Physical Review A*, **68**, 023810.

[79] Capogrosso-Sansone, B., ProkofEv, N. V., and Svistunov, B. V. 2007. Phase diagram and thermodynamics of the three-dimensional Bose-Hubbard model. *Physical Review B*, **75**(13), 134302.

[80] Carmichael, Howard. 2009. *An open systems approach to quantum optics.* Lecture presented at the Université Libre de Bruxelles, October 28 to November 4, 1991. Vol. 18. Springer Science and Business Media.

[81] Carnal, O., and Mlynek, J. 1991. Young's double-slit experiment with atoms: a simple atom interferometer. *Physical Review Letters*, **66**, 2689–2692.

[82] Carr, Lincoln D., DeMille, David, Krems, Roman V., and Ye, Jun. 2009. Cold and ultracold molecules: science, technology and applications. *New Journal of Physics*, **11**(5), 055049.

[83] Caspers, Willem Jan. 1989. *Spin Systems*. World Scientific.

[84] Cassettari, Donatella, Hessmo, Björn, Folman, Ron, Maier, Thomas, and Schmiedmayer, Jörg. 2000. Beam splitter for guided atoms. *Physical Review Letters*, **85**(26), 5483.

[85] Cataliotti, F. S., Burger, S., Fort, C., Maddaloni, P., Minardi, F., Trombettoni, A., Smerzi, A., and Inguscio, M. 2001. Josephson junction arrays with Bose–Einstein condensates. *Science*, **293**, 843–846.

[86] Caves, C. M. 1981. Quantum-mechanical noise in an interferometer. *Physical Review D*, **23**, 1693–1708.

[87] Chabé, Julien, Lemarié, Gabriel, Grémaud, Benoît, Delande, Dominique, Szriftgiser, Pascal, and Garreau, Jean Claude. 2008. Experimental observation of the Anderson metal-insulator transition with atomic matter waves. *Physical Review Letters*, **101**(25), 255702.

[88] Chen, Qijin, Stajic, Jelena, Tan, Shina, and Levin, Kathryn. 2005. BCS–BEC crossover: from high temperature superconductors to ultracold superfluids. *Physics Reports*, **412**(1), 1–88.

[89] Cheuk, Lawrence W., Nichols, Matthew A., Okan, Melih, Gersdorf, Thomas, Ramasesh, Vinay V., Bakr, Waseem S., Lompe, Thomas, and Zwierlein, Martin W. 2015. Quantum-gas microscope for fermionic atoms. *Physical Review Letters*, **114**(19), 193001.

[90] Chin, Cheng, Grimm, Rudolf, Julienne, Paul, and Tiesinga, Eite. 2010. Feshbach resonances in ultracold gases. *Reviews of Modern Physics*, **82**(2), 1225.

[91] Cirac, J. Ignacio, and Zoller, Peter. 2012. Goals and opportunities in quantum simulation. *Nature Physics*, **8**(4), 264.

[92] Cleve, Richard, Ekert, Artur, Macchiavello, Chiara, and Mosca, Michele. 1998. Quantum algorithms revisited. *Proceedings of the Royal Society of London A: Mathematical, Physical and Engineering Sciences*, **454**(1969), 339–354.

[93] Cohen-Tannoudji, C. 1992. Laser cooling and trapping of neutral atoms: theory. *Physics Reports*, **219**(3–6), 153–164.

[94] Cohen-Tannoudji, C., Diu, B., and Laloë, F. 1977. *Quantum Mechanics*. New York, NY: Wiley.

[95] Cohen-Tannoudji, C., Dupont-Roc, J., and Grynberg, G. 1998. *Atom-Photon Interactions: Basic Process and Applications*. Wiley.

[96] Cooper, Nigel R. 2008. Rapidly rotating atomic gases. *Advances in Physics*, **57**(6), 539–616.

[97] Cornish, Simon L., Claussen, Neil R., Roberts, Jacob L., Cornell, Eric A., and Wieman, Carl E. 2000. Stable ^{85}Rb Bose–Einstein condensates with widely tunable interactions. *Physical Review Letters*, **85**(9), 1795.

[98] Courteille, P., Freeland, R. S., Heinzen, Daniel J., Van Abeelen, F. A., and Verhaar, B. J. 1998. Observation of a Feshbach resonance in cold atom scattering. *Physical Review Letters*, **81**(1), 69.

[99] Cronin, A. D., Schmiedmayer, J., and Pritchard, D. E. 2009. Optics and interferometry with atoms and molecules. *Reviews of Modern Physics*, **81**, 1051.

[100] Dalfovo, F., Giorgini, S., Pitaevskii, L. P., and Stringari, S. 1999a. Ground-state properties of magnetically trapped Bose-condensed rubidium gas. *Reviews of Modern Physics*, **71**, 463–512.

[101] Dalfovo, F., Giorgini, S., Pitaevskii, L. P., and Stringari, S. 1999b. Theory of Bose–Einstein condensation in trapped gases. *Reviews of Modern Physics*, **71**(3), 463.

[102] Dalibard, Jean. 1999. Collisional dynamics of ultra-cold atomic gases. Page 14 of: *Proceedings of the International School of Physics–Enrico Fermi*, Vol. 321.

[103] Dalibard, Jean, Castin, Yvan, and Mølmer, Klaus. 1992. Wave-function approach to dissipative processes in quantum optics. *Physical Review Letters*, **68**(5), 580.

[104] Dalibard, Jean, and Cohen-Tannoudji, Claude. 1985. Dressed-atom approach to atomic motion in laser light: the dipole force revisited. *Journal of the Optical Society of America B*, **2**, 1707.

[105] Dalibard, Jean, Gerbier, Fabrice, Juzeliūnas, Gediminas, and Öhberg, Patrik. 2011. Colloquium: artificial gauge potentials for neutral atoms. *Reviews of Modern Physics*, **83**(4), 1523.

[106] Dalton, B. J., Goold, John, Garraway, B. M., and Reid, M. D. 2017. Quantum entanglement for systems of identical bosons: II. Spin squeezing and other entanglement tests. *Physica Scripta*, **92**(2), 023005.

[107] Damski, B., Santos, L., Tiemann, E., Lewenstein, M., Kotochigova, S., Julienne, P., and Zoller, P. 2003a. Creation of a dipolar superfluid in optical lattices. *Physical Review Letters*, **90**(11), 110401.

[108] Damski, Bogdan, Zakrzewski, Jakub, Santos, Luis, Zoller, Peter, and Lewenstein, Maciej. 2003b. Atomic Bose and Anderson glasses in optical lattices. *Physical Review Letters*, **91**(8), 080403.

[109] Das, Arnab, and Chakrabarti, Bikas K. 2008. Colloquium: Quantum annealing and analog quantum computation. *Reviews of Modern Physics*, **80**(Sep), 1061–1081.

[110] Davis, Kendall B., Mewes, M.-O., Andrews, Michael R., van Druten, Nicolaas J., Durfee, Dallin S., Kurn, D. M., and Ketterle, Wolfgang. 1995. Bose–Einstein condensation in a gas of sodium atoms. *Physical Review Letters*, **75**(22), 3969.

[111] De Paz, Aurelie, Sharma, Arijit, Chotia, Amodsen, Marechal, Etienne, Huckans, J. H., Pedri, Paolo, Santos, Luis, Gorceix, Olivier, Vernac, Laurent, and Laburthe-Tolra, Bruno. 2013. Nonequilibrium quantum magnetism in a dipolar lattice gas. *Physical Review Letters*, **111**(18), 185305.

[112] Deng, L., Hagley, Edward W., Wen, J., Trippenbach, M., Band, Y., Julienne, Paul S., Simsarian, J. E., Helmerson, Kristian, Rolston, S. L., and Phillips, William D. 1999. Four-wave mixing with matter waves. *Nature*, **398**(6724), 218.

[113] Deng, Shujin, Shi, Zhe-Yu, Diao, Pengpeng, Yu, Qianli, Zhai, Hui, Ran, Qi, and Wu, Haibin. 2016. Observation of the Efimovian expansion in scale-invariant Fermi gases. *Science*, **353**(6297), 371–374.

[114] Denschlag, J., Simsarian, J. E., Feder, D. L., Clark, Charles W., Collins, L. A., Cubizolles, J., Deng, L., Hagley, Edward W., Helmerson, Kristian, Reinhardt, William P., et al. 2000. Generating solitons by phase engineering of a Bose–Einstein condensate. *Science*, **287**(5450), 97–101.

[115] Deutsch, David, and Jozsa, Richard. 1992. Rapid solution of problems by quantum computation. *Proceedings of the Royal Society of London A: Mathematical and Physical Sciences*, **439**(1907), 553–558.

[116] Dirac, Paul Adrien Maurice. 1927. The quantum theory of the emission and absorption of radiation. *Proceedings of the Royal Society of London A: Containing Papers of a Mathematical and Physical Character*, **114**(767), 243–265.

[117] Donley, Elizabeth A., Claussen, Neil R., Cornish, Simon L., Roberts, Jacob L., Cornell, Eric A., and Wieman, Carl E. 2001. Dynamics of collapsing and exploding Bose–Einstein condensates. *Nature*, **412**(6844), 295.

[118] Dowling, Jonathan P. 2008. Quantum optical metrology – the lowdown on high-N00N states. *Contemporary Physics*, **49**, 125–143.

[119] Dowling, Jonathan P. 1998. Correlated input-port, matter-wave interferometer: quantum-noise limits to the atom-laser gyroscope. *Physical Review A*, **57**(6), 4736.

[120] Dowling, Jonathan P., and Milburn, Gerard J. 2003. Quantum technology: the second quantum revolution. *Philosophical Transactions of the Royal Society of London A: Mathematical, Physical and Engineering Sciences*, **361**(1809), 1655–1674.

[121] Dowling, Jonathan P., Agarwal, Girish S., and Schleich, Wolfgang P. 1994. Wigner distribution of a general angular-momentum state: applications to a collection of two-level atoms. *Physical Review A*, **49**(5), 4101.

[122] Drummond, P. D., Kheruntsyan, K. V., and He, H. 1998. Coherent molecular solitons in Bose–Einstein condensates. *Physical Review Letters*, **81**(15), 3055.

[123] Duan, L.-M., Cirac, J. Ignacio, and Zoller, P. 2002. Quantum entanglement in spinor Bose–Einstein condensates. *Physical Review A*, **65**(3), 033619.

[124] Duan, L.-M., Giedke, G., Cirac, J. I., and Zoller, P. 2000. Inseparability criterion for continuous variable systems. *Physical Review Letters*, **84**(12), 2722.

[125] Dudarev, Artem M., Diener, Roberto B., Carusotto, Iacopo, and Niu, Qian. 2004. Spin-orbit coupling and Berry phase with ultracold atoms in 2D optical lattices. *Physical Review Letters*, **92**(15), 153005.

[126] Dulieu, Olivier, and Gabbanini, Carlo. 2009. The formation and interactions of cold and ultracold molecules: new challenges for interdisciplinary physics. *Reports on Progress in Physics*, **72**(8), 086401.

[127] Dum, R., and Olshanii, M. 1996. Gauge structures in atom-laser interaction: Bloch oscillations in a dark lattice. *Physical Review Letters*, **76**(11), 1788.

[128] Dum, R., Zoller, P., and Ritsch, H. 1992. Monte Carlo simulation of the atomic master equation for spontaneous emission. *Physical Review A*, **45**(7), 4879.

[129] Dumke, R., Müther, T., Volk, M., Ertmer, Wolfgang, and Birkl, G. 2002. Interferometer-type structures for guided atoms. *Physical Review Letters*, **89**(22), 220402.

[130] Dürr, Stephan, Volz, Thomas, Marte, Andreas, and Rempe, Gerhard. 2004. Observation of molecules produced from a Bose–Einstein condensate. *Physical Review Letters*, **92**(2), 020406.

[131] Ebert, M., Kwon, M., Walker, T. G., and Saffman, M. 2015. Coherence and Rydberg blockade of atomic ensemble qubits. *Physical Review Letters*, **115**(9), 093601.

[132] Efremidis, Nikolaos K., Hudock, Jared, Christodoulides, Demetrios N., Fleischer, Jason W., Cohen, Oren, and Segev, Mordechai. 2003. Two-dimensional optical lattice solitons. *Physical Review Letters*, **91**(21), 213906.

[133] Eiermann, B., Anker, T., Albiez, M., Taglieber, M., Treutlein, Philipp, Marzlin, K.-P., and Oberthaler, M. K. 2004. Bright Bose–Einstein gap solitons of atoms with repulsive interaction. *Physical Review Letters*, **92**(23), 230401.

[134] Einstein, Albert. 1924. Quantentheorie des einatomigen idealen Gases. *SB Preuss. Akad. Wiss. phys.-math. Klasse.*

[135] Engels, Peter, Coddington, I., Haljan, P. C., and Cornell, Eric A. 2002. Nonequilibrium effects of anisotropic compression applied to vortex lattices in Bose–Einstein condensates. *Physical Review Letters*, **89**(10), 100403.

[136] Erdős, László, Schlein, Benjamin, and Yau, Horng-Tzer. 2007. Rigorous derivation of the Gross–Pitaevskii equation. *Physical Review Letters*, **98**(4), 040404.

[137] Erdős, László, Schlein, Benjamin, and Yau, Horng-Tzer. 2010. Derivation of the Gross–Pitaevskii equation for the dynamics of Bose–Einstein condensate. *Annals of Mathematics*, 291–370.

[138] Estève, J., Gross, C., Weller, A., Giovanazzi, S., and Oberthaler, M. K. 2008. Squeezing and entanglement in a Bose–Einstein condensate. *Nature*, **455**, 1216–1219.

[139] Eto, Y., Ikeda, H., Suzuki, H., Hasegawa, S., Tomiyama, Y., Sekine, S., Sadgrove, M., and Hirano, T. 2013. Spin-echo-based magnetometry with spinor Bose–Einstein condensates. *Physical Review A*, **88**, 031602.

[140] Eto, Y., Shibayama, H., Saito, H., and Hirano, T. 2018. Spinor dynamics in a mixture of spin-1 and spin-2 Bose–Einstein condensates. *Physical Review A*, **97**, 021602.

[141] Fadel, Matteo, Zibold, Tilman, Décamps, Boris, and Treutlein, Philipp. 2018. Spatial entanglement patterns and Einstein–Podolsky–Rosen steering in Bose–Einstein condensates. *Science*, **360**(6387), 409–413.

[142] Farhi, Edward, Goldstone, Jeffrey, Gutmann, Sam, Lapan, Joshua, Lundgren, Andrew, and Preda, Daniel. 2001. A quantum adiabatic evolution algorithm applied to random instances of an NP-complete problem. *Science*, **292**(5516), 472–475.

[143] Fetter, A. L. 1969. Page 351 of: Mahanthappa, K. T., and Brittin, W. E. (eds.), *Lectures in Theoretical Physics*. Vol. XIB. Gordon and Breach.

[144] Fetter, Alexander L. 2009. Rotating trapped Bose–Einstein condensates. *Reviews of Modern Physics*, **81**(2), 647.

[145] Fetter, Alexander L., and Svidzinsky, Anatoly A. 2001. Vortices in a trapped dilute Bose–Einstein condensate. *Journal of Physics: Condensed Matter*, **13**(12), R135.

[146] Feynman, Richard P. 1982. Simulating physics with computers. *International Journal of Theoretical Physics*, **21**(6–7), 467–488.

[147] Folman, Ron, Krüger, Peter, Cassettari, Donatella, Hessmo, Björn, Maier, Thomas, and Schmiedmayer, Jörg. 2000. Controlling cold atoms using nanofabricated surfaces: atom chips. *Physical Review Letters*, **84**(20), 4749.

[148] Foot, Christopher J. 2005. *Atomic Physics*. Vol. 7. Oxford University Press.

[149] Fortágh, József, and Zimmermann, Claus. 2007. Magnetic microtraps for ultracold atoms. *Reviews of Modern Physics*, **79**(1), 235.

[150] Foulkes, W. M. C., Mitas, L., Needs, R. J., and Rajagopal, G. 2001. Quantum Monte Carlo simulations of solids. *Reviews of Modern Physics*, **73**(1), 33.

[151] Fox, Mark. 2006. *Quantum Optics: An Introduction*. Vol. 15. Oxford University Press.

[152] Fray, Sebastian, Diez, Cristina Alvarez, Hänsch, Theodor W., and Weitz, Martin. 2004. Atomic interferometer with amplitude gratings of light and its applications to atom based tests of the equivalence principle. *Physical Review Letters*, **93**(24), 240404.

[153] Fried, Dale G., Killian, Thomas C., Willmann, Lorenz, Landhuis, David, Moss, Stephen C., Kleppner, Daniel, and Greytak, Thomas J. 1998. Bose–Einstein condensation of atomic hydrogen. *Physical Review Letters*, **81**(18), 3811.

[154] Friis, Nicolai, Vitagliano, Giuseppe, Malik, Mehul, and Huber, Marcus. 2019. Entanglement certification from theory to experiment. *Nature Reviews Physics*, **1**(1), 72–87.

[155] Furusawa, Akira, Sørensen, Jens Lykke, Braunstein, Samuel L., Fuchs, Christopher A., Kimble, H. Jeff, and Polzik, Eugene S. 1998. Unconditional quantum teleportation. *Science*, **282**(5389), 706–709.

[156] Furusawa, Akira, and Takei, Nobuyuki. 2007. Quantum teleportation for continuous variables and related quantum information processing. *Physics Reports*, **443**(3), 97–119.

[157] Furusawa, Akira, and van Loock, Peter. 2011. *Quantum Teleportation and Entanglement: A Hybrid Approach to Optical Quantum Information Processing*. Wiley.

[158] Gadway, Bryce, Pertot, Daniel, Reimann, René, and Schneble, Dominik. 2010. Superfluidity of interacting bosonic mixtures in optical lattices. *Physical Review Letters*, **105**(4), 045303.

[159] Gaëtan, Alpha, Miroshnychenko, Yevhen, Wilk, Tatjana, Chotia, Amodsen, Viteau, Matthieu, Comparat, Daniel, Pillet, Pierre, Browaeys, Antoine, and Grangier, Philippe. 2009. Observation of collective excitation of two individual atoms in the Rydberg blockade regime. *Nature Physics*, **5**(2), 115.

[160] Galitski, Victor, and Spielman, Ian B. 2013. Spin-orbit coupling in quantum gases. *Nature*, **494**(7435), 49–54.

[161] Garcia, O., Deissler, B., Hughes, K. J., Reeves, J. M., and Sackett, C. A. 2006. Bose–Einstein condensate interferometer with macroscopic arm separation. *Physical Review A*, **74**, 031601.

[162] Gardiner, C. W., Anglin, J. R., and Fudge, T. I. A. 2002. The stochastic Gross–Pitaevskii equation. *Journal of Physics B: Atomic, Molecular and Optical Physics*, **35**(6), 1555.

[163] Gardiner, C. W., and Zoller, P. 2004. *Quantum Noise: A Handbook of Markovian and Non-Markovian Quantum Stochastic Methods with Applications to Quantum Optics*. Springer-Verlag.

[164] Georges, Antoine, Kotliar, Gabriel, Krauth, Werner, and Rozenberg, Marcelo J. 1996. Dynamical mean-field theory of strongly correlated fermion

systems and the limit of infinite dimensions. *Reviews of Modern Physics*, **68**(1), 13.

[165] Georgescu, I. M., Ashhab, Sahel, and Nori, Franco. 2014. Quantum simulation. *Reviews of Modern Physics*, **86**(1), 153.

[166] Gerd, Leuchs, et al. 2007. *Quantum Information with Continuous Variables of Atoms and Light*. World Scientific.

[167] Gericke, Tatjana, Würtz, Peter, Reitz, Daniel, Langen, Tim, and Ott, Herwig. 2008. High-resolution scanning electron microscopy of an ultracold quantum gas. *Nature Physics*, **4**(12), 949.

[168] Gerry, Christopher, and Knight, Peter. 2005. *Introductory Quantum Optics*. Cambridge University Press.

[169] Gibble, Kurt, Chang, Seongsik, and Legere, Ronald. 1995. Direct observation of s-wave atomic collisions. *Physical Review Letters*, **75**(14), 2666.

[170] Ginzburg, V. L., and Pitaevskii, L. P. 1958. On the theory of superfluidity. *Soviet Physics JETP*, **7**(5), 858–861.

[171] Giovannetti, Vittorio, Mancini, Stefano, Vitali, David, and Tombesi, Paolo. 2003. Characterizing the entanglement of bipartite quantum systems. *Physical Review A*, **67**(2), 022320.

[172] Gleyzes, Sebastien, Kuhr, Stefan, Guerlin, Christine, Bernu, Julien, Deleglise, Samuel, Hoff, Ulrich Busk, Brune, Michel, Raimond, Jean-Michel, and Haroche, Serge. 2007. Quantum jumps of light recording the birth and death of a photon in a cavity. *Nature*, **446**(7133), 297.

[173] Godun, R. M., D'Arcy, M. B., Summy, G. S., and Burnett, K. 2001. Prospects for atom interferometry. *Contemporary Physics*, **42**, 77–95.

[174] Goldman, Nathan, Juzeliūnas, G., Öhberg, Patrik, and Spielman, Ian B. 2014. Light-induced gauge fields for ultracold atoms. *Reports on Progress in Physics*, **77**(12), 126401.

[175] Gong, Ming, Tewari, Sumanta, and Zhang, Chuanwei. 2011. BCS–BEC crossover and topological phase transition in 3D spin-orbit coupled degenerate Fermi gases. *Physical Review Letters*, **107**(19), 195303.

[176] Goodman, J. W. 1996. *Introduction to Fourier Optics*. 2nd ed. McGraw-Hill.

[177] Greiner, Markus, Mandel, Olaf, Esslinger, Tilman, Hänsch, Theodor W., and Bloch, Immanuel. 2002. Quantum phase transition from a superfluid to a Mott insulator in a gas of ultracold atoms. *Nature*, **415**(6867), 39.

[178] Greiner, Markus, Regal, Cindy A., and Jin, Deborah S. 2005. Probing the excitation spectrum of a Fermi gas in the BCS–BEC crossover regime. *Physical Review Letters*, **94**(7), 070403.

[179] Griesmaier, Axel, Werner, Jörg, Hensler, Sven, Stuhler, Jürgen, and Pfau, Tilman. 2005. Bose–Einstein condensation of chromium. *Physical Review Letters*, **94**(16), 160401.

[180] Griffin, Allan, Snoke, David W., and Stringari, Sandro. 1996. *Bose–Einstein condensation*. Cambridge University Press.

[181] Griffiths, David J., and Schroeter, Darrell F. 2018. *Introduction to Quantum Mechanics*. Cambridge University Press.

[182] Grimm, Rudolf, Weidemüller, Matthias, and Ovchinnikov, Yurii B. 2000. Optical dipole traps for neutral atoms. Pages 95–170 of: *Advances in Atomic, Molecular, and Optical Physics*. Vol. 42. Elsevier.

[183] Gross, C. 2012. Spin squeezing, entanglement and quantum metrology with Bose–Einstein condensates. *Journal of Physics B: Atomic, Molecular and Optical Physics*, **45**, 103001.

[184] Gross, C., and Bloch, I. 2017. Quantum simulations with ultracold atoms in optical lattices. *Science*, **357**(6355), 995–1001.

[185] Gross, C., Zibold, T., Nicklas, E., Estève, J., and Oberthaler, M. K. 2010. Nonlinear atom interferometer surpasses classical precision limit. *Nature*, **464**, 1165–1169.

[186] Gross, Eugene P. 1961. Structure of a quantized vortex in boson systems. *Il Nuovo Cimento (1955–1965)*, **20**(3), 454–477.

[187] Grynberg, G., Lounis, B., Verkerk, P., Courtois, J.-Y., and Salomon, C. 1993. Quantized motion of cold cesium atoms in two- and three-dimensional optical potentials. *Physical Review Letters*, **70**(15), 2249.

[188] Gühne, Otfried, and Tóth, Géza. 2009. Entanglement detection. *Physics Reports*, **474**(1–6), 1–75.

[189] Gupta, S., Leanhardt, A. E., Cronin, A. D., and Pritchard, D. E. 2001. Coherent manipulation of atoms with standing lightwaves. *Comptes Rendus de l'Académie des Sciences Series IV: Physics*, **2**, 479–495.

[190] Gustavson, T. L., Bouyer, P., and Kasevich, M. A. 1997. Precision rotation measurements with an atom interferometer gyroscope. *Physical Review Letters*, **78**(11), 2046.

[191] Gustavson, T. L., Landragin, A., and Kasevich, M. A. 2000. Rotation sensing with a dual atom-interferometer Sagnac gyroscope. *Classical and Quantum Gravity*, **17**(12), 2385.

[192] Hald, J., Sørensen, J. L., Schori, Christian, and Polzik, E. S. 1999. Spin squeezed atoms: a macroscopic entangled ensemble created by light. *Physical Review Letters*, **83**(7), 1319.

[193] Hall, B. V., Whitlock, S., Anderson, Russell, Hannaford, P., and Sidorov, A. I. 2007. Condensate splitting in an asymmetric double well for atom chip based sensors. *Physical Review Letters*, **98**(3), 030402.

[194] Hall, D. S., Matthews, M. R., Ensher, J. R., Wieman, C. E., and Cornell, E. A. 1998a. Dynamics of component separation in a binary mixture of Bose–Einstein condensates. *Physical Review Letters*, **81**, 1539–1542.

[195] Hall, D. S., Matthews, M. R., Wieman, C. E., and Cornell, E. A. 1998b. Measurements of relative phase in two-component Bose–Einstein condensates. *Physical Review Letters*, **81**, 1543–1546.

[196] Haller, Elmar, Hudson, James, Kelly, Andrew, Cotta, Dylan A., Peaudecerf, Bruno, Bruce, Graham D., and Kuhr, Stefan. 2015. Single-atom imaging of fermions in a quantum-gas microscope. *Nature Physics*, **11**(9), 738.

[197] Halliday, D., Walker, J., and Resnick, R. 2013. *Fundamentals of Physics*. 10th ed. Wiley.

[198] Hamer, C. J., and Barber, M. N. 1981. Finite-lattice methods in quantum Hamiltonian field theory: I. O(2) and O(3) Heisenberg models. *Journal of Physics A: Mathematical and General*, **14**(1), 259.

[199] Hammerer, Klemens, Sørensen, Anders S., and Polzik, Eugene S. 2010. Quantum interface between light and atomic ensembles. *Reviews of Modern Physics*, **82**(2), 1041.

[200] Hänsel, Wolfgang, Hommelhoff, Peter, Hänsch, T. W., and Reichel, Jakob. 2001. Bose–Einstein condensation on a microelectronic chip. *Nature*, **413**(6855), 498–501.

[201] Happer, W., and Mathur, B. S. 1967. Off-resonant light as a probe of optically pumped alkali vapors. *Physical Review Letters*, **18**(15), 577.

[202] Harber, D. M., Lewandowski, H. J., McGuirk, J. M., and Cornell, Eric A. 2002. Effect of cold collisions on spin coherence and resonance shifts in a magnetically trapped ultracold gas. *Physical Review A*, **66**(5), 053616.

[203] Hart, Russell A., Duarte, Pedro M., Yang, Tsung-Lin, Liu, Xinxing, Paiva, Thereza, Khatami, Ehsan, Scalettar, Richard T., Trivedi, Nandini, Huse, David A., and Hulet, Randall G. 2015. Observation of antiferromagnetic correlations in the Hubbard model with ultracold atoms. *Nature*, **519**(7542), 211.

[204] Hasan, M. Zahid, and Kane, Charles L. 2010. Colloquium: topological insulators. *Reviews of Modern Physics*, **82**(4), 3045.

[205] Hau, Lene Vestergaard, Harris, Stephen E., Dutton, Zachary, and Behroozi, Cyrus H. 1999. Light speed reduction to 17 metres per second in an ultracold atomic gas. *Nature*, **397**(6720), 594.

[206] He, Q. Y., Reid, M. D., Vaughan, T. G., Gross, C., Oberthaler, M., and Drummond, P. D. 2011. Einstein–Podolsky–Rosen entanglement strategies in two-well Bose–Einstein condensates. *Physical Review Letters*, **106**(12), 120405.

[207] Hecht, E. 2002. *Optics*. 4th ed. Addison-Wesley.

[208] Hegerfeldt, Gerhard C., and Wilser, S. 1993. Ensemble or individual system, collapse or no collapse: a description of a single radiating atom. *Classical and Quantum Systems*, 104.

[209] Heisenberg, Werner. 1985. Über den anschaulichen Inhalt der quantentheoretischen Kinematik und Mechanik. Pages 478–504 of: *Original Scientific Papers Wissenschaftliche Originalarbeiten*. Springer.

[210] Helmerson, Kristian, and You, Li. 2001. Creating massive entanglement of Bose–Einstein condensed atoms. *Physical Review Letters*, **87**(17), 170402.

[211] Helstrom, C. W., and Kennedy, R. S. 1974. Noncommuting observables in quantum detection and estimation theory. *IEEE Transactions on Information Theory*, **20**, 16–24.

[212] Hemmerich, A., and Hänsch, T. W. 1993. Two-dimesional atomic crystal bound by light. *Physical Review Letters*, **70**(4), 410.

[213] Hemmerich, A., Zimmermann, C., and Hänsch, T. W. 1993. Sub-kHz Rayleigh resonance in a cubic atomic crystal. *EPL (Europhysics Letters)*, **22**(2), 89.

[214] Hensinger, Winfried K., Häffner, Hartmut, Browaeys, Antoine, Heckenberg, Norman R., Helmerson, Kris, McKenzie, Callum, Milburn, Gerard J., Phillips, William D., Rolston, Steve L., Rubinsztein-Dunlop, Halina, et al. 2001. Dynamical tunnelling of ultracold atoms. *Nature*, **412**(6842), 52.

[215] Hillery, Mark, and Zubairy, M. Suhail. 2006. Entanglement conditions for two-mode states. *Physical Review Letters*, **96**(5), 050503.

[216] Hinds, E. A., Vale, C. J., and Boshier, M. G. 2001. Two-wire waveguide and interferometer for cold atoms. *Physical Review Letters*, **86**(8), 1462.

[217] Hines, Andrew P., McKenzie, Ross H., and Milburn, Gerard J. 2003. Entanglement of two-mode Bose–Einstein condensates. *Physical Review A*, **67**(1), 013609.

[218] Ho, T.-L. 1998. Spinor Bose condensates in optical traps. *Physical Review Letters*, **81**, 742–745.

[219] Ho, T.-L., and Shenoy, V. B. 1996. Binary mixtures of Bose condensates of alkali atoms. *Physical Review Letters*, **77**, 3276–3279.

[220] Ho, T.-L., and Zhang, S. 2011. Bose–Einstein condensates with spin-orbit interaction. *Physical Review Letters*, **107**(15), 150403.

[221] Hodby, E., Hechenblaikner, G., Hopkins, S. A., Marago, O. M., and Foot, C. J. 2001. Vortex nucleation in Bose–Einstein condensates in an oblate, purely magnetic potential. *Physical Review Letters*, **88**(1), 010405.

[222] Hofmann, Holger F., and Takeuchi, Shigeki. 2003. Violation of local uncertainty relations as a signature of entanglement. *Physical Review A*, **68**(3), 032103.

[223] Holland, M. J., and Burnett, K. 1993. Interferometric detection of optical phase shifts at Heisenberg Limit. *Physical Review Letters*, **71**, 1355–1358.

[224] Holstein, T., and Primakoff, H. 1940. Field dependence of the intrinsic domain magnetization of a ferromagnet. *Physical Review*, **58**(12), 1098.

[225] Holtz, R., and Hanus, J. 1974. On coherent spin states. *Journal of Physics A: Mathematical, Nuclear and General*, **7**(4), L37.

[226] Horedecki, M., Horodecki, P., and Horodecki, R. 1996. Separability of mixed states: necessary and sufficient conditions. *Physics Letters A*, **223**, 1–8.

[227] Horikoshi, M., and Nakagawa, K. 2006. Dephasing due to atom-atom interaction in a waveguide interferometer using a Bose–Einstein condensate. *Physical Review A*, **74**, 031602.

[228] Horikoshi, M., and Nakagawa, K. 2007. Suppression of dephasing due to a trapping potential and atom-atom interactions in a trapped-condensate interferometer. *Physical Review Letters*, **99**, 180401.

[229] Horodecki, M., Horodecki, P., and Horodecki, R. 1996. Separability of mixed states: necessary and sufficient conditions. *Physics Letters A*, **223**(1–2), 1–8.

[230] Horodecki, Paweł. 1997. Separability criterion and inseparable mixed states with positive partial transposition. *Physics Letters A*, **232**(5), 333–339.

[231] Horodecki, Ryszard, Horodecki, Paweł, Horodecki, Michał, and Horodecki, Karol. 2009. Quantum entanglement. *Reviews of Modern Physics*, **81**(2), 865.

[232] Hu, Hui, Jiang, Lei, Liu, Xia-Ji, and Pu, Han. 2011. Probing anisotropic superfluidity in atomic Fermi gases with Rashba spin-orbit coupling. *Physical Review Letters*, **107**(19), 195304.

[233] Hughes, K. J., Deissler, B., Burke, J. H. T., and Sackett, C. A. 2007. High-fidelity manipulation of a Bose–Einstein condensate using optical standing wave. *Physical Review A*, **76**, 035601.

[234] Ilo-Okeke, E. O., and Byrnes, T. 2014. Theory of single-shot phase contrast imaging in spinor Bose–Einstein condensates. *Physical Review Letters*, **112**(23), 233602.

[235] Ilo-Okeke, E. O., and Byrnes, T. 2016. Information and backaction due to phase-contrast-imaging measurements of cold atomic gases: beyond Gaussian states. *Physical Review A*, **94**, 013617.

[236] Ilo-Okeke, E. O., and Zozulya, A. A. 2010. Atomic population distribution in the output ports of cold-atom interferometers with optical splitting and recombination. *Physical Review A*, **82**, 053603.

[237] Inouye, S., Andrews, M. R., Stenger, J., Miesner, H.-J., Stamper-Kurn, D. M., and Ketterle, W. 1998. Observation of Feshbach resonances in a Bose–Einstein condensate. *Nature*, **392**(6672), 151.

[238] Jackiw, Roman W. 1988. Berry's phase: topological ideas from atomic, molecular and optical physics. *Comments on Atomic and Molecular Physics*, **21**, 71.

[239] Jaksch, Dieter, Bruder, Christoph, Cirac, Juan Ignacio, Gardiner, Crispin W., and Zoller, Peter. 1998. Cold bosonic atoms in optical lattices. *Physical Review Letters*, **81**(15), 3108.

[240] James, D. F. V., and Jerke, J. 2000. Effective Hamiltonian theory and its applications in quantum information. *Fortschritte der Physik*, **48**, 823.

[241] Javanainen, J., and Ivanov, M. Y. 1999. Splitting a trap containing a Bose–Einstein condensate: atom number fluctuations. *Physical Review A*, **60**, 2351–2359.

[242] Javanainen, J., and Wilkens, M. 1997. Phase and phase diffusion of a split Bose–Einstein condensate. *Physical Review Letters*, **78**, 4675–4678.

[243] Jeltes, Tom, McNamara, John M., Hogervorst, Wim, Vassen, Wim, Krachmalnicoff, Valentina, Schellekens, Martijn, Perrin, Aurélien, Chang, Hong, Boiron, Denis, Aspect, Alain, et al. 2007. Comparison of the Hanbury Brown–Twiss effect for bosons and fermions. *Nature*, **445**(7126), 402.

[244] Jendrzejewski, Fred, Bernard, Alain, Mueller, Killian, Cheinet, Patrick, Josse, Vincent, Piraud, Marie, Pezzé, Luca, Sanchez-Palencia, Laurent, Aspect, Alain, and Bouyer, Philippe. 2012. Three-dimensional localization of ultracold atoms in an optical disordered potential. *Nature Physics*, **8**(5), 398.

[245] Jessen, Poul S., Gerz, C., Lett, Paul D., Phillips, William D., Rolston, S. L., Spreeuw, R. J. C., and Westbrook, C. I. 1992. Observation of quantized motion of Rb atoms in an optical field. *Physical Review Letters*, **69**(1), 49.

[246] Jing, Yumang, Fadel, Matteo, Ivannikov, Valentin, and Byrnes, Tim. 2019. Split spin-squeezed Bose–Einstein condensates. *New Journal of Physics*, **21**(9), 093038.

[247] Jo, G. B., Shin, Y., Will, S., Pasquini, T. A., Saba, M., Ketterle, W., and Pritchard, D. E. 2007. Long phase coherence time and number squeezing of two Bose–Einstein condensates on an atom chip. *Physical Review Letters*, **98**, 030407.

[248] Jochim, Selim, Bartenstein, Markus, Altmeyer, Alexander, Hendl, Gerhard, Riedl, Stefan, Chin, Cheng, Denschlag, J. Hecker, and Grimm, Rudolf. 2003. Bose–Einstein condensation of molecules. *Science*, **302**(5653), 2101–2103.

[249] Jones, Robert O. 2015. Density functional theory: its origins, rise to prominence, and future. *Reviews of Modern Physics*, **87**(3), 897.

[250] Jordan, P. 1935. Der Zusammenhang der symmetrischen und linearen Gruppen und das Mehrkörperproblem. *Zeitschrift für Physik*, **94**(7–8), 531–535.

[251] Julienne, P. S., Mies, F. H., Tiesinga, E., and Williams, C. J. 1997. Collisional stability of double Bose condensates. *Physical Review Letters*, **78**, 1880–1883.

[252] Julsgaard, Brian, Kozhekin, Alexander, and Polzik, Eugene S. 2001. Experimental long-lived entanglement of two macroscopic objects. *Nature*, **413**(6854), 400.

[253] Kafle, R. P., Anderson, D. Z., and Zozulya, A. A. 2011. Analysis of a free oscillation atom interferometer. *Physical Review A*, **84**, 033639.

[254] Kasamatsu, K., Tsubota, M., and Ueda, M. 2005. Vortices in multicomponent Bose–Einstein condensates. *International Journal of Modern Physics B*, **19**, 1835–1904.

[255] Kasevich, M., and Chu, S. 1991. Atomic interferometry using stimulated Raman transition. *Physical Review Letters*, **67**, 181.

[256] Kastberg, A., Phillips, William D., Rolston, S. L., Spreeuw, R. J. C., and Jessen, Poul S. 1995. Adiabatic cooling of cesium to 700 nK in an optical lattice. *Physical Review Letters*, **74**(9), 1542.

[257] Kawaguchi, Y., and Ueda, M. 2012. Spinor Bose-Einstein condensates. arXiv:1001.2072.

[258] Kazantsev, A., Surdutovich, G., and Yakovlev, V. 1990. *Mechanical Action of Light on Atoms*. World Scientific.

[259] Keil, Mark, Amit, Omer, Zhou, Shuyu, Groswasser, David, Japha, Yonathan, and Folman, Ron. 2016. Fifteen years of cold matter on the atom chip: promise, realizations, and prospects. *Journal of Modern Optics*, **63**(18), 1840–1885.

[260] Khalatnikov, Isaac M. 2018. *An Introduction to the Theory of Superfluidity*. CRC Press.

[261] Khaykovich, L., Schreck, F., Ferrari, G., Bourdel, Thomas, Cubizolles, Julien, Carr, Lincoln D., Castin, Yvan, and Salomon, Christophe. 2002. Formation of a matter-wave bright soliton. *Science*, **296**(5571), 1290–1293.

[262] Kitagawa, M., and Ueda, M. 1993. Squeezed spin states. *Physical Review A*, **47**(6), 5138.

[263] Kittel, Charles. 1987. *Quantum Theory of Solids*. Wiley.

[264] Kitzinger, Jonas, Chaudhary, Manish, Kondappan, Manikandan, Ivannikov, Valentin, and Byrnes, Tim. 2020. Two-axis two-spin squeezed spin states. *Phys. Rev. Research*, **2**, 033504.

[265] Klausen, Nille N., Bohn, John L., and Greene, Chris H. 2001. Nature of spinor Bose–Einstein condensates in rubidium. *Physical Review A*, **64**(5), 053602.

[266] Köhler, Thorsten, Góral, Krzysztof, and Julienne, Paul S. 2006. Production of cold molecules via magnetically tunable Feshbach resonances. *Reviews of Modern Physics*, **78**(4), 1311.

[267] Kolomeisky, Eugene B., Newman, T. J., Straley, Joseph P., and Qi, Xiaoya. 2000. Low-dimensional Bose liquids: beyond the Gross–Pitaevskii approximation. *Physical Review Letters*, **85**(6), 1146.

[268] Korbicz, J. K., Cirac, J. Ignacio, and Lewenstein, M. 2005. Spin squeezing inequalities and entanglement of N qubit states. *Physical Review Letters*, **95**(12), 120502.

[269] Kordas, George, Witthaut, D., Buonsante, P., Vezzani, A., Burioni, R., Karanikas, A. I., and Wimberger, S. 2015. The dissipative Bose-Hubbard model. *European Physical Journal Special Topics*, **224**(11), 2127–2171.

[270] Kozuma, M. M., Deng, L., Hagley, E. W., Wen, J., Lutwak, R., Helmerson, K., Rolston, S. L., and Phillips, W. D. 1999. Coherent splitting of Bose–Einstein condensed atoms with optically induced Bragg diffraction. *Physical Review Letters*, **82**, 871–875.

[271] Kraft, Sebastian, Vogt, Felix, Appel, Oliver, Riehle, Fritz, and Sterr, Uwe. 2009. Bose-Einstein condensation of alkaline earth atoms: Ca 40. *Physical Review Letters*, **103**(13), 130401.

[272] Krane, Kenneth S., Halliday, David, et al. 1987. *Introductory Nuclear Physics*. Wiley.

[273] Krauter, H., Muschik, C. A., Jensen, K., Wasilewski, W., Petersen, J. M., Cirac, J. I., and Polzik, E. S. 2011. Entanglement generated by dissipation and steady state entanglement of two macroscopic objects. *Physical Review Letters*, **107**(8), 080503.

[274] Krauter, H., Salart, D., Muschik, C. A., Petersen, Jonas Meyer, Shen, Heng, Fernholz, Thomas, and Polzik, Eugene Simon. 2013. Deterministic quantum teleportation between distant atomic objects. *Nature Physics*, **9**(7), 400.

[275] Kuhr, Stefan. 2016. Quantum-gas microscopes: a new tool for cold-atom quantum simulators. *National Science Review*, **3**(2), 170–172.

[276] Kunkel, Philipp, Prüfer, Maximilian, Strobel, Helmut, Linnemann, Daniel, Frölian, Anika, Gasenzer, Thomas, Gärttner, Martin, and Oberthaler, Markus K. 2018. Spatially distributed multipartite entanglement enables EPR steering of atomic clouds. *Science*, **360**(6387), 413–416.

[277] Kurkjian, Hadrien, Pawłowski, Krzysztof, Sinatra, Alice, and Treutlein, Philipp. 2013. Spin squeezing and Einstein–Podolsky–Rosen entanglement of two bimodal condensates in state-dependent potentials. *Physical Review A*, **88**(4), 043605.

[278] Kuzmich, A., Bigelow, N. P., and Mandel, L. 1998. Atomic quantum non-demolition measurements and squeezing. *EPL (Europhysics Letters)*, **42**(5), 481.

[279] Kuzmich, A., Mandel, L., and Bigelow, N. P. 2000. Generation of spin squeezing via continuous quantum nondemolition measurement. *Physical Review Letters*, **85**(8), 1594.

[280] Kuzmich, A., Mølmer, K., and Polzik, E. S. 1997. Spin squeezing in an ensemble of atoms illuminated with squeezed light. *Physical Review Letters*, **79**(24), 4782.

[281] Kuzmich, A., and Polzik, E. S. 2000. Atomic quantum state teleportation and swapping. *Physical Review Letters*, **85**(26), 5639.

[282] Labeyrie, Guillaume, Vaujour, E., Mueller, Cord A., Delande, Dominique, Miniatura, Christian, Wilkowski, David, and Kaiser, Robin. 2003. Slow diffusion of light in a cold atomic cloud. *Physical Review Letters*, **91**(22), 223904.

[283] Labuhn, Henning, Barredo, Daniel, Ravets, Sylvain, De Léséleuc, Sylvain, Macrì, Tommaso, Lahaye, Thierry, and Browaeys, Antoine. 2016. Tunable two-dimensional arrays of single Rydberg atoms for realizing quantum Ising models. *Nature*, **534**(7609), 667.

[284] Lambrecht, A., Coudreau, Thomas, Steinberg, A. M., and Giacobino, E. 1996. Squeezing with cold atoms. *EPL (Europhysics Letters)*, **36**(2), 93.

[285] Landau, L. D. 1941. Theory of the superfluidity of helium II. *Physical Review*, **60**(4), 356.

[286] Landau, L. D., and Lifshitz, E. M. 2013. *Quantum Mechanics: Non-relativistic Theory*. Vol. 3. Elsevier.

[287] Lange, Karsten, Peise, Jan, Lücke, Bernd, Kruse, Ilka, Vitagliano, Giuseppe, Apellaniz, Iagoba, Kleinmann, Matthias, Tóth, Géza, and Klempt, Carsten. 2018. Entanglement between two spatially separated atomic modes. *Science*, **360**(6387), 416–418.

[288] Langen, Tim, Erne, Sebastian, Geiger, Remi, Rauer, Bernhard, Schweigler, Thomas, Kuhnert, Maximilian, Rohringer, Wolfgang, Mazets, Igor E., Gasenzer, Thomas, and Schmiedmayer, Jörg. 2015. Experimental observation of a generalized Gibbs ensemble. *Science*, **348**(6231), 207–211.

[289] Lanyon, Ben P., Hempel, Cornelius, Nigg, Daniel, Müller, Markus, Gerritsma, Rene, Zähringer, F., Schindler, Philipp, Barreiro, Julio T., Rambach, Markus, Kirchmair, Gerhard, et al. 2011. Universal digital quantum simulation with trapped ions. *Science*, **334**(6052), 57–61.

[290] Las Heras, U., Mezzacapo, A., Lamata, L., Filipp, S., Wallraff, A., and Solano, E. 2014. Digital quantum simulation of spin systems in superconducting circuits. *Physical Review Letters*, **112**(20), 200501.

[291] Laudat, T., Durgrain, V., Mazzoni, T., Huang, M. Z., Alzar, C. L. G., Sinatra, A., Rosenbush, P., and Reichel, J. 2018. Spontaneous spin squeezing in a rubidium BEC. *New Journal of Physics*, **20**, 073018.

[292] LeBlanc, Lindsay J., Beeler, M. C., Jimenez-Garcia, Karina, Perry, Abigail R., Sugawa, Seiji, Williams, R. A., and Spielman, Ian B. 2013. Direct observation of zitterbewegung in a Bose–Einstein condensate. *New Journal of Physics*, **15**(7), 073011.

[293] Lee, C. T. 1984. Q representation of the atomic coherent states and the origin of fluctuations in superfluorescence. *Physical Review A*, **30**(6), 3308.

[294] Leggett, Anthony J. 1999. Superfluidity. *Reviews of Modern Physics*, **71**(2), S318.

[295] Leggett, Anthony J. 2001. Bose–Einstein condensation in the alkali gases: some fundamental concepts. *Reviews of Modern Physics*, **73**(2), 307.

[296] Lemke, Nathan D., Ludlow, Andrew D., Barber, Z. W., Fortier, Tara M., Diddams, Scott A., Jiang, Yanyi, Jefferts, Steven R., Heavner, Thomas P., Parker, Thomas E., and Oates, Christopher W. 2009. Spin-1/2 optical lattice clock. *Physical Review Letters*, **103**(6), 063001.

[297] Leroux, Ian D., Schleier-Smith, Monika H., and Vuletić, Vladan. 2010. Implementation of cavity squeezing of a collective atomic spin. *Physical Review Letters*, **104**(7), 073602.

[298] Lesanovsky, Igor, and von Klitzing, Wolf. 2007. Time-averaged adiabatic potentials: versatile matter-wave guides and atom traps. *Physical Review Letters*, **99**(8), 083001.

[299] Lett, Paul D., Watts, Richard N., Westbrook, Christoph I., Phillips, William D., Gould, Phillip L., and Metcalf, Harold J. 1988. Observation of atoms laser cooled below the Doppler limit. *Physical Review Letters*, **61**(2), 169.

[300] Levine, Mindy, Heath, A., et al. 1991. *Quantum Chemistry*. Citeseer.

[301] Lewenstein, M., Kraus, B., Cirac, J. I., and Horodecki, P. 2000. Optimization of entanglement witnesses. *Physical Review A*, **62**(5), 052310.

[302] Lewenstein, M., Sanpera, A., Ahufinger, V., Damski, B., Sen, A., and Sen, U. 2007. Ultracold atomic gases in optical lattices: mimicking condensed matter physics and beyond. *Advances in Physics*, **56**(2), 243–379.

[303] Lewenstein, M., Santos, L., Baranov, M. A., and Fehrmann, H. 2004. Atomic Bose-Fermi mixtures in an optical lattice. *Physical Review Letters*, **92**(5), 050401.

[304] Li, Jun-Ru, Lee, Jeongwon, Huang, Wujie, Burchesky, Sean, Shteynas, Boris, Top, Furkan Çağrı, Jamison, Alan O., and Ketterle, Wolfgang. 2017. A stripe phase with supersolid properties in spin-orbit-coupled Bose–Einstein condensates. *Nature*, **543**(7643), 91–94.

[305] Li, Tongcang, Kheifets, Simon, Medellin, David, and Raizen, Mark G. 2010. Measurement of the instantaneous velocity of a Brownian particle. *Science*, **328**(5986), 1673–1675.

[306] Lin, Y.-J., Compton, Robert L., Jiménez-García, Karina, Phillips, William D., Porto, James V., and Spielman, Ian B. 2011b. A synthetic electric force acting on neutral atoms. *Nature Physics*, **7**(7), 531–534.

[307] Lin, Y.-J., Compton, Robert L., Jiménez-García, Karina, Porto, James V., and Spielman, Ian B. 2009. Synthetic magnetic fields for ultracold neutral atoms. *Nature*, **462**(7273), 628–632.

[308] Lin, Y.-J., Jiménez-García, K., and Spielman, Ian B. 2011a. Spin-orbit-coupled Bose–Einstein condensates. *Nature*, **471**(7336), 83–86.

[309] Liu, Chien, Dutton, Zachary, Behroozi, Cyrus H., and Hau, Lene Vestergaard. 2001. Observation of coherent optical information storage in an atomic medium using halted light pulses. *Nature*, **409**(6819), 490.

[310] Liu, Y. C., Xu, Z. F., Jin, G. R., and You, L. 2011. Spin squeezing: transforming one-axis twisting into two-axis twisting. *Physical Review Letters*, **107**(1), 013601.

[311] Lloyd, Seth. 1995. Almost any quantum logic gate is universal. *Physical Review Letters*, **75**(2), 346.

[312] Lloyd, Seth. 1996. Universal quantum simulators. *Science*, 1073–1078.

[313] Longuet-Higgins, Hugh Christopher, Öpik, U., Pryce, Maurice Henry Lecorney, and Sack, R. A. 1958. Studies of the Jahn-Teller effect. II. The dynamical problem. *Proceedings of the Royal Society of London A: Mathematical and Physical Sciences*, **244**(1236), 1–16.

[314] Lukin, M. D., Fleischhauer, M., Cote, R., Duan, L. M., Jaksch, D., Cirac, J. I., and Zoller, P. 2001. Dipole blockade and quantum information processing in mesoscopic atomic ensembles. *Physical Review Letters*, **87**(3), 037901.

[315] Ma, Jian, Wang, Xiaoguang, Sun, Chang-Pu, and Nori, Franco. 2011. Quantum spin squeezing. *Physics Reports*, **509**(2–3), 89–165.

[316] Madison, K. W., Chevy, F., Wohlleben, W., and Dalibard, J. 2000. Vortex formation in a stirred Bose–Einstein condensate. *Physical Review Letters*, **84**(5), 806.

[317] Mandel, L., and Wolf, E. 1995. *Optical Coherence and Quantum Optics*. Cambridge University Press.

[318] Mandel, Olaf, Greiner, Markus, Widera, Artur, Rom, Tim, Hänsch, Theodor W., and Bloch, Immanuel. 2003. Coherent transport of neutral atoms in spin-dependent optical lattice potentials. *Physical Review Letters*, **91**(1), 010407.

[319] Marago, O. M., Hopkins, S. A., Arlt, J., Hodby, E., Hechenblaikner, G., and Foot, C. J. 2000. Observation of the scissors mode and evidence for superfluidity of a trapped Bose–Einstein condensed gas. *Physical Review Letters*, **84**(10), 2056.

[320] Marchant, A. L., Billam, T. P., Wiles, T. P., Yu, M. M. H., Gardiner, S. A., and Cornish, S. L. 2013. Controlled formation and reflection of a bright solitary matter-wave. *Nature Communications*, **4**, 1865.

[321] Marte, M., and Stenholm, S. 1992. Multiphoton resonance in atomic Bragg scattering. *Applied Physics B*, **54**, 443–450.

[322] Martin, P. J., Oldaker, B. G., Miklich, H., and Pritchard, D. E. 1988. Bragg scattering of atoms from a standing wave light. *Physical Review Letters*, **60**, 515–518.

[323] Matthews, M. R., Anderson, B. P., Haljan, P. C., Hall, D. S., Wieman, C. E., and Cornell, E. A. 1999. Vortices in a Bose–Einstein condensate. *Physical Review Letters*, **83**(13), 2498.

[324] Matthews, M. R., Hall, D. S., Jin, D. S., Ensher, J. R., Wieman, C. E., Cornell, E. A., Dalfovo, F., Minniti, C., and Stringari, S. 1998. Dynamical response of a Bose–Einstein condensate to a discontinuous change in internal state. *Physical Review Letters*, **81**, 243–247.

[325] McArdle, Sam, Endo, Suguru, Aspuru-Guzik, Alan, Benjamin, Simon, and Yuan, Xiao. 2018. Quantum computational chemistry. arXiv:1808.10402.

[326] Mead, C. Alden, and Truhlar, Donald G. 1979. On the determination of Born–Oppenheimer nuclear motion wave functions including complications due to conical intersections and identical nuclei. *Journal of Chemical Physics*, **70**(5), 2284–2296.

[327] Mekhov, Igor B., Maschler, Christoph, and Ritsch, Helmut. 2007. Probing quantum phases of ultracold atoms in optical lattices by transmission spectra in cavity quantum electrodynamics. *Nature Physics*, **3**(5), 319.

[328] Melko, R. G., Paramekanti, A., Burkov, A. A., Vishwanath, A., Sheng, D. N., and Balents, Leon. 2005. Supersolid order from disorder: hard-core bosons on the triangular lattice. *Physical Review Letters*, **95**(12), 127207.

[329] Mertes, K. M., Merrillc, J. W., Carretero-González, R., Frantzeskakis, D. J., Kevrekidis, P. G., and Hall, D. S. 2007. Nonequilibrium dynamics and superfluid ring excitations in binary Bose–Einstein condensates. *Physical Review Letters*, **99**, 190402.

[330] Metcalf, Harold J., and der Straten, P. V. 1999. *Laser Cooling and Trapping*. Springer-Verlag.

[331] Metcalf, Harold, and van der Straten, Peter. 1994. Cooling and trapping of neutral atoms. *Physics Reports*, **244**(4–5), 203–286.

[332] Metcalf, Harold J., and van der Straten, Peter. 2007. Laser cooling and trapping of neutral atoms. *The Optics Encyclopedia: Basic Foundations and Practical Applications*. Wiley.

[333] Meystre, Pierre. 2001. *Atom Optics*. Vol. 33. Springer Science and Business Media.

[334] Micheli, Andrea, Jaksch, D., Cirac, J. Ignacio, and Zoller, P. 2003. Many-particle entanglement in two-component Bose–Einstein condensates. *Physical Review A*, **67**(1), 013607.

[335] Milburn, G. J., Corney, J., Wright, E. M., and Walls, D. F. 1997. Quantum dynamics of an atomic Bose-Einstein condensate in a double-well potential. *Physical Review A*, **55**(6), 4318–4324.

[336] Miroshnychenko, Yevhen, Alt, Wolfgang, Dotsenko, Igor, Förster, Leonid, Khudaverdyan, Mkrtych, Meschede, Dieter, Schrader, Dominik, and Rauschenbeutel, Arno. 2006. Quantum engineering: an atom-sorting machine. *Nature*, **442**(7099), 151.

[337] Mizel, Ari, Lidar, Daniel A., and Mitchell, Morgan. 2007. Simple proof of equivalence between adiabatic quantum computation and the circuit model. *Physical Review Letters*, **99**(7), 070502.

[338] Mohseni, Naeimeh, Narozniak, Marek, Pyrkov, Alexey N., Ivannikov, Valentin, Dowling, Jonathan P., and Byrnes, Tim. 2019. Error suppression in adiabatic quantum computing with qubit ensembles. npj Quantum Information 7, 71 (2021)

[339] Moler, K., Weiss, D. S., Kasevich, K., and Chu, S. 1992. Theoretical analysis of velocity-selective Raman transitions. *Physical Review A*, **45**, 342–348.

[340] Mølmer, Klaus. 1999. Twin-correlations in atoms. *European Physical Journal D. Atomic, Molecular, Optical and Plasma Physics*, **5**(2), 301–305.

[341] Mølmer, Klaus, Castin, Yvan, and Dalibard, Jean. 1993. Monte Carlo wave-function method in quantum optics. *Journal of the Optical Society of America B*, **10**(3), 524–538.

[342] Mueller, E. J., and Ho, T. L. 2002. Two-component Bose–Einstein condensates with large number of vortices. *Physical Review Letters*, **88**, 180403.

[343] Muruganandam, Paulsamy, and Adhikari, Sadhan K. 2009. Fortran programs for the time-dependent Gross–Pitaevskii equation in a fully anisotropic trap. *Computer Physics Communications*, **180**(10), 1888–1912.

[344] Myatt, C. J., Burt, E. A., Ghrist, R. W., Cornell, E. A., and Wieman, C. E. 1997. Production of two overlapping Bose–Einstein condensates by sympathetic cooling. *Physical Review Letters*, **78**, 586–589.

[345] Nielsen, M. A., and Chuang, I. L. 2000. *Quantum Computation and Quantum Information*. Cambridge University Press.

[346] Oberthaler, M. K., Abfalterer, R., Bernet, S., Keller, C., Schmiedmayer, J., and Zeilinger, A. 1999. Dynamical diffraction of atomic matter waves by crystals of light. *Physical Review A*, **60**, 456–472.

[347] Obrecht, John Michael, Wild, R. J., Antezza, M., Pitaevskii, L. P., Stringari, S., and Cornell, Eric A. 2007. Measurement of the temperature dependence of the Casimir–Polder force. *Physical Review Letters*, **98**(6), 063201.

[348] Ockeloen, C. F., Schmied, R., Riedel, M. F., and Treutlein, P. 2013. Quantum metrology with a scanning probe atom interferometer. *Physical Review Letters*, **111**, 143001.

[349] ODell, Duncan H. J., Giovanazzi, Stefano, and Eberlein, Claudia. 2004. Exact hydrodynamics of a trapped dipolar Bose–Einstein condensate. *Physical Review Letters*, **92**(25), 250401.

[350] O'hara, K. M., Hemmer, S. L., Gehm, M. E., Granade, S. R., and Thomas, J. E. 2002. Observation of a strongly interacting degenerate Fermi gas of atoms. *Science*, **298**(5601), 2179–2182.

[351] Ohashi, Yoji, and Griffin, A. 2002. BCS–BEC crossover in a gas of Fermi atoms with a Feshbach resonance. *Physical Review Letters*, **89**(13), 130402.

[352] Ohmi, T., and Machida, K. 1998. Spinor Bose condensates in optical traps. *Journal of the Physical Society of Japan*, **67**, 1822–1825.

[353] Oitmaa, Jaan, Hamer, Chris, and Zheng, Weihong. 2006. *Series Expansion Methods for Strongly Interacting Lattice Models*. Cambridge University Press.

[354] Olshanii, M., and Dunjko, V. 2005. Interferometry in dense nonlinear media and interaction-induced loss of contrast in microfabricated atom interferometers. arXiv:cond-mat/0505358.

[355] Orzel, Chad, Tuchman, A. K., Fenselau, M. L., Yasuda, M., and Kasevich, M. A. 2001. Squeezed states in a Bose–Einstein condensate. *Science*, **291**(5512), 2386–2389.

[356] Osterloh, K., Baig, M., Santos, L., Zoller, P., and Lewenstein, M. 2005. Cold atoms in non-Abelian gauge potentials: from the Hofstadter "moth" to lattice gauge theory. *Physical Review Letters*, **95**(1), 010403.

[357] Ostrovskaya, Elena A., and Kivshar, Yuri S. 2003. Matter-wave gap solitons in atomic band-gap structures. *Physical Review Letters*, **90**(16), 160407.

[358] Paivi, Torma, and Klaus, Sengstock. 2014. *Quantum Gas Experiments: Exploring Many-Body States*. Vol. 3. World Scientific.

[359] Pancharatnam, Shivaramakrishnan. 1956. Generalized theory of interference and its applications. Pages 398–417 of: *Proceedings of the Indian Academy of Sciences Section A*. Vol. 44. Springer.

[360] Paredes, Belén, Widera, Artur, Murg, Valentin, Mandel, Olaf, Fölling, Simon, Cirac, Ignacio, Shlyapnikov, Gora V., Hänsch, Theodor W., and Bloch, Immanuel. 2004. Tonks–Girardeau gas of ultracold atoms in an optical lattice. *Nature*, **429**(6989), 277.

[361] Paris, M. G. 2009. Quantum estimation for quantum technology. *International Journal of Quantum Information*, **7**, 125–137.

[362] Parker, N. G., and Adams, C. S. 2005. Emergence and decay of turbulence in stirred atomic Bose–Einstein condensates. *Physical Review Letters*, **95**(14), 145301.

[363] Perelomov, Askold. 2012. *Generalized Coherent States and Their Applications*. Springer Science and Business Media.

[364] Peres, Asher. 1996. Separability criterion for density matrices. *Physical Review Letters*, **77**(Aug), 1413–1415.

[365] Perez-Garcia, Victor M., Michinel, Humberto, Cirac, J. I., Lewenstein, M., and Zoller, P. 1997. Dynamics of Bose–Einstein condensates: variational solutions of the Gross–Pitaevskii equations. *Physical Review A*, **56**(2), 1424.

[366] Perez-Garcia, Victor M., Michinel, Humberto, and Herrero, Henar. 1998. Bose–Einstein solitons in highly asymmetric traps. *Physical Review A*, **57**(5), 3837.

[367] Peters, A., Chung, K. Y., and Chu, S. 1999. Measurement of gravitational acceleration by dropping atoms. *Nature*, **400**, 849.

[368] Peters, A., Chung, K. Y., and Chu, S. 2001. High-precision gravity measurements using atom interferometry. *Metrologia*, **38**, 25.

[369] Pethick, C. J., and Smith, H. 2002. *Bose–Einstein Condensation in Dilute Gases*. Cambridge University Press.

[370] Pezzè, L., and Smerzi, A. 2014. Quantum theory of phase estimation. arXiv:1411.5164.

[371] Pfeuty, Pierre. 1970. The one-dimensional Ising model with a transverse field. *Annals of Physics*, **57**(1), 79–90.

[372] Phillips, William D. 1998. Nobel lecture: laser cooling and trapping of neutral atoms. *Reviews of Modern Physics*, **70**(3), 721.

[373] Phillips, William D., Prodan, John V., and Metcalf, Harold J. 1985. Laser cooling and electromagnetic trapping of neutral atoms. *Journal of the Optical Society of America B*, **2**(11), 1751–1767.

[374] Pitaevskii, L. P. 1961. Vortex lines in an imperfect Bose gas. *Soviet Physics JETP*, **13**(2), 451–454.

[375] Pitaevskii, L. P., and Stringari, S. 2016. *Bose–Einstein Condensation and Superfluidity*. Vol. 164. Oxford University Press.

[376] Plenio, M. B. 2005. Logarithmic negativity: a full entanglement monotone that is not convex. *Physical Review Letters*, **95**(Aug), 090503.

[377] Plenio, M. B., and Knight, P. L. 1998. The quantum-jump approach to dissipative dynamics in quantum optics. *Reviews of Modern Physics*, **70**(1), 101.

[378] Plenio, M. B., and Virmani, S. S. 2014. An introduction to entanglement theory. Pages 173–209 of: *Quantum Information and Coherence*. Springer.

[379] Preiss, Philipp M., Ma, Ruichao, Tai, M. Eric, Lukin, Alexander, Rispoli, Matthew, Zupancic, Philip, Lahini, Yoav, Islam, Rajibul, and Greiner, Markus. 2015. Strongly correlated quantum walks in optical lattices. *Science*, **347**(6227), 1229–1233.

[380] Pu, H., and Meystre, Pierre. 2000. Creating macroscopic atomic Einstein–Podolsky–Rosen states from Bose–Einstein condensates. *Physical Review Letters*, **85**(19), 3987.

[381] Pyrkov, Alexey N., and Byrnes, Tim. 2013. Entanglement generation in quantum networks of Bose–Einstein condensates. *New Journal of Physics*, **15**(9), 093019.

[382] Raab, E. L., Prentiss, M., Cable, Alex, Chu, Steven, and Pritchard, David E. 1987. Trapping of neutral sodium atoms with radiation pressure. *Physical Review Letters*, **59**(23), 2631.

[383] Rabl, P., DeMille, D., Doyle, John M., Lukin, Mikhail D., Schoelkopf, R. J., and Zoller, P. 2006. Hybrid quantum processors: molecular ensembles

as quantum memory for solid state circuits. *Physical Review Letters*, **97**(3), 033003.

[384] Radcliffe, J. M. 1971. Some properties of coherent spin states. *Journal of Physics A: General Physics*, **4**(3), 313.

[385] Raghavan, S., Smerzi, A., Fantoni, S., and Shenoy, S. R. 1999. Coherent oscillations between two weakly coupled Bose–Einstein condensates: Josephson effects, π oscillations, and macroscopic quantum self-trapping. *Physical Review A*, **59**, 620–633.

[386] Raman, C., Abo-Shaeer, J. R., Vogels, J. M., Xu, K., and Ketterle, W. 2001. Vortex nucleation in a stirred Bose–Einstein condensate. *Physical Review Letters*, **87**(Nov), 210402.

[387] Ramsey, N. F. 1950. A molecular beam resonance method with separated oscillating fields. *Physical Review*, **78**(6), 695–699.

[388] Ramsey, N. F. 1985. *Molecular Beams*. Oxford University Press.

[389] Rasel, E. M., Oberthaler, M. K., Batelaan, H., Schmiedmayer, J., and Zeilinger, A. 1995. Atom wave interferometry with diffraction gratings of Light. *Physical Review Letters*, **75**, 2633–2637.

[390] Ratti, Claudia. 2018. Lattice QCD and heavy ion collisions: a review of recent progress. *Reports on Progress in Physics*, **81**(8), 084301.

[391] Regal, C. A., Greiner, M., and Jin, D. S. 2004. Observation of resonance condensation of fermionic atom pairs. *Physical Review Letters*, **92**(4), 040403.

[392] Regal, C. A., Ticknor, C., Bohn, J. L., and Jin, D. S. 2003. Creation of ultracold molecules from a Fermi gas of atoms. *Nature*, **424**(6944), 47.

[393] Reichel, Jakob, and Vuletic, Vladan. 2011. *Atom Chips*. Wiley.

[394] Riedel, M. F., Böhi, P., L. i., Y., Hänsch, T. W., Sinatra, A., and Treutlein, P. 2010. Atom-chip-based generation of entanglement for quantum metrology. *Nature*, **464**, 1170.

[395] Roach, T. M., Abele, H., Boshier, M. G., Grossman, H. L., Zetie, K. P., and Hinds, E. A. 1995. Realization of a magnetic mirror for cold atoms. *Physical Review Letters*, **75**(4), 629.

[396] Roati, G., De Mirandes, E., Ferlaino, F., Ott, H., Modugno, G., and Inguscio, M. 2004. Atom interferometry with trapped Fermi gases. *Physical Review Letters*, **92**(23), 230402.

[397] Robertson, H. P. 1929. The uncertainty principle. *Physical Review*, **34**(1), 163.

[398] Robertson, H. P. 1934. An indeterminacy relation for several observables and its classical interpretation. *Physical Review*, **46**(9), 794.

[399] Rogel-Salazar, Jesus. 2013. The Gross–Pitaevskii equation and Bose–Einstein condensates. *European Journal of Physics*, **34**(2), 247.

[400] Romans, M. W. J., Duine, R. A., Sachdev, Subir, and Stoof, H. T. C. 2004. Quantum phase transition in an atomic Bose gas with a Feshbach resonance. *Physical Review Letters*, **93**(2), 020405.

[401] Rosi, G., Sorrentino, F., Cacciapuoti, L., Prevedelli, M., and Tino, G. M. 2014. Precision measurement of the Newtonian gravitational constant using cold atoms. *Nature*, **510**(7506), 518.

[402] Rosseau, Daniel, Ha, Qianqian, and Byrnes, Tim. 2014. Entanglement generation between two spinor Bose–Einstein condensates with cavity QED. *Physical Review A*, **90**(5), 052315.

[403] Ruostekoski, J., and Anglin, J. R. 2001. Creating vortex rings and three-dimensional skyrmions in Bose–Einstein condensates. *Physical Review Letters*, **86**(18), 3934.

[404] Ruseckas, Julius, Juzeliūnas, G., Öhberg, Patrik, and Fleischhauer, Michael. 2005. Non-Abelian gauge potentials for ultracold atoms with degenerate dark states. *Physical Review Letters*, **95**(1), 010404.

[405] Sadgrove, M., Horikoshi, M., Sekimura, T., and Nakagawa, K. 2007. Rectified momentum transport for a kicked Bose–Einstein condensate. *Physical Review Letters*, **99**, 043002.

[406] Saffman, Mark, Walker, Thad G., and Mølmer, Klaus. 2010. Quantum information with Rydberg atoms. *Reviews of Modern Physics*, **82**(3), 2313.

[407] Sakurai, Jun John, and Commins, Eugene D. 1995. *Modern Quantum Mechanics, Revised Edition*. Cambridge University Press.

[408] Salasnich, L., Parola, A., and Reatto, L. 2002. Effective wave equations for the dynamics of cigar-shaped and disk-shaped Bose condensates. *Physical Review A*, **65**(4), 043614.

[409] Sauter, T., Neuhauser, W., Blatt, R., and Toschek, P. E. 1986. Observation of quantum jumps. *Physical Review Letters*, **57**(14), 1696.

[410] Scarola, Vito W., and Sarma, S. Das. 2005. Quantum phases of the extended Bose-Hubbard Hamiltonian: Possibility of a supersolid state of cold atoms in optical lattices. *Physical Review Letters*, **95**(3), 033003.

[411] Schaetz, Tobias, Monroe, Chris R., and Esslinger, Tilman. 2013. Focus on quantum simulation. *New Journal of Physics*, **15**(8), 085009.

[412] Schellekens, Martijn, Hoppeler, Rodolphe, Perrin, Aurélien, Gomes, J. Viana, Boiron, Denis, Aspect, Alain, and Westbrook, Christoph I. 2005. Hanbury Brown Twiss effect for ultracold quantum gases. *Science*, **310**(5748), 648–651.

[413] Schindler, Philipp, Müller, Markus, Nigg, Daniel, Barreiro, Julio T., Martinez, Esteban A., Hennrich, Markus, Monz, T., Diehl, Sebastian, Zoller, Peter, and Blatt, Rainer. 2013. Quantum simulation of dynamical maps with trapped ions. *Nature Physics*, **9**(6), 361.

[414] Schlosshauer, Maximilian A. 2007. *Decoherence and the Quantum-to-Classical Transition*. Springer Science and Business Media.

[415] Schneider, U., Hackermüller, L., Will, S., Best, T., Bloch, Immanuel, Costi, T. A., Helmes, R. W., Rasch, D., and Rosch, A. 2008. Metallic and insulating phases of repulsively interacting fermions in a 3D optical lattice. *Science*, **322**(5907), 1520–1525.

[416] Schollwöck, Ulrich. 2011. The density-matrix renormalization group in the age of matrix product states. *Annals of Physics*, **326**(1), 96–192.

[417] Schreiber, Michael, Hodgman, Sean S., Bordia, Pranjal, Lüschen, Henrik P., Fischer, Mark H., Vosk, Ronen, Altman, Ehud, Schneider, Ulrich, and Bloch, Immanuel. 2015. Observation of many-body localization of interacting fermions in a quasirandom optical lattice. *Science*, **349**(6250), 842–845.

[418] Schumm, T., Hofferberth, S., Andersson, L. M., Wildermuth, S., Groth, S., Bar-Joseph, I., Schmiedmayer, J., and Krüger, P. 2005. Matter-wave interferometry in a double well on an atom chip. *Nature Physics*, **1**, 57–62.

[419] Schweikhard, Volker, Coddington, I., Engels, Peter, Tung, Shihkuang, and Cornell, Eric A. 2004. Vortex-lattice dynamics in rotating spinor Bose–Einstein condensates. *Physical Review Letters*, **93**(21), 210403.

[420] Schwinger, J. 1965. *Quantum Mechanics of Angular Momentum. A Collection of Reprints and Original Papers.* Ed. L. C. Biedenharn and H. van Dam. Academic Press.

[421] Scully, Marlan O., and Zubairy, M. Suhail. 1999. *Quantum Optics*. Cambridge University Press.

[422] Scully, M. O., and Wódkiewicz, K. 1994. Spin quasi-distribution functions. *Foundations of Physics*, **24**(1), 85–107.

[423] Seaman, B. T., Krämer, M., Anderson, D. Z., and Holland, M. J. 2007. Atomtronics: ultracold-atom analogs of electronic devices. *Physical Review A*, **75**(2), 023615.

[424] Semenenko, Henry, and Byrnes, Tim. 2016. Implementing the Deutsch–Jozsa algorithm with macroscopic ensembles. *Physical Review A*, **93**(5), 052302.

[425] Sherson, Jacob F., Krauter, Hanna, Olsson, Rasmus K., Julsgaard, Brian, Hammerer, Klemens, Cirac, Ignacio, and Polzik, Eugene S. 2006. Quantum teleportation between light and matter. *Nature*, **443**(7111), 557.

[426] Sherson, Jacob F., Weitenberg, Christof, Endres, Manuel, Cheneau, Marc, Bloch, Immanuel, and Kuhr, Stefan. 2010. Single-atom-resolved fluorescence imaging of an atomic mott insulator. *Nature*, **467**(7311), 68.

[427] Shin, Y., Saba, M., Pasquini, T. A., Ketterle, W., Pritchard, D. E., and Leanhardt, A. E. 2004. Atom interferometry with Bose–Einstein condensates in a double-well potential. *Physical Review Letters*, **92**, 050405.

[428] Shore, B. W. 1990. *The Theory of Coherent Atomic Excitation. Vol. 2*. Wiley-Interscience.

[429] Simon, R. 2000. Peres–Horodecki separability criterion for continuous variable systems. *Physical Review Letters*, **84**(Mar), 2726–2729.

[430] Simula, T. P., and Blakie, P. B. 2006. Thermal activation of vortex-antivortex pairs in quasi-two-dimensional Bose–Einstein condensates. *Physical Review Letters*, **96**(2), 020404.

[431] Slusher, R. E., Hollberg, L. W., Yurke, Bernard, Mertz, J. C., and Valley, J. F. 1985. Observation of squeezed states generated by four-wave mixing in an optical cavity. *Physical Review Letters*, **55**(22), 2409.

[432] Smerzi, A., Fantoni, S., Giovanazzi, S., and Shenoy, S. R. 1997. Quantum coherent atomic tunneling between two trapped Bose–Einstein condensates. *Physical Review Letters*, **79**, 4950–4953.

[433] Söding, J., Guéry-Odelin, D., Desbiolles, P., Chevy, F., Inamori, H., and Dalibard, J. 1999. Three-body decay of a rubidium Bose–Einstein condensate. *Applied Physics B*, **69**(4), 257–261.

[434] Sørensen, A. S., Duan, L.-M., Cirac, J. I., and Zoller, Peter. 2001. Many-particle entanglement with Bose–Einstein condensates. *Nature*, **409**(6816), 63.

[435] Sørensen, A. S., and Mølmer, K. 2001. Entanglement and extreme spin squeezing. *Physical Review Letters*, **86**(20), 4431.

[436] Stamper-Kurn, D. M., Andrews, M. R., Chikkatur, A. P., Inouye, S., Miesner, H.-J., Stenger, J., and Ketterle, W. 1998. Optical confinement of a Bose–Einstein condensate. *Physical Review Letters*, **80**, 2027–2030.

[437] Stanescu, Tudor D., Anderson, Brandon, and Galitski, Victor. 2008. Spin-orbit coupled Bose–Einstein condensates. *Physical Review A*, **78**(2), 023616.

[438] Steck, Daniel A. 2001. *Rubidium 87 D Line Data*.

[439] Steck, Daniel A. 2007. Quantum and atom optics. *Oregon Center for Optics and Department of Physics, University of Oregon*, **47**.

[440] Steel, M. J., and Collett, M. J. 1998. Quantum state of two trapped Bose–Einstein condensates with a Josephson coupling. *Physical Review A*, **57**, 2920–2930.

[441] Steinhauer, J., Ozeri, R., Katz, N., and Davidson, N. 2002. Excitation spectrum of a Bose–Einstein condensate. *Physical Review Letters*, **88**(12), 120407.

[442] Stickney, J. A., Anderson, D. Z., and Zozulya, A. A. 2007. Increasing the coherence tome of Bose–Einstein condensate interferometers with optical control of dynamics. *Physical Review A*, **75**, 063603.

[443] Stöferle, Thilo, Moritz, Henning, Schori, Christian, Köhl, Michael, and Esslinger, Tilman. 2004. Transition from a strongly interacting 1D superfluid to a Mott insulator. *Physical Review Letters*, **92**(13), 130403.

[444] Stöferle, Thilo, Moritz, Henning, Günter, Kenneth, Köhl, Michael, and Esslinger, Tilman. 2006. Molecules of fermionic atoms in an optical lattice. *Physical Review Letters*, **96**(3), 030401.

[445] Strecker, Kevin E., Partridge, Guthrie B., Truscott, Andrew G., and Hulet, Randall G. 2002. Formation and propagation of matter-wave soliton trains. *Nature*, **417**(6885), 150.

[446] Strinati, Giancarlo Calvanese, Pieri, Pierbiagio, Röpke, Gerd, Schuck, Peter, and Urban, Michael. 2018. The BCS–BEC crossover: from ultra-cold Fermi gases to nuclear systems. *Physics Reports*, **738**, 1–76.

[447] Suzuki, Masuo. 1976a. Generalized Trotter's formula and systematic approximants of exponential operators and inner derivations with applications to many-body problems. *Communications in Mathematical Physics*, **51**(2), 183–190.

[448] Suzuki, Masuo. 1976b. Relationship between d-dimensional quantal spin systems and (d + 1)-dimensional Ising systems: Equivalence, critical exponents and systematic approximants of the partition function and spin correlations. *Progress of Theoretical Physics*, **56**(5), 1454–1469.

[449] Szangolies, Jochen, Kampermann, Hermann, and Bruß, Dagmar. 2015. Detecting entanglement of unknown quantum states with random measurements. *New Journal of Physics*, **17**(11), 113051.

[450] Takamoto, Masao, Hong, Feng-Lei, Higashi, Ryoichi, and Katori, Hidetoshi. 2005. An optical lattice clock. *Nature*, **435**(7040), 321–324.

[451] Takano, T., Fuyama, M., Namiki, R., and Takahashi, Y. 2009. Spin squeezing of a cold atomic ensemble with the nuclear spin of one-half. *Physical Review Letters*, **102**(3), 033601.

[452] Tang, Yijun, Kao, Wil, L. i., Kuan-Yu, Seo, Sangwon, Mallayya, Krishnanand, Rigol, Marcos, Gopalakrishnan, Sarang, and Lev, Benjamin L. 2018. Thermalization near integrability in a dipolar quantum Newtons cradle. *Physical Review X*, **8**(2), 021030.

[453] Tarruell, Leticia, and Sanchez-Palencia, Laurent. 2018. Quantum simulation of the Hubbard model with ultracold fermions in optical lattices. *Comptes Rendus Physique*, **19**(6), 365–393.

[454] Terhal, Barbara M. 2002. Detecting quantum entanglement. *Theoretical Computer Science*, **287**(1), 313–335.

[455] Theocharis, G., Schmelcher, P., Kevrekidis, P. G., and Frantzeskakis, D. J. 2005. Matter-wave solitons of collisionally inhomogeneous condensates. *Physical Review A*, **72**(3), 033614.

[456] Thompson, William J. 2008. *Angular Momentum: An Illustrated Guide to Rotational Symmetries for Physical Systems*. Wiley.

[457] Thomsen, L. K., Mancini, Stefano, and Wiseman, Howard Mark. 2002. Spin squeezing via quantum feedback. *Physical Review A*, **65**(6), 061801.

[458] Tilley, David R. 2019. *Superfluidity and Superconductivity*. Routledge.

[459] Tilma, Todd, Everitt, Mark J., Samson, John H., Munro, William J., and Nemoto, Kae. 2016. Wigner functions for arbitrary quantum systems. *Physical Review Letters*, **117**(18), 180401.

[460] Timmermans, Eddy, Tommasini, Paolo, Hussein, Mahir, and Kerman, Arthur. 1999. Feshbach resonances in atomic Bose–Einstein condensates. *Physics Reports*, **315**(1–3), 199–230.

[461] Torii, Y., Suzuki, Y., Kozuma, M., Sugiura, T., Kuga, T., Deng, L., and Hagley, E. W. 2000. Mach–Zehnder Bragg interferometer for a Bose–Einstein condensate. *Physical Review A*, **61**, 041602.

[462] Tóth, Géza, and Apellaniz, Iagoba. 2014. Quantum metrology from a quantum information science perspective. *Journal of Physics A: Mathematical and Theoretical*, **47**(42), 424006.

[463] Tóth, Géza, Knapp, Christian, Gühne, Otfried, and Briegel, Hans J. 2007. Optimal spin squeezing inequalities detect bound entanglement in spin models. *Physical Review Letters*, **99**(25), 250405.

[464] Tóth, Géza, Knapp, Christian, Gühne, Otfried, and Briegel, Hans J. 2009. Spin squeezing and entanglement. *Physical Review A*, **79**(4), 042334.

[465] Treutlein, Philipp. 2008. *Coherent Manipulation of Ultracold Atoms on Atom Chips*. Ph.D. thesis, LMU Munich.

[466] Treutlein, Philipp, Hänsch, Theodor W., Reichel, Jakob, Negretti, Antonio, Cirone, Markus A., and Calarco, Tommaso. 2006. Microwave potentials and optimal control for robust quantum gates on an atom chip. *Physical Review A*, **74**(2), 022312.

[467] Treutlein, Philipp, Hommelhoff, Peter, Steinmetz, Tilo, Hänsch, Theodor W., and Reichel, Jakob. 2004. Coherence in microchip traps. *Physical Review Letters*, **92**(20), 203005.

[468] Tripathi, Vinay, Radhakrishnan, Chandrashekar, and Byrnes, Tim. 2020. Covariance matrix entanglement criterion for an arbitrary set of operators. *New Journal of Physics*, **22**(7), 073055.

[469] Trotter, Hale F. 1959. On the product of semi-groups of operators. *Proceedings of the American Mathematical Society*, **10**(4), 545–551.

[470] Urban, E., Johnson, Todd A., Henage, T., Isenhower, L., Yavuz, D. D., Walker, T. G., and Saffman, M. 2009. Observation of Rydberg blockade between two atoms. *Nature Physics*, **5**(2), 110–114.

[471] Ushijima, Ichiro, Takamoto, Masao, Das, Manoj, Ohkubo, Takuya, and Katori, Hidetoshi. 2015. Cryogenic optical lattice clocks. *Nature Photonics*, **9**(3), 185–189.

[472] Vaidman, Lev. 1994. Teleportation of quantum states. *Physical Review A*, **49**(2), 1473.

[473] Verkerk, P., Lounis, B., Salomon, C., Cohen-Tannoudji, C., Courtois, J.-Y., and Grynberg, G. 1992. Dynamics and spatial order of cold cesium atoms in a periodic optical potential. *Physical Review Letters*, **68**(26), 3861.

[474] Vidal, Guifré, and Werner, Reinhard F. 2002. Computable measure of entanglement. *Physical Review A*, **65**(3), 032314.

[475] von Neumann, J. 1927. Thermodynamik quantenmechanischer Grossen. *Gott. Nachr.*, 273.

[476] Walls, Daniel F. 1983. Squeezed states of light. *Nature*, **306**(5939), 141.

[477] Walls, Daniel F., and Milburn, Gerard J. 2007. *Quantum Optics*. Springer Science and Business Media.

[478] Wang, Chunji, Gao, Chao, Jian, Chao-Ming, and Zhai, Hui. 2010. Spin-orbit coupled spinor Bose–Einstein condensates. *Physical Review Letters*, **105**(16), 160403.

[479] Wang, Xiaoguang, and Sanders, Barry C. 2003. Spin squeezing and pairwise entanglement for symmetric multiqubit states. *Physical Review A*, **68**(1), 012101.

[480] Wang, Y. J., Anderson, D. Z., Bright, V. M., Cornell, E. A., Diot, Q., Kishimoto, T., Prentiss, M., Saravanan, R. A., Segal, S. R., and Wu, S. 2005. Atom Michelson interferometer on a chip using a Bose–Einstein condensate. *Physical Review Letters*, **94**, 090405.

[481] Weedbrook, Christian, Pirandola, Stefano, García-Patrón, Raúl, Cerf, Nicolas J., Ralph, Timothy C., Shapiro, Jeffrey H., and Lloyd, Seth. 2012. Gaussian quantum information. *Reviews of Modern Physics*, **84**(2), 621.

[482] Wehrl, Alfred. 1978. General properties of entropy. *Reviews of Modern Physics*, **50**(2), 221.

[483] Weiler, Chad N., Neely, Tyler W., Scherer, David R., Bradley, Ashton S., Davis, Matthew J., and Anderson, Brian P. 2008. Spontaneous vortices in the formation of Bose–Einstein condensates. *Nature*, **455**(7215), 948.

[484] Weimer, Hendrik, Müller, Markus, Lesanovsky, Igor, Zoller, Peter, and Büchler, Hans Peter. 2010. A Rydberg quantum simulator. *Nature Physics*, **6**(5), 382.

[485] Weisskopf, V. F., and Wigner, E. 1930. *Z. Physik*, **63**, 54.

[486] Weld, David M., Medley, Patrick, Miyake, Hirokazu, Hucul, David, Pritchard, David E., and Ketterle, Wolfgang. 2009. Spin gradient thermometry for ultracold atoms in optical lattices. *Physical Review Letters*, **103**(24), 245301.

[487] White, M., Pasienski, M., McKay, D., Zhou, S. Q., Ceperley, D., and DeMarco, B. 2009. Strongly interacting bosons in a disordered optical lattice. *Physical Review Letters*, **102**(5), 055301.

[488] White, Steven R. 1992. Density matrix formulation for quantum renormalization groups. *Physical Review Letters*, **69**(19), 2863.

[489] Wieman, Carl E., Pritchard, David E., and Wineland, David J. 1999. Atom cooling, trapping, and quantum manipulation. *Reviews of Modern Physics*, **71**(2), S253.

[490] Wiesner, Stephen. 1996. Simulations of many-body quantum systems by a quantum computer. arXiv:quant-ph/9603028.

[491] Wineland, David J., Bollinger, John J., Itano, Wayne M., and Heinzen, D. J. 1994. Squeezed atomic states and projection noise in spectroscopy. *Physical Review A*, **50**(1), 67.

[492] Wineland, David J., Bollinger, John J., Itano, Wayne M., Moore, F. L., and Heinzen, Daniel J. 1992. Spin squeezing and reduced quantum noise in spectroscopy. *Physical Review A*, **46**(11), R6797.

[493] Wiseman, Howard M., and Milburn, Gerard J. 2009. *Quantum Measurement and Control*. Cambridge University Press.

[494] Wu, S., Wang, Y. J., Diot, Q., and Prentis, M. 2005. Splitting matter waves using an optimized standing-wave light-pulse sequence. *Physical Review A*, **71**, 043602.

[495] Wu, Zhan, Zhang, Long, Sun, Wei, Xu, Xiao-Tian, Wang, Bao-Zong, Ji, Si-Cong, Deng, Youjin, Chen, Shuai, Liu, Xiong-Jun, and Pan, Jian-Wei. 2016. Realization of two-dimensional spin-orbit coupling for Bose–Einstein condensates. *Science*, **354**(6308), 83–88.

[496] Yamamoto, Ryuta, Kobayashi, Jun, Kuno, Takuma, Kato, Kohei, and Takahashi, Yoshiro. 2016. An ytterbium quantum gas microscope with narrow-line laser cooling. *New Journal of Physics*, **18**(2), 023016.

[497] Yamamoto, Yoshihisa, and Imamoglu, Atac. 1999. Mesoscopic quantum optics. *Mesoscopic Quantum Optics* Wiley.

[498] Yan, Bo, Moses, Steven A., Gadway, Bryce, Covey, Jacob P., Hazzard, Kaden R. A., Rey, Ana Maria, Jin, Deborah S., and Ye, Jun. 2013. Observation of dipolar spin-exchange interactions with lattice-confined polar molecules. *Nature*, **501**(7468), 521.

[499] Yang, Chen Ning. 1962. Concept of off-diagonal long-range order and the quantum phases of liquid He and of superconductors. *Reviews of Modern Physics*, **34**(4), 694.

[500] Yang, Jianke, and Musslimani, Ziad H. 2003. Fundamental and vortex solitons in a two-dimensional optical lattice. *Optics Letters*, **28**(21), 2094–2096.

[501] Yavin, I., Weel, M., Andreyuk, A., and Kumarakrishnan, A. 2002. A calculation of the time-of-flight distribution of trapped atoms. *American Journal of Physics*, **70**(2), 149–152.

[502] Young, H. D., and Freedman, R. A. 2012. *Sears and Zemansky's University Physics*. Addison-Wesley.

[503] Yurke, B., McCall, S. L., and Klauder, J. R. 1986. SU(2) and SU(1,1) interferometers. *Physical Review A*, **33**, 4033–4054.

[504] Zhai, Hui. 2012. Spin-orbit coupled quantum gases. *International Journal of Modern Physics B*, **26**(01), 1230001.

[505] Zhang, Long, and Liu, Xiong-Jun. 2018. Spin-orbit coupling and topological phases for ultracold atoms. arXiv:1806.05628.

[506] Zirbel, J. J., Ni, K.-K., Ospelkaus, S., DIncao, J. P., Wieman, C. E., Ye, J., and Jin, DS. 2008. Collisional stability of fermionic Feshbach molecules. *Physical Review Letters*, **100**(14), 143201.

[507] Zwerger, Wilhelm. 2011. *The BCS–BEC Crossover and the Unitary Fermi Gas*. Vol. 836. Springer Science and Business Media.

[508] Zwierlein, Martin W., Stan, Claudiu A., Schunck, Christian H., Raupach, Sebastian M. F., Gupta, Subhadeep, Hadzibabic, Zoran, and Ketterle, Wolfgang. 2003. Observation of Bose–Einstein condensation of molecules. *Physical Review Letters*, **91**(25), 250401.

[509] Zwierlein, M. W., Stan, C. A., Schunck, C. H., Raupach, S. M. F., Kerman, A. J., and Ketterle, W. 2004. Condensation of pairs of fermionic atoms near a Feshbach resonance. *Physical Review Letters*, **92**(12), 120403.

Index

ac Stark shift, 53, 54, 159
adiabatic approximation, 161
adiabatic elimination, 112
adiabatic quantum computing, 199, 211
anticommutation relations, 2, 3, 5
artificial gauge field, 160
atom interferometry, 130
atom wave, 117
atom-light interactions, 102
Avogadro's number, 152

Bardeen–Cooper–Schrieffer state, 160
basis transformation, 2
bath, 56
Bell state, 176, 189
Berry phase, 160, 162
binding energy, 49, 60
binomial, 119, 120
bipartite state, 177
bipartite system, 176, 177
Bloch function, 169
Bloch sphere, 69, 72–74, 78, 87, 88, 93, 140, 187, 200, 207
Bloch state, 172
Bloch vector, 141
Bloch's theorem, 168
Bogoliubov approximation, 30
Bogoliubov distribution, 23
Bogoliubov operators, 23
Bogoliubov transformation, 21
bogoliubovon, 21
bogoliubovons, 22
Bohr magneton, 16
Bohr radius, 50
Boltzmann
  constant, 10, 13
  distribution, 10, 155
  factor, 13
  statistics, 12–14
Bose–Hubbard Hamiltonian, 172
Bose–Hubbard model, 153, 167, 170, 171
bosonic enhancement, 59
Bragg condition, 107
Bragg diffraction, 107, 109
Bragg's law, 99

Casimir operator, 70
Cauchy–Schwarz inequality, 75, 180
channel, 47, 54
  closed, 54
  open, 54
chemical potential, 14–18, 32, 36, 128, 131, 135, 171, 172, 191
Clebsch–Gordan coefficient, 48, 88
closed system, 55, 56
coheren state, 187
coherence, 58, 139
coherent process, 55, 61
coherent state, 75, 94
  optical, 65, 66, 76, 77, 94, 200
  spin, 64–66, 69, 70, 72, 76–78, 81, 82, 85–89, 91–95, 187–189, 200, 204
commutation matrix, 182
commutation relations, 2, 7, 21, 45, 67, 77, 181
condensate wavefunction, 131
continuity equation, 37
convection currents, 38
cooling of atoms, 105
covariance matrix, 182–186
Cramer–Rao inequality, 145
creation and destruction operators, 4
critical temperature for BEC, 19

dark state, 108
de Broglie wavelength, 19, 101
detuning, 104
  blue-detuned, 104, 106
  red-detuned, 104, 106
Deutsch's algorithm, 199, 207, 208, 210
Deutsch–Jozsa algorithm, 210
devil's crevasse, 191
Dicke states, 88
diffraction, 99, 101, 139
  Atom diffraction, 106
  Bragg diffraction, 106
  Raman–Nath diffraction, 106, 107
digital quantum simulation, 153
dipole approximation, 102
dispersion, 22, 24, 25
  Bogoliubov, 10, 22, 24, 25, 27
  energy, 22, 24
  linear, 22
  parabolic, 22
  relation, 20, 23–25
dissipative force, 105
disspationless flow, 36
distinguishable particles, 11–14, 85
Dresselhaus effect, 163
Duan–Giedke–Cirac–Zoller criterion, 180

Ehrenfest equations, 105
electric dipole
  approximation, 49
  Hamiltonian, 50–52
  moment, 52
electron beam diffraction, 99
ensemble, 86, 177, 186, 199, 201, 206
ensembles, 78, 82, 84, 85
entanglement generation, 54
entanglement witness, 182
environment, 56
error propagation formula, 140, 142

Feshbach resonance, 44, 54, 55, 159, 160
filling factor, 171
Fisher information, 144–147
fluorescence imaging, 166, 167
Fock state, 93, 208
Fock states, 4–6, 12, 14, 21, 59, 64–68, 73, 74, 83, 84, 86, 88–95, 181, 189, 206
fractal, 190, 191

Galilean transformation, 24, 26
grand canonical distribution, 15
grand canonical ensemble, 17, 18
Gross–Pitaevskii equation, 118, 131

Hadamard gate, 205
harmonic trap, 117, 127
Haussdorff dimension, 190, 191
healing length, 22, 28, 35, 36, 39
Heisenberg equation, 31
Heisenberg equation of motion, 104, 141
Heisenberg limit, 81
Heisenberg picture, 202, 203
Heisenberg uncertainty relation, 74
Hillery–Zubairy criteria, 181
Hillery–Zubairy criterion, 186
Holstein–Primakoff transformation, 84, 180, 181, 199–201, 204, 208
homodyne measurement, 200
hydrodynamic equation, 28, 36
hydrogen atom, 38, 39

incoherent process, 55–57, 60, 61
index picture, 51
indistinguishable particles, 10–13, 84
interference, 101, 117, 124, 139
interferometer, 115
  Fabry–Perot interferometer, 115
  guided-wave atom interferometer, 117
  Mach–Zehnder interferometer, 115, 116
  Michelson interferometer, 115
irrotational flow, 37
Ising model, 154–158, 191

Jordan map, 86
Josephson oscillations, 131

ladder operators, 67
Landau's criterion, 23, 25, 27
Landé g-factor, 45
Laplacian, 1
Legendre polynomials, 90
Levi–Civita antisymmetric tensor, 67
light potential, 104
  reactive force, 105
Lindblad operator, 57, 58, 61
logarithmic negativity, 178, 179
loss, 44, 59
  one-body, 59
  three-body, 60
  two-body, 60
lowest common multiple, 190

Markovian approximation, 56
master equation, 56–58, 60
Mathieu equations, 107
matter wave, 99, 101
Maxwell–Boltzmann distribution, 166
mean-field approximation, 59
Mercator projection, 89
mixed state, 177, 179
momentum operator, 66
Mott insulating phase, 171
Mott insulator, 171, 172
Mott insulator phase, 172

negativity, 178, 179
nonclassical state, 94, 130
  twin-Fock state, 130
  N00N state, 130
  squeezed state, 130
nondestructive measurement, 54
number operators, 6

occupation number, 17, 19
on-site interaction, 170
one-axis one-spin squeezed states, 78
one-axis twisting Hamiltonian, 78, 80, 81
one-axis two-spin squeezed states, 186
open system, 55, 56
optical lattice, 158, 159, 168, 172
order parameter, 28–37, 41, 154, 158

P-function, 64, 87
Padé approximants, 39
parity effect, 52
partition function, 14
Pauli exclusion principle, 5
Pauli operators, 67, 70, 71, 85, 154, 164, 165, 212
Peres–Horodecki criterion, 177, 178
perturbation theory, 53
phase diffusion, 117, 129, 132
phase distribution, 39
phase estimation, 158
phase estimation algorithm
  quantum, 157
phase singularity, 41
phase transition, 154
photon counting, 200
population, 139
position operator, 66
PPT criterion, 178, 179

Probability, 121, 123–125
covariance, 121
mean, 122–126
variance, 122–126
product state, 175
pure state, 175–177

Q-function, 64, 87–94, 136, 141, 143, 187, 204, 208
quantum algorithm, 206
quantum gas microscope, 166
quantum information processing, 199
quantum jump method, 60, 61
quantum metrology, 199
quantum phase transition, 154
quantum simulation, 152, 167
analogue, 153, 154, 157, 158, 167
digital, 153, 155, 157, 158
quantum teleportation, 201, 202
quasiprobability distribution, 86

Rabi frequency, 51, 159
radiation gauge, 49
raising and lowering operators, 67
Raman transition, 99, 110
Ramsey fringes, 138
Ramsey interferometry, 137
Ramsey spectroscopy, 139
Rashba spin-orbit coupling, 163
Robertson uncertainty relation, 75
rotating frame, 139
rotating wave approximation, 52, 103, 112, 138
rotational flow, 37
Rydberg atom, 130

s-wave scattering, 47
s-wave scattering interaction, 31, 48
s-wave scattering length, 131
scalar function, 49
scalar potential, 49, 50, 162
scattering length, 8, 54, 131
Schrodinger cat state, 93, 189
Schrodinger picture, 51
Schrodinger uncertainty relation, 75, 76, 183
Schwinger boson operator, 66
Schwinger boson operators, 67–69, 71, 83, 86
second quantization, 1
selection rule, 52
separable state, 177, 181, 184, 185
single particle limit, 7
singlet channel, 47
singlet state, 48
soliton, 38, 41, 43
bright, 42
dark, 42
solitons, 28
spin coherent state, 143
spin squeezed state, 143
spin-exchange, 60
spin-orbit coupling, 162
spin-orbit interaction, 164
spinor Bose–Einstein condensates, 46, 130
spontaneous emission, 44, 55, 57, 58, 102, 104, 105, 140
decay, 105
squeezed state, 64, 77, 78, 80, 85, 92, 186, 203, 204
one-axis twisting, 81, 90, 91
one-axis two-spin, 186
optical, 77
optical squeezed vacuum, 77
two-axis countertwisting , 81, 82
two-axis two-spin, 191
standard quantum limit, 80, 130
stationary solutions, 32
statistical mixture, 62
Stirling's approximation, 121
superfluid, 23, 37, 38, 171, 172
superfluid phase, 171
superfluidity, 10, 20, 23–25
superradiance, 59
Suzuki–Trotter expansion, 81, 156–158, 206

Taylor series, 73, 96, 180
thermodynamic limit, 154
Thomas–Fermi approximation, 34, 35, 40, 127
Thomas–Fermi limit, 35
three-level atom, 134
time-of-flight measurement, 165–168, 172
topological charge, 41
topological insulator, 162
translational invariance, 9
triplet channel, 47
triplet state, 48
Trotter number, 156
two-axis countertwisting Hamiltonian, 81, 82
two-axis one-spin squeezed states, 80
two-axis two-spin squeezed states, 191
two-component Bose–Einstein condensates, 130–132, 134, 148
Two-mode model, 131

uncertainty relation, 74

vacuum, 3, 5, 131
variance, 142–144, 146, 147
vector norm, 162
vector potential, 49, 160, 162
velocity, 37
visibility, 116, 117, 126
von Neumann entropy, 176–179, 187
vortex core, 39, 41
vortices, 28, 38

Wannier functions, 169
weak link, 131
Wigner function, 64, 87–94, 200, 204, 208
X bit flip gate, 205

Young's double slit experiment, 99

Z phase flip gate, 205
Zeeman effect, 45, 54, 72
Zeeman energy shift, 44
Zeeman shift, 48, 49